SPRINGER
LAB MANUAL

M.D. Luque de Castro   M. Valcárcel   M.T. Tena

Analytical Supercritical Fluid Extraction

With 180 Figures

Springer-Verlag
Berlin Heidelberg GmbH

Professor MARIA DOLORES LUQUE DE CASTRO
Professor Dr. MIGUEL VALCÁRCEL
Dr. MARIA TERESA TENA

University of Cordoba
Department of Analytical Chemistry
Avda. San Alberto Magno, s/n
E-14004 Cordoba
Spain

ISBN 978-3-642-78675-4   ISBN 978-3-642-78673-0 (eBook)
DOI 10.1007/978-3-642-78673-0

Library of Congress Cataloging-in-Publication Data. Valcárcel Cases, Miguel. Analytical supercritical fluid extraction / M. Valcarcel, M.D. Luque de Castro, M.T. Tena. p.   cm. — (Springer laboratory) Includes bibliographical references. ISBN 978-3-642-78675-4 1. Supercritical fluid extraction. I. Luque de Castro, M.D. II. Tena, M.T. (Maria Teresa) III. Title. IV. Series. TP156.E8V28 1994 543'.089—dc20 94-7781

© Springer-Verlag Berlin Heidelberg 1994
Softcover reprint of the hardcover 1st edition 1994

Typesetting: Best-set Typesetter Ltd., Hong Kong
52/3130-5 4 3 2 1– Printed on acid-free paper

Preface

Recent advances in analytical chemistry have turned it into a virtually unrecognizable science compared to a few decades ago, when it lagged behind other sciences and techniques. However, advances in analytical science have been far from universal: while innovations in instrumentation and data acquisition and processing systems have reached unprecedented levels thanks to parallel breakthroughs in computer science and chemometrics, progress in preliminary operations has been much slower despite their importance to analytical results. Thus, such clear trends in analytical process development as automation and miniaturization have not reached preliminary operations to the same extent, even though this area is probably in the greatest need. Improvement in preliminary operations is thus an urgent goal of analytical chemistry on the verge of the twenty first century. Increased R&D endeavours and manufacture of commercially available automatic equipment for implementation of the wide variety of operations that separate the uncollected, unmeasured, untreated sample from the signal measuring step are thus crucial on account of the wide variability of such operations, which precludes development of all-purpose equipment, and the complexity of some, particularly relating to solid samples.

Supercritical fluid extraction opens up interesting prospects in this context and is no doubt an effective approach to automation and miniaturization in the preliminary steps of the analytical process. The dramatic developments achieved in its short life are atypical in many respects. Thus, unlike other analytical alternatives, commercialization of SFE equipment has not been preceded by exhaustive R&D work; its chromatographic counterpart was developed in advance. The advantages of SFE over other solid–liquid extraction techniques make it particularly attractive for routine analyses of large numbers of solid samples.

This book provides an overview of the state of the art in analytical-scale supercritical fluid extraction and its potential. Chapter 1 reviews the preliminary operations of the analytical process and relates supercritical fluid extraction to them. The physico–chemical properties of the supercritical state (Chapter 2) and the fundamentals of supercritical fluid extraction (Chapter 3) make obvious stepping-stones to a comprehensive description of the SF extractor (Chapter 4) and selected applications and trends of SFE techniques (Chapter 5). Supercritical fluid extraction is doubtless of interest to all analytical laboratories concerned with the analysis of solid samples, as well as to teachers of chemical analysis. No doubt, the SFE technique has a promising future that will come with the spread of its high potential for solving analytical problems, to which this book is essentially devoted.

The authors should like to thank Springer-Verlag for their warm welcome for the book, Antonio Losada for translation of the Spanish manuscript, and José Manuel Membrives and Francisco Doctor for producing the artwork.

Córdoba, May 1994   M.D. LUQUE DE CASTRO
M. VALCÁRCEL
M.T. TENA

Contents

1 Preliminary Operations of the Analytical Process

1.1 Analytical Chemistry Today

Information is central to today's society and modern economy. Inasmuch as analytical chemistry is the chemical metrological science, it obviously plays a prominent role in this respect [1]. Delivery of analytical results and their interpretation make two cornerstones on which correct, well-founded decisions should rest. Some estimates suggest that between 5 and 6% of the gross domestic product of developed countries is expended on chemical measurements. Denying the significance of analytical chemistry can only be the result of a manifest lack of knowledge or unscientific interests.

Analytical chemistry must meet the challenges posed by society on the verge of the twenty first century, a continuously developing society requiring this science to adjust itself to advances in other scientific and technical areas. The primary goals of current analytical chemistry can be summed up as the obtaining of greater amounts of chemical information of a higher quality on any type of material or system studied by using increasingly less material, time and human resources, taking the lowest hazards and incurring the lowest expenses. For analytical chemistry to accomplish these generic goals, it must include three major constituents, namely (a) research and development, (b) the suite of analytical methods and techniques that make up what was formerly called *chemical analysis*; and (c) education. This in turn entails the existence of two branches, *basic* and *applied* analytical chemistry, which must bear appropriate relationships to each other and the rest of the diverse scientific and technical areas [2].

Automation and miniaturization [3] are two solidly established trends in science and technology with an undeniable effect on current breakthroughs in analytical chemistry. Few analytical innovations involve neither as they respond to decreased material, time and human resource expenditure and diminished hazards. On the other hand, the impressive development of analytical instrumentation and massive use of (micro)-computers in the laboratory has facilitated the obtaining of more analytical information of a higher quality. **Automation and miniaturization**

1.2 The Analytical Process

As can be seen in Fig. 1.1, the analytical process can be defined as the suite of operations separating the uncollected, untreated, unmeasured sample from the desired results. Even though such operations can be classified according to various criteria, the soundest of all relies on a distinction to between the measurement step and the rest. Thus, one can distinguish between three primary groups of operations, namely:

a) *Preliminary operations*, which make the raw sample ready for analytical measurement. This involves a wide variety of procedures essentially related to the sample and its treatment which are dealt with throughout this book.

Difference between instrument and apparatus

b) *Measurement and transducing of the analytical signal*. This entails recording a signal that can be related to the presence or absence (qualitative detection), concentration (quantitative determination) or structure of the analyte(s), i.e. using an instrument. A clear distinction must be made at this point between an *instrument* and an *apparatus* (the latter may perform a given, major function, but provides no

ANALYTICAL PROCESS

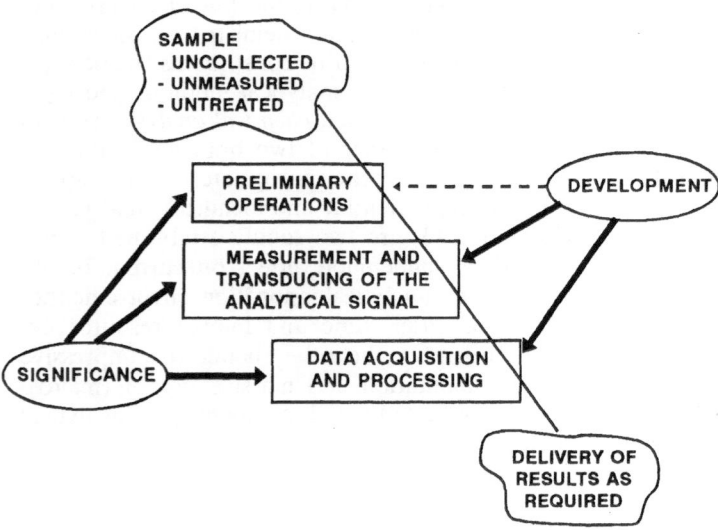

Fig. 1.1. Significance and degree of development of each of the three stages of the analytical process

analytical information). Thus, a centrifuge or even a supercritical fluid extractor is an apparatus, whereas a chromatograph is an instrument because it is fitted on-line to a detector and is different from a chromatographic fraction collection system, which can be considered an apparatus. An analytical instrument (e.g. a balance, spectrophotometer, polarograph) is the materialization of an analytical technique (e.g. gravimetry, photometry, differential pulse polarography). Nor should *technique* and *method* be confused. A method is the specific implementation of a given technique in order to develop a specific analytical procedure to be applied to a definite type of sample and analyte(s). Finally, a clear distinction should also be made between *determination*, which affects the analyte, and *measurement*, which involves a chemical, physical or physicochemical property, of the analyte or a reaction product.

Differential between technique and method

c) *Data acquisition and processing.* Data produced at the second stage of the analytical process can be collected in a number of ways that range from simply recording them in a digital or analogical readout notebook to using a straightforward chart recorder to acquiring them automatically by means of a (micro)computer interfaced to the measuring instrument. Data can also be processed by previously inputing them to a calculator or a computer, or having them handled directly and immediately if either is connected to the instrument. In this context, analytical chemistry plays a prominent role [4,5], both in increasing the amount and quality of the information drawn from raw data and in optimizing and simulating analytical processes.

1.3 Preliminary Operations of the Analytical Process

There has been a clear trend in the last few decades to placing special emphasis on the significance of instrumentation and computers to the analytical process. Objectively, one must acknowledge the deep changes in the substances managed and procedures used in analytical laboratories in the intervening period. All this has led to underestimating the importance of preliminary operations, which are actually as significant as or even more so than the operations involved in the other two stages of the analytical process. Figure 1.1 is a simplified depiction of the state of the art in this context: preliminary operations have developed to a markedly smaller extent than have instrumentation and data acquisition and processing systems, so they continue to be a pending goal of analytical chemistry [6].

Instruments and computers

As noted earlier, the preliminary operations of the analytical process can be defined as the set of steps that make the uncollected, untreated, unmeasured sample ready for instrumental measurement at the second, subsequent stage. There follows a brief description of their most salient features, the steps they most frequently involve and the latest advances in this context.

1.3.1 Basic Features

The generic features of the first stage of the analytical process can be summed up as follows (see Fig. 1.2).

Variability a) They are *widely variable*. In fact, each analytical process has some peculiarities that call for a specific approach in terms of a number of factors, namely:

1. The objectives of the analytical problem concerned. For example, the operations to be performed will vary depending on whether one is to determine toxins/additives or the basic composition (proteins, fats, carbohydrates) in a given type of foodstuff. Surface analysis and speciation call for unique preliminary operations. Also, sampling will vary widely depending on the required degree of representativeness, not only of the sample itself, but also of the decisions to be made on the results obtained.
2. The state of aggregation and nature of the sample. Naturally, the experimental approach to be adopted will depend on whether the sample is a gas, rock, biological fluid, yoghurt, toothpaste, etc.

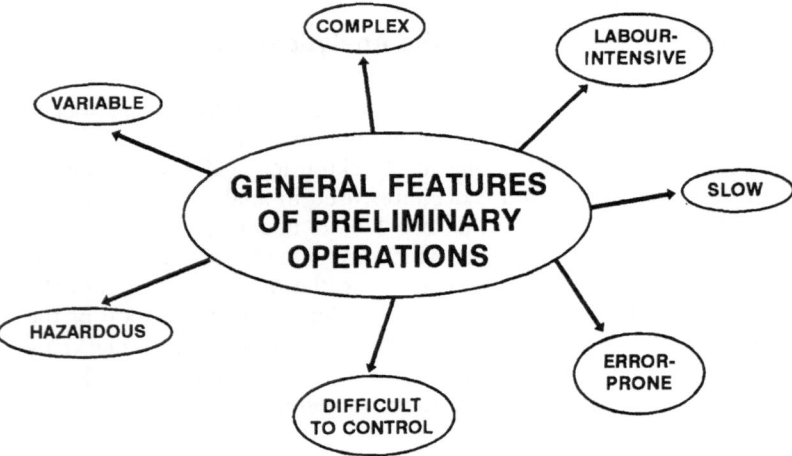

Fig. 1.2. Features of the first stage of the analytical process

3. The availability/affordability of the sample. Preliminary operations will be rather different depending on whether one is to deal with earth or moon rocks, blood or dioxin, adult's or newborn's blood, a work of art or a building material, sea water or water in a suspectedly contaminated or poisoned bottle, etc.

4. The nature and concentration of the analytes. The preliminary operations to be implemented will also be different if the analytes to be determined are metals, pesticides or aminoacids, for example. They will also vary broadly depending on whether a given analyte is present as a major component or in trace amounts in the sample (or at similar concentration levels in various matrices).

5. The analytical technique required or available. The preliminary operations of choice will differ if one is to identify/quantify an organic compound (IR, MS, NMR) or a metal (atomic, x-ray spectroscopy). The choice will also be dictated by the available instrumental equipment of the laboratory concerned. For example, if only a UV-visible photometer is available, the preliminary operations to be implemented for the determination of metal traces in biological fluids will be rather different from those one would use if an atomic absorption spectrophotometer were accessible.

b) They entail *intensive human labour* in clear contrast with the other two stages of the analytical process. Because of their inherent complexity, they are rather difficult to automate. In addition, because of their variability, analytical instrumentation manufacturers have only undertaken the automation of preliminary operations in those cases where a wide market was anticipated. Such is the case with clinical chemistry, where dramatic endeavours to replace human labour are being made and endless developments achieved thanks to increasing entrepreneurial competition. It should be noted that the advertising for many commercially available continuous [7] and batch analysers [3] boasts a high degree of automation and sampling throughout, which is in clear contrast with the facts as it ignores the manual preliminary operations required to condition samples for treatment by an autosampler. This is also the case with automatic systems for introduction of samples into an electrothermal–vaporization atomic absorption spectrophotometer. In any case, there is a growing trend towards developing sample processing modules for off-line or on-line coupling to traditional instruments via very simple interfaces.

c) They are rather *time-consuming* or even tedious. As a rule, preliminary operations take up a high fraction (70–95%) of the overall time devoted to the analytical process, which is particularly outstanding when the fast instruments and computers recently available are used. Samples calling for complex treatments (dissolution, disaggregation, evaporation, centrifugation, solid–liquid or liquid–liquid extraction, etc.) compel the operator to repeat a long sequence of operations each day, so weariness can have a significant effect on human performance in this context.

Errors   d) They are the potential *source of extremely significant bias or accidental errors* that can decisively influence the quality of the results. One should bear in mind that the risk of making serious errors is much higher in preliminary operations than it is at the other two stages of the analytical process. Inappropriate sampling, incomplete dissolution or disaggregation, poor sample preservation, inefficient removal of interferents, etc., can result in very much larger errors – occasionally over 100% – than those normally made in measuring or transducing the analytical signal by using even a fairly unsophisticated instrument. Such errors arise from the technical complexity of preliminary operations and the so-called "human factor", which in turn originates in the high involvement of human operators. Consequently, as clearly shown in Fig. 1.1, preliminary operations are at least as significant as the other two stages of the analytical process.

e) They *defy systematic control.* In contrast with the ease with which an instrument can be calibrated for checking purposes, controlling preliminary operations entails thorough, continuous, systematic verification of all the apparatuses, calibrated glassware and instruments involved in the process. Such a tedious task can be addressed in three ways: first, one can rely on the quality assurance principles set in the so-called Quality Manual and performing external or internal checkups on the laboratory work [8]; second, one can assess the overall analytical process by subjecting blind samples of known composition or certified reference materials (CEMs) resembling as far as possible the constituent materials of the matrices of the real samples concerned; finally, one can use a combination of the previous two approaches – this is the most comprehensive and hence the most desirable of the three choices.

f)) They are major *sources of potential hazards* to both human operators and the environment. Some preliminary operations are rather complex and require the use of acids, solvents, pressurized gases, digestors, evaporators, distillers, etc., which may be harmful to the operator in the act (e.g. accidents) or in the medium-to-long run. Reducing human involvement in this respect is therefore essential in order to augment laboratory safety. The disposal of toxic laboratory waste poses further problems since preliminary operations typically use large volumes of reagents.

1.3.2 Most Common Steps

First of all, it should be noted that not all of the common steps described below are included in every analytical process and that the extremely wide variety of possible analytical problems precludes a systematic approach in this respect. Each process calls for some unavoidable preliminary operations in order to adapt the unknown sample to the instrumental technique to be used and ensure that adequate representativeness, integrity preservation, sensitivity [minimum detectable analyte concentration(s)], selectivity (the lack of response to potential interferents) and precision are achieved. In some cases, the preliminary operations required are minimal (e.g. with in vivo sensors and in the continuous monitoring of evolving processes). **Adapting sample to technique**

The most commonly employed analytical preliminary operations are as follows:

Sampling includes a wide assortment of operations aimed at obtaining a sub-sample portion that is representative of the starting material and consistent with the definition of the analytical problem for subjection to the analytical process [9–11].

Weight/volume measurements are indispensable in order to express the results, except in direct analytical monitoring systems [e.g. (bio)chemical sensors]. Implementing this operation calls for an instrument, which is in contradiction with the scheme for the analytical process shown in Fig. 1.1. Thus, the instrument is used in two steps: first, to measure a sample property (weight or volume), and then to measure an analyte property. This step can be simultaneous with sampling (e.g. measurement of the gas volume that passes through a filter in atmospheric pollutant analyses) or, subsequently, at the laboratory, after a large sample portion has been taken and – optionally – homogenized (e.g. weighing of a sub-sample subjected to all other operations in the analysis of rocks, coal, butter, etc., or measurement of a volume of liquid sample taken from a river).

Moisture in solid samples (e.g. soils) must previously be removed in order that the measured weight can be assigned to the dry sample, thereby avoiding major errors in the results potentially arising from the high variability of the amount of retained water. If water is an analyte in the analytical process concerned, then this operation will obviously be included in its second stage. **Drying**

Solid samples of variable particle size must be ground in order to obtain even grain sizes and hence avoid the variability potentially arising from compositional differences between portions containing different sizes. In addition, grinding and sieving facilitate subsequent dissolution/ **Grinding/ sieving** disaggregation of the sample as they increase its specific surface area, thereby substantially diminishing the time required for the attack.

Homogenization The aliquot of solid or liquid sample that is finally subjected to the other operations of the analytical process must be representative of the entire population if an average composition is to be determined. This step can be included in the initial sampling or, alternatively, implemented on arrival of the samples at the laboratory.

Preservation of collected sample Chemical and physico-chemical alterations of the sample resulting from an instable character or an interaction with its container should be avoided by taking preventive measures such as adding a suitable preservative, freezing, using a container material that can neither contaminate the sample nor retain the analytes to be determined, and closing the container tight so as to avoid interactions with atmospheric components (water, oxygen, carbon dioxide, metal and organic pollutants). Alternatively, one can use a device for retaining the analytes directly (e.g. a filter or a suitable support); in this case, preservation refers to the stability of the analytes on the support used.

Dissolution/ disaggregation Inorganic samples are most often dissolved by attack with an appropriate acid or flux.

Removal of organic matter The determination of inorganic analytes in organic or biological samples requires a dry or wet attack of the sample in order to decompose and volatilize organic matter (i.e. to release water, carbon dioxide, sulphur dioxide, nitrogen, etc.), which poses a serious constraint on the analytical process. This step can be regarded as an analytical separation inasmuch as it involves the release of gases by oxidation and subsequent volatilization of the reaction products.

Analytical separation Analytical separation techniques play a central role in many analytical processes since, among other advantages, they enhance two capital analytical properties, viz. selectivity (through interference removal) and sensitivity (through preconcentration) [12]. Since supercritical fluid extraction is a fairly recent separation technique, it is dealt with in detail in a special section.

Analytical reaction is intended to convert the analytes into products with physico-chemical features that allow their determination by the analytical technique to be used in the second stage of the process.

Other chemical reaction are used for a variety of purposes such as facilitating a separation (e.g. derivatization prior to column chromatography) or removing interferences (e.g. by use of masking reagents), among others.

Transport to and introduction of the treated sample into the instrument **Link between** involves establishing a manual, semi-automatic or fully automatic link **process stages** between the first and second stage of the analytical process. Most often, this operation is performed by the operator, whether sequentially (sample by sample) or in batches, using a sampler as inerface provided the instrument is furnished with an automatic sample introduction system. Occasionally, some preliminary operations are integrated with detection (e.g. in gas and liquid chromatographs, where the separation step is linked on-line with detection). In a few special cases, all preliminary operations are integrated with detection (e.g. in sensors), so the analytical process consists of two distinct stages only.

The above description provides a broad view of the basic features (complexity, varitety) of the preliminary operations of the analytical process [13,14].

1.3.3 Recent Developments

The many pitfalls of the preliminary operations of the analytical process, particularly in those cases where they involve complex procedures, has fostered R&D activity aimed at circumventing them. It should be noted that most recent advances, whether or not commercially available, rely on current scientific and technological trends such as automation, miniaturization and the use of new technologies and materials. Figure 1.3 shows some recently reported approaches, most of which are commercially available; to the authors' minds, they make an advance of the developments one can anticipate for the very near future.

Robotic stations [15,16] make the most comprehensive choice for the **Robots** automation of laboratory processes. Their foremost asset is their ability to undertake most of the operations forbidden to analysers (e.g. weighing, dissolution/disaggregation, solid–liquid extraction, evaporation, solvent changeover, centrifugation, preparation of chromatographic vials) without human participation. As with computers, the current high cost of robotic systems is bound to fall as their performance is enhanced and their scope of application expanded. No doubt they make a promising alternative, particularly in relation to analytical processes involving a large number of complex preliminary operations.

Non-chromatographic continuous separation techniques (sorption, extraction, dialysis, gas diffusion, etc.) [26] have helped to solve a variety of problems encountered in the treatment – and, occasionally, collection – of samples. These techniques are quite simple and candidates for automation, yet can only handle liquid samples; a few continuous systems, though, can process gas or solid samples. This is notably the most appealing feature of supercritical fluid extraction.

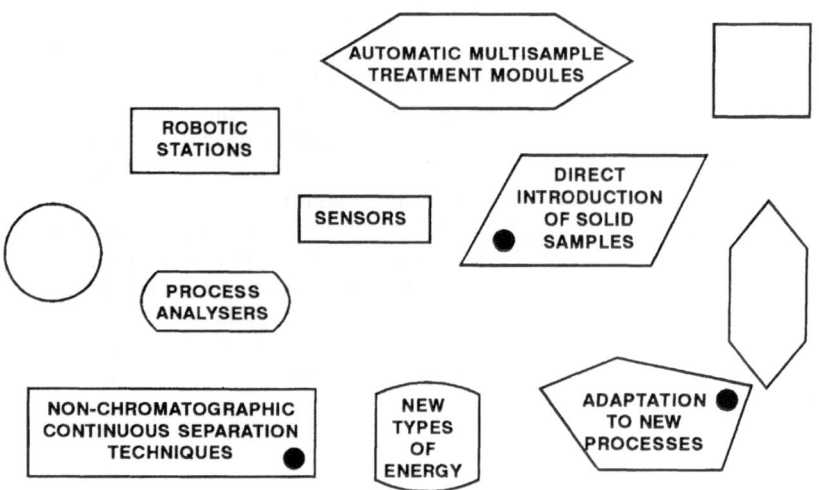

Fig. 1.3. Recent developments aimed at enhancing the features of preliminary operations of the analytical process. (●) Supercritical fluid extraction

The development of physical and (bio)chemical sensors responding in a rapid, continuous, reversible and precise way to changes in the analyte concentrations is of special interest in this context as it can help to minimize preliminary operations [17–19]. Notwithstanding the thousands of publications on the topic, only a few instances have proved to resolve real problems effectively and have thus materialized in commercially available alternatives (e.g. solid-state gas sensors). Much R&D endeavour must still be made, though, in order to make sensors usable for routine applications. One major hurdle in this respect lies in their understandable limitations for direct analysis of solid samples.

Process analysers   The development of process analysers (on-line, in-line and non-invasive) is a subject of great, not only industrial, but also environmental and biotechnological interest [20]. This type of analyser is very easy to automate and involves a minimum of preliminary operations. Because such operations must inevitably be quite simple, they occasionally fail to provide reliable results, which also rely on proper calibration.

A number of modules for automatic treatment (heating, digestion, agitation, evaporation, extraction) of several samples (4 to 24) simultaneously emerged in the 1980s. While such modules cannot deal with every possible preliminary operation, they are being increasingly frequently used in routine laboratories as they substantially shorten the time expended in these operations.

Attempts at simplifying preliminary operations involving solid samples are usually aimed at avoiding dissolution, which can be achieved by direct introduction of the samples into the analytical system used (e.g. direct introduction of solid samples into atomic spectrometers [21] or indirect introduction of slurries). Supercritical fluid extraction enables rapid, efficient extraction of analytes from a solid that is directly introduced into the extractor, though not into the measuring instrument.

The use of uncommon sources of energy (ultrasound [22], microwave [23] or laser) in automatic modules allows samples to be attacked (dissolved, digested, etc.) in a faster, more efficient way that with ordinary systems. **Uncommon energy sources**

Some formerly innovative industrial and preparative procedures have been transferred to the analytical field with the development of labware of proper technical performance. Such is the case with analytical lyophilizers for preparing certified reference materials or attacking biological samples [24], and with supercritical fluid extractors for direct treatment of solid analytes, which are the subject matter of this book.

1.4 Analytical Separation Techniques

Mass transfer between two phases is rather commonplace in two central areas of chemistry: synthesis and analysis. An analytical separation involves splitting a sample into at least two parts (one per phase) that are enriched or impoverished with one or more species of interest (analytes, interferents). This operation entails the transfer or relative motion of the parts (phases) involved [25]. As noted earlier, mass transfer makes an essential element of the preliminary operations of the analytical process as some of its features respond to enhanced basic analytical properties. A significant fraction of the most outstanding analytical chemical developments of the last few years relies on the use of separation techniques [12]. In fact, few analytical processes include no separation.

1.4.1 Objectives

Where does the significance of analytical separation techniques lie? The answer is very simple: they enhance such capital analytical properties as sensitivity (through preconcentration) and selectivity (through interference removal). These two, in addition to precision, are central to assuring quality (accurate) results from an analytical process (Fig. 1.4).

Analytical separation techniques offer a number of other, more specific, but also interesting advantages, namely:

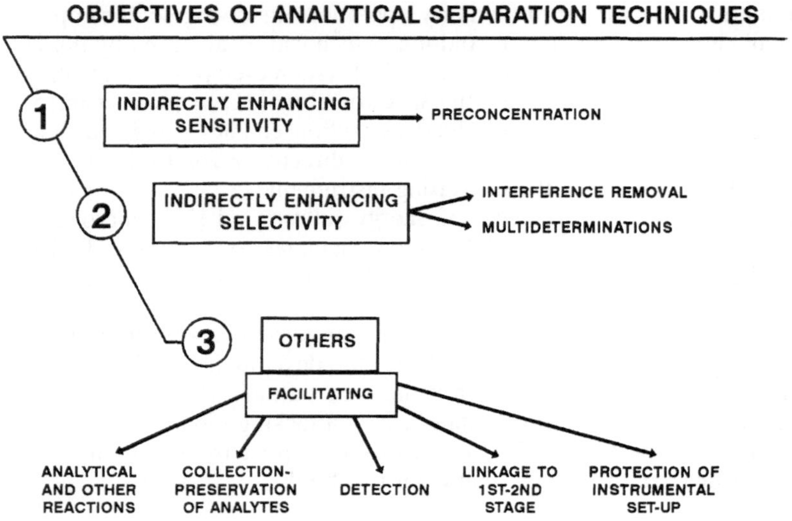

OBJECTIVES OF ANALYTICAL SEPARATION TECHNIQUES

Fig. 1.4. Principal advantages of analytical separation techniques incorporated into preliminary operations of the analytical process

Advantages   They facilitate or improve the analytical reactions on which determinations rely. If the reaction in question takes place simultaneously with separation, the complexity of the preliminary operations involved is substantially decreased.

They enable easier collection and preservation of the analytes on direct application to the sample. By using a large enough mass, the sample can be efficiently preconcentrated; also, the collecting phase can help to preserve the identity of the retained analytes.

They facilitate detection. For example, water analyses by gas chromatography involve solvent changeovers, which are routinely dealt with by liquid-liquid extraction.

They aid transfer of the analyte(s) or reaction product(s) to the measuring instrument, particularly in the case of continuous or dynamic separation techniques.

They protect the instrumental set-up from potential aggressions (by species or particles) that might result in reversible or irreversible damage of potentially very serious economic consequences.

1.4.2 Classifications

The large number and variety of analytical separation techniques available calls for appropriate classification according to various criteria such as the

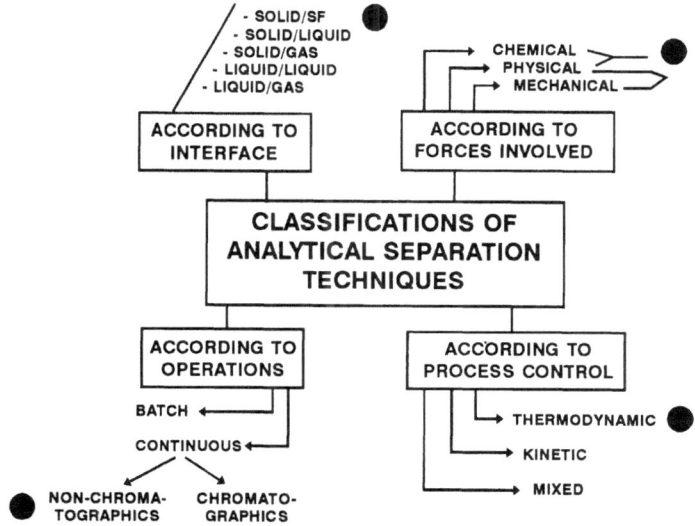

Fig. 1.5. General criteria used to classify the wide variety of available analytical separation techniques. (●) Supercritical fluid extraction

nature of the phases involved, the mechanism by which mass transfer takes place, and the way the process is controlled or implemented [12]. Figure 1.5 shows a scheme of such classifications, which are commented on briefly below.

One distinction is based on the nature of the interface across which mass transfer takes place. Of all the possible binary combinations of the four types of phase (solid, liquid, gas and supercritical fluid), whether of two different or the same phase (e.g. liquid–liquid), only five are technically feasible and employed in practice: liquid–gas (e.g. distillation, gas diffusion, gas chromatography) [7], liquid–liquid (e.g. extraction, dialysis, high performance liquid chromatography), solid–gas (e.g. sublimation) [7], solid–liquid (e.g. ion exchange, precipitation, high performance liquid chromatography) and solid–supercritical fluid (e.g. supercritical fluid extraction and chromatography). The phase acting as or containing the sample is taken as the first phase and the second can be either produced in situ (e.g. in distillation or precipitation) or added externally (e.g. in extraction, chromatography or dialysis).

Mass transfer interface

The forces involved in the mass transfer process can be of different nature, namely: mechanical (e.g. filtration, centrifugation); physical, based on state changes (e.g. distillation zone melting); adsorption (e.g. gas–liquid chromatography, solid–liquid extraction); dissolution (e.g. liquid–liquid extraction); electric (e.g. electrophoresis); or chemical (e.g.

precipitation, ion exchange). In any case, most separation techniques involve more than one type of force.

Foundations of analytical separation techniques
Analytical separation techniques can have a thermodynamic, kinetic or mixed foundation. Separation techniques based on the establishment of an equilibrium involve the usual thermodynamic relations (e.g. the phase rule, the Gibbs and Clausius–Clapeyron equations, and distribution constants). Such is the case with liquid–liquid, solid–liquid and solid–supercritical fluid extraction, precipitation and distillation, where the time needed for equilibrium to be reached is assumed to be negligible. On the other hand, kinetic separation techniques rely on differences in the speed the different components of matter acquire on establishment of a concentration (dialysis), electric potential (electrophoresis), temperature (thermal diffusion) or density (centrifugation) gradient; the migration processes involved obey the laws of diffusion. Other separation techniques such as chromatography and ion exchange have a twofold (thermodynamic and kinetic) foundation. Thus, chromatography is ruled by the coefficients of partitioning of the analytes between the phases on the one hand, and the diffusion resulting from longitudinal and axial concentration gradients in the chromatographic bed, on the other. Ion exchange is an equilibrium process governed by a selectivity coefficient, even though film and particle diffusion also play a prominent role in the separation.

Separation techniques can also be classified according to various criteria based on dynamic and operational factors (e.g. the presence or absence of a well-defined interface or the number of times a partitioning equilibrium is reached). However, the most interesting classification of this type is that based on the use or not of a flow of one or both phases to facilitate differential transfer of the mixture of analytes and other sample components. Discrete separation techniques (foremost of which are precipitation and classical electrophoresis) are characterized by the absence of

Stationary phase
motion of one of the phases and the fact that kinetic plays no significant role except in a few cases; the link with detection (the second stage of the analytical process) is established off-line. On the other hand, continuous separation techniques are characterized by the flowing moting of one or both phases, whether in the same or the opposite direction. These

Moving phase
techniques can be classified into two broad categories, viz. chromatographic and non-chromatographic, depending on the efficiency of the separation process, which usually depends on the number of times equilibrium is reached. Detection can be done on-line by monitoring the mobile phase, or off-line by collecting fractions of such a phase. A given separation technique can be implemented in various operational modes. Thus, liquid–liquid extraction can be carried out batchwise (e.g. in separatory funnels) or continuously (whether chromatographically, as in distillation/condensation systems and continuous segmented organic/aqueous flow

analysers, or non-chromatographically, as in partitioning liquid chromatography and HPLC). Sorption processes (ion exchange included) are usually implemented in a continuous, chromatographic or non-chromatographic mode – the batch mode is of very little analytical use. Finally, the supercritical fluid separation technique has a continuous non-chromatographic variant (extraction), the subject matter of this book, and a chromatographic one.

1.4.3 Continuous Separation Techniques

Non-chromatographic continuous separation techniques lie between batch (manual) and chromatographic separation techniques [26]. In fact, they resemble the former in their separation efficiency and the latter in their operational foundation. On the other hand, the greatest differences with the batch techniques lies in their ready automation and miniaturization, while those with the chromatographic ones are related to their efficiency and purpose (an overall vs an individual separation).

They dynamic character of continuous separation techniques relies on the use of a propulsion system to continuously propel a liquid, gas or supercritical fluid through a continuous system under strictly controlled operational conditions (pressure and flow-rate, basically). A liquid mobile phase can be propelled by means of a peristaltic or another type of pump, a gas pressure system or even gravity. Gases are introduced by using commercially available high-pressure cylinders. Supercritical fluids must be kept in pressurized cylinders and require a syringe or piston pump for raising the pressure and a heating system for establishing the critical conditions.

Propulsion system

Continuous separation systems involve a variety of interfaces including gas–liquid (e.g. introduction of gaseous samples into hydrodynamic systems, distillation, gas diffusion, hydride generation), liquid–liquid (e.g. dialysis, extraction), liquid–solid (e.g. extraction, sorption, precipitation, electrochemical stripping) and sold–supercritical fluid (extraction).

Continuous systems accept both solid, liquid and gaseous samples, which can be introduced by injection (insertion) with the aid of an appropriate valve or by continuous aspiration in the case of liquids and gases. Solid samples can only be introduced directly in a few cases (e.g. by automatic leaching using electric energy to facilitate dissolution or ultrasonic energy to favour extraction). The most valuable asset of supercritical fluid extraction is its ability to handle solid samples directly.

Sample introduction

The second phase, intended to collect the analytes (or their reaction products) or the interfaces, can either be directly introduced into the separation system (e.g. in liquid–liquid and supercritical fluid extraction, dialysis, gas diffusion), be a permanent part of the system (e.g. sorbent material packed in a minicolumn), originate from a chemical reaction (e.g. precipitation) or result from a combination of two or more of these.

Most often, analytical separation systems are connected on-line with a continuous (optical or electroanalytical) system through which the sample to be monitored is passed – some systems can also be connected to chromatographic (HPLC, GC, SFC) instruments as well. The automation of this operation is self-evident. In some cases (e.g. with supercritical fluids), the off-line mode excels the on-line mode. Separation and detection can also be integrated in the same microzone [27]; also, the two operations can take place sequentially (e.g. in anodic and cathodic stripping) or simultaneously (e.g. with continuous-flow sensors using a photometric or fluorimetric flow-cell packed with sorbent material) [28].

Analyte derivatization and any accessory reactions involved can be developed prior to (e.g. formation of volatile species in gas diffusion or scarcely polar derivatives in supercritical fluid extraction), during (e.g. precipitation) or after the continuous separation process (e.g. formation of detectable derivatives).

Most continuous separation techniques are usually implemented in a batch, manual fashion. However, in some special instances including dialysis, gas diffusion, ion exchange and supercritical fluid extraction, the batch (static) alternative is rarely used for analytical purposes, even though it is highly useful for physico-chemical studies.

In addition to the basic assets derived from the use of analytical separation techniques, continuous separation systems offer several added advantages, namely: (a) they facilitate the automation of preliminary operations; (b) they increase precision by minimizing errors associated with the so-called "human factor"; (c) they boost throughput; (d) they result in dramatic savings in sample, reagents and solvents; (e) they augment operator safety and decrease environmental pollution; and (f) they reduce analytical costs.

1.5   Extraction Systems in Analytical Chemistry

The Oxford English Dictionary defines *extraction* as "the action or process of extracting" and the verb *to extract* as "to take from something of which the thing taken was a part", or, in its chemical sense, "to obtain from a thing or substance by any chemical or mechanical operation". There is thus little conceptual difference between extraction and separation, the latter relying on the establishment of a partitioning equilibrium, since both involve a chemical mass transfer (of the analytes or other components) from one phase to the other. In fact, extraction is hardly different from many separation techniques in many respects. Thus, ion exchange can be

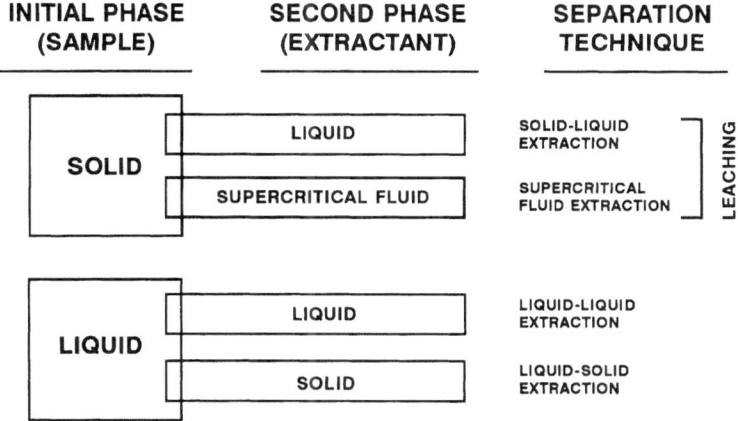

INITIAL PHASE (SAMPLE) | **SECOND PHASE (EXTRACTANT)** | **SEPARATION TECHNIQUE**

Fig. 1.6. Extraction techniques most frequently used in analytical chemistry

defined as the "extraction" of ions from a liquid phase into a solid one (the exchanger). Precipitation is similar in this regard. Gas diffusion can also be considered the "extraction" of volatile analytes from a liquid phase into a gaseous phase. The question therefore arises as to what warrants using the word "extraction" to refer to some separation processes. To the authors' minds, it is scarcely scientific and the result of tradition usage rather than a founded choice. However, there are indeed some exceptions. Thus, "extraction" is rarely used in connection with kinetic separation processes (e.g. chromatography or electrophoresis). Nor is it applied to separation systems involving gases, whether as a phase (e.g. gas chromatography) or as analytes (e.g. gas diffusion).

Figure 1.6 shows a simplified scheme of extraction techniques used in analytical processes. For conceptual clarity, a distinction is made between the state of aggregation of the sample phase (the first or initial phase) and that of the second, added phase, which can be called the "extracting" phase as it collects the analytes (or interferents, depending on the particular extraction purpose).

When a solid sample is directly subjected to the separation process known as leaching, the analytes are the species involved in the mass transfer. The process consists of a single step (extraction–elution) and the analytes are carried by the extracting mobile phase. There are two alternatives depending on the state of aggregation of the second phase: (a) solid–liquid extraction, in which a liquid (an acid or organic solvent) thoroughly "wets" a portion of a solid sample to directly extract the solutes; and (b) supercritical fluid extraction, the subject matter of this book.

Solid sample leaching

Liquid sample   If the initial sample is a liquid (even if it is obtained by dissolution/ disaggregation) or a solid sample, then the species involved in the mass transfer can be either the analytes or the interferents. There are two general manners of developing an extraction process. First, liquid–liquid extraction [29], which is based on the virtually complete immiscibility of the two phases, can be carried out in at least four ways, namely: (a) in separatory funnels; (b) in a closed circulation system with solvent distillation/condensation; (c) in a Craig upstream extraction tube; and (d) in a continuous segmented organic solvent/aqueous solution flow system. Second, solid–liquid extraction, which involves passing the liquid sample phase through a minicolumn containing a sorbent of widely variable nature (e.g. carbon, silica, alumina, an ion exchanger, covalently bonded silica) that selectively retains the analytes or interferents. If the analytes are retained, a second step is required to elute them with another solvent (e.g. one of different polarity) for their subsequent determination. This separation process has gained popularity in the last few years and fostered commercialization of a host of disposable cartridges of sorbent material for sample clean-up. This process can also include analyte preconcentration. In fact, it is frequently used to condition complex samples for subsequent chromatographic (HPLC, GC) determination of the analytes. The very slight differences between solid–liquid extraction (solid sample) and liquid–solid extraction (liquid sample) frequently lead to terminological confusion. Taking into account that the latter is much more frequently used than the former, it is hardly surprising that both denominations are used indifferently in the industrial field to refer to the same phenomenon. The term "leaching", which originated in industrial chemistry, is more distinct; however, the recent advent of supercritical fluid extraction has rendered it inappropriate for denoting solid–liquid extraction exclusively.

Gaseous sample   In a few analytical non-chromatographic separation processes, extraction is done by retaining the analytes or interferents present in an initial gaseous phase by using a suitable solid or liquid. It is worth emphasizing the great similarity with extracting processes; however, these separation processes rarely warrant the denomination "extraction". If the second phase is a solid, separation can be achieved by simple filtration (retention of particulates in a gas), adsorption (a physico–chemical interaction between the solid and the retained species, or a combination of both); this separation strategy is widely used for the determination of atmospheric pollutants in environmental analysis. If the collecting phase is liquid and a gaseous sample is passed through it, the process is essentially one

Atmospheric pollutants   of absorption by dissolution and a chemical reaction can be optionally employed to facilitate retention; this process is also often applied to environmental samples.

It should be noted that most extraction processes have a chromatographic counterpart with the same physico–chemical foundation but a much higher efficiency resulting from the fact that analytes (or interferents) are isolated individually rather than jointly. Obviously, there are no chromatographic analogues for extractions involving solid samples. On the other hand, liquid chromatographic (particle adsorption) modes do have some extraction counterpart. In fact, some partitioning chromatographic modes used to be referred to as "extraction chromatography" in the 1970s [30].

1.6 Analytical Leaching Methodologies

Leaching is the extraction of substances from a solid by using a second, usually liquid phase. It thus differs from dissolution in the fact that a substantial portion of the solid does not disappear by virtue of the process. Leaching is rather commonplace in the chemical industry, but is also present in some everyday actions (e.g. doing the washing or making coffee). Analytically, leaching is a preliminary operation of the analytical process by which a solid sample (dried or lyophilized, if necessary) is treated with a second phase (a liquid or supercritical fluid) into which the analytes or, occasionally, the matrix, are transferred with the aid of external (thermal, ultrasonic, electric) energy.

As noted earlier, the word "leaching" is rarely used in analytical chemistry; its synonym "solid–liquid extraction" is much more commonplace. The problem arising from usage of this latter is a potential confusion with liquid–solid extraction (see Fig. 1.6) since the mere order of the two words is insufficient to set a clear-cut, distinct denomination emphasizing the fact that the (initial) sample is solid or liquid. Therefore, the word "leaching" and the verb "to leach" should be used in the analytical field in the strict sense defined in reputable dictionaries: "subject to the action of percolating water, etc., with the view of removing the soluble constituents" (Oxford English Dictionary), or "dissolve out by the action of a percolating liquid" (Webster's Third New International Dictionary), for example. Some authors use the term "selective extraction" to describe analytical leaching [31].

Solid–liquid vs liquid–solid extraction

Essentially, leaching involves separating (extracting) a solid component by means of another phase (a liquid or supercritical fluid), the efficiency of which depends on several factors, namely:

a) The properties of the solid sample, particularly its degree of dispersion (specific surface area) and the physico-chemical properties of the matrix (the presence or absence or active surface groups).
b) The physico-chemical properties of the extracted analyte-solute, which determine its interaction with the matrix and its solubility in the extracting phase.

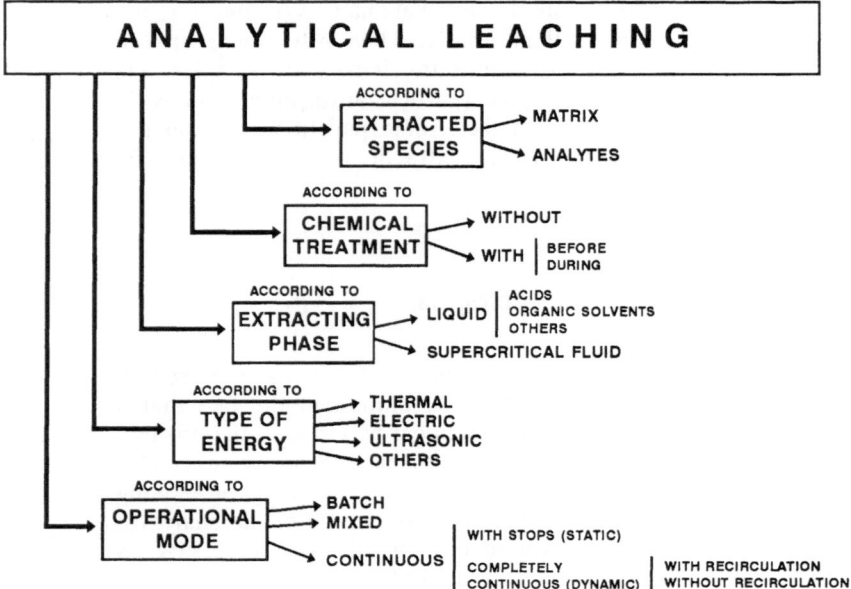

Fig. 1.7. Classification of solid–liquid extraction systems for analytical purposes according to various criteria

c) The physical (e.g. viscosity, diffusivity) and physico-chemical properties (e.g. polarity) of the extracting phase (usually the mobile
   phase), which determine both the solubility of the solute and its ability
   to "navigate" across the active sites of the matrix.
d) The operational conditions (temperature, way of bringing the phases
   into contact, type of energy used to the accelerate the process, etc.).

All these factors influence two generic properties of analytical leaching
that are of a great practical interest: the duration and yield (or efficiency)
of the separation process.

Figure 1.7 shows a scheme of the most relevant analytical leaching
alternatives, classified according to several criteria. Some specific instances
(e.g. extraction of volatiles in vegetable material by use of a steam flow)
have been excluded for simplicity.

One possible distinction is based on which part of the solid sample is
dissolved into the extracting phase. Analyses of inorganic materials (e.g.
metals, alloys) often involve dissolving (extracting) the matrix to leave a

residue (oxides, nitrides, carbides) containing the analytes. However, it is the analytes which are normally extracted.

Leaching can be directly applied to untreated samples. Extraction of the analytes can be facilitated or improved by a prior derivatization. The chemical reaction used for this purpose can be introduced before the extraction (e.g. formation of scarcely polar derivatives from highly polar analytes for supercritical fluid extraction) or during it (e.g. conversion of a silicon matrix into silica by heating with NaOH at 360 °C and 350 atm, and determination of boron in the resulting liquid phase). Derivatization can thus act on both the matrix and the analytes.

The extracting phase can be liquid or a supercritical fluid, depending on the nature of the analytes and matrix of the solid sample. Acidic aqueous solutions are often used to extract ionic species from a variety of samples, both inorganic, organic and biological; in many cases, an additional reagent (e.g. an oxidant or complex-forming agent) is employed to aid dissolution of the analytes. Organic solvents, whether pure or mixed, are also extensively used to extract analytes of widely variable polarity from diverse materials. The use of supercritical fluids in this context has resulted in great developments based on the excellent advantages they offer and has revitalized the significance and use of solid–liquid extraction for analytical purposes.

Choice of extracting phase

The rate at which physical (e.g. transport) and physico-chemical phenomena involved in the leaching process occur is determined by such factors as the availability of extracting phase to all analyte species, which are not always present at the surface, the ionic displacement or indicator required to replace one analyte species with one or more molecules of extracting phase, the way in which a given matrix is related to some analytes, the solubility of the analytes in the extracting phase, etc. This is clearly seen in studying the variation of the leaching efficiency as a function of time. As a rule, the efficiency increases substantially at the beginning of the process, which shows that the rate of leaching depends on the above-mentioned factors. Accelerating the process entails increasing the pressure to favour penetration and transport, the temperature to augment the solubility and diffusivity, or boosting transport phenomena to facilitate displacement of the partitioning equilibrium. The use of external energy to aid the kinetics of solid–liquid separation is virtually unavoidable. Most often, the temperature is raised by ordinary or microwave heating for this purpose. On the other hand, the pressure can be increased by using a heated closed circuit, ultrasound, or a supercritical fluid as extracting phase. Transport phenomena are determined by the operational modes described below. Ultrasonic probes, of consistent, readily controllable functioning, are highly useful in this context as they provide instantaneous local pressure and temperature increases. Electric energy has occasionally been employed [32,33] to aid selective dissolution of some

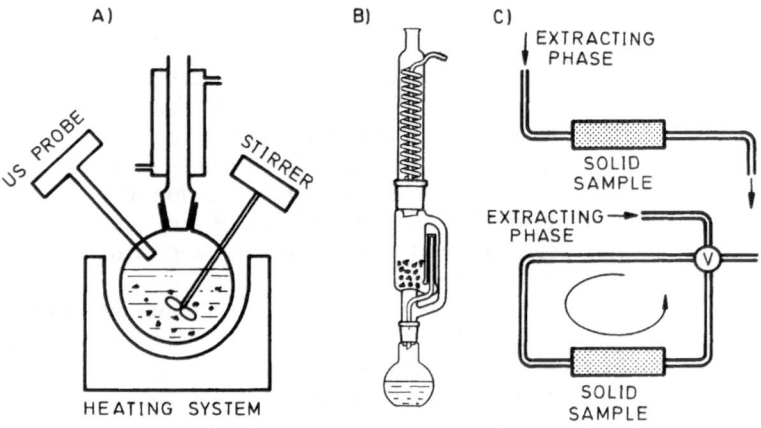

Fig. 1.8A–C. Scheme of the most common operational modes of analytical leaching. A Batch. **B** Mixed (batch–continuous), with a Soxhlet extractor. **C** Continuous, with and without recirculation

steel components by electrolysis (oxidation). Finally, mechanical energy is frequently used in connection with the batch operational mode to effect vigorous shaking of the solid suspended in the extracting phase.

Analytical leaching is typically carried out in the discrete or batch mode (Fig. 1.8A). The solid sample is placed in a vessel containing the extracting phase, where it is immersed. Most frequently, the vessel is open and furnished with a condensation system in order to avoid solvent losses through vaporization. The process is usually accelerated by heating, even though ultrasonic probes are much more efficient and convenient for this purpose as they provide precisely controllable pressure and temperature increases. Efficient agitation also favours development of the extraction process, yet is not strictly necessary provided the extracting, liquid phase is kept boiling and stirred (e.g. with a magnetic bar). Because the extracting phase is not renewed during the process, a displacement of the separation equilibrium to the liquid phase is not seen, so the extraction efficiency may be rather low if the analytes are not very soluble and the liquid molecules do not compete advantageously with those of the analyte for the active sites of the matrix – hence dynamic or mixed modes are in principle more likely to provide yields approaching 100%. This shortcoming can be circumvented by using large solvent volumes (500 ml to 2 l), which, however, has serious drawbacks such as the need for a subsequent preconcentration, a high cost per analysed sample and the hazards involved in handling inflammable and/or toxic substances. One other technical pitfall lies in the implicit need to separate the phases (e.g. by

filtration) after leaching, which is further complicated if the whole extracting phase is to be collected. Occasionally (e.g. in the determination of
available elements in soils), leaching is carried out without heating or
shaking, particularly when only a specific fraction of the analytes present
need be extracted.

Figure 1.8B depicts a Soxhlet solid–liquid reactor where the initial solid
phase (or a solid–liquid mixture) is placed in a cavity that is gradually
filled with extracting liquid phase by condensation of vapours from a
distillation flask. When the liquid reaches a preset level, a syphon aspirates
the whole contents of the cavity and unloads it back into the distillation
flask, but carrying the extracted analytes in the bulk liquid. This operation
is repeated many times until virtually complete separation is achieved and
the analytes are all in the flask. Inasmuch as the solvent acts stepwise, the
assembly can be considered a batch system; however, since the solvent
is recirculated through the sample, the system also bears a continuous
character. Hence this is regarded as a mixed operational mode (see Fig.
1.7). The most salient advantage of continuous leaching systems lies in the
fact that the sample phase is repeatedly brought into contact with fresh
portions of the solvent, thereby aiding displacement of the separation
equilibrium. One technical drawback in this context is the inability to
provide agitation, which would be of help to accelerate the process. On
the other hand, the temperature of the system is kept quite high since the
heat applied to the distillation matrix reaches the extraction cavity to some
extent. One further advantage is that no filtration is required. This procedure is widely used in a variety of applications including the determination of fats in foods and that of organic pollutants in soils and
environmental filters.

Continuous leaching systems are characterized by the fact that the
extracting phase is flowed through the solid sample, which can be heated
or subjected to ultrasound. For various reasons, the separation is occasionally performed while the flow of dissolving phase is stopped; more
often than not, though, the extracting phase is kept in motion throughout
the process. As can be seen in Fig. 1.8C, there are two general ways of
performing continuous leaching; (a) by having fresh extracting phase permanently pass through the sample; and (b) by recirculating a fixed volume
of extracting phase in a closed–open circuit furnished with suitable propulsion system and valves. The former choice has the advantage that
separation equilibria always occur under favourable conditions, even
though the phase contact time is shorter than in batch or mixed modes;
the chief shortcoming here is that the analytes are collected in rather a
large phase volume that must almost always be preconcentrated before the
determinative analytical technique in question is applied. The recirculation
assembly is intended to overcome this drawback by extracting the analytes
into a much lower final volume of extracting phase; on the other hand, the

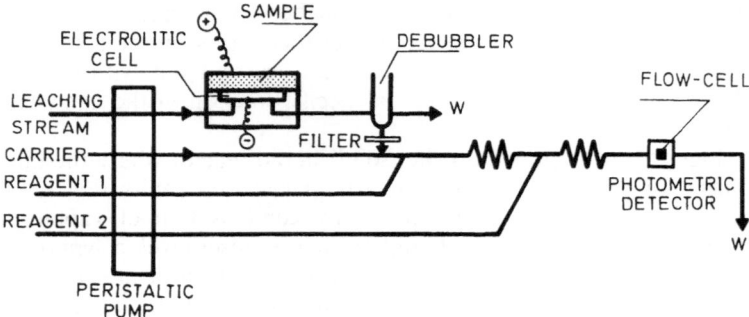

Fig. 1.9. Automatic continuous leaching of metallurgical samples by means of an electrolytic cell coupled on-line to an analyte derivatization system, and photometric detection of the reaction products. For further details, see [31,32]. (Reproduced with permission of Elsevier Science Publishers)

efficiency may suffer as a result since the separation is not effected with fresh portions of the extracting mobile phase.

Continuous solid–liquid extraction systems were scarcely used until fairly recently, when supercritical fluid chromatography emerged with previously unseen strength in analytical chemistry as a result of its major assets, which are appraised throughout this book. In this way, leaching was given a place among non-chromatographic continuous separation techniques [26]. There have been some other very promising developments, which, however, have not reached the maturity required for commercialization. Such is the case with automatic continuous unsegmented configurations, in which the sample is not liquid – and hence introduced by injection or aspiration – but solid. Leaching is effected by a solvent that is circulated through the sample under the action of some type of energy. For example, the determination of aluminium [32] and molybdenum [33] **Metallurgical** in metallurgical samples can be carried out by placing the sample in **analysis** question in a continuous electrolytic cell accommodated in a PTFE vessel where the sample acts as cathode and another electrode as anode, an acid stream being circulated through the system (Fig. 1.9). As an electric current is applied, the surface of the metal die undergoes partial oxidation/dissolution and the leaching phase drives the dissolved analytes to a derivatizing system where they are conditioned for subsequent photometric determination. A degassing unit is needed to separate the hydrogen released in the process. The determination of available boron in soils [34] and iron in vegetable material [35] can be carried out by directly introducing an amount of sample of 3–10 mg into a sonicated minireactor through which a leaching stream of hydrochloric acid is circulated (Fig. 1.10). A switching valve enables opening and closing of a circuit through which

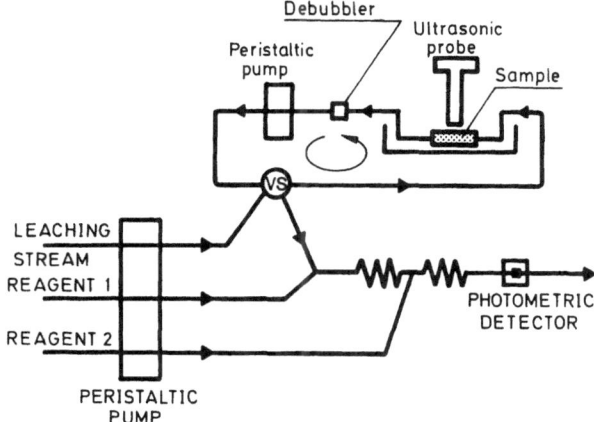

Fig. 1.10. Automatic leaching of soil and vegetable samples with recirculation and ultrasonic irradiation, coupled on-line to continuous derivatization/detection. SV: switching valve. For further details, see [33,34]. (Reproduced with permission of Elsevier Science Publishers)

200–300 µl of extracting solution is circulated with the aid of an additional peristaltic pump. In the leaching position, the valve disconnects the circuit from the remaining continuous assembly. Once leaching is finished (after a few seconds or minutes), the valve is switched to have the leaching stream sweep the circuit contents to the derivatization/detection zone. These systems have been validated by comparing their performance with that of conventional manual batch methods.

Switching valves

1.7 Ideal Features of an Analytical Leaching System

For a solid–liquid (leaching) system to function as intended in the analytical process and contribute to the quality of an analytical laboratory (both in the results it produces and the work it performs [8]) it must meet a number of requirements as strictly as possible. Such requirements are shown schematically in Fig. 1.11 and commented on only briefly below on account of their self-evidence.

Design and operational simplicity are decisively influential on analytical costs and laboratory work. Automation in general and systems for simultaneous treatment of several samples in particular aid fulfillment of these two requisites.

Simplicity

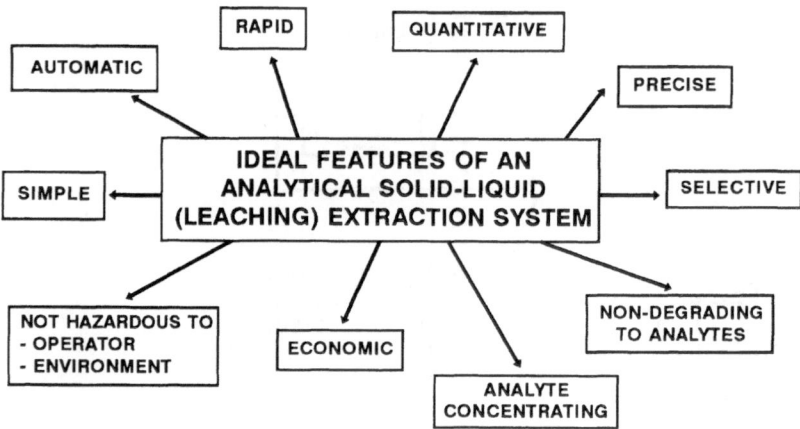

Fig. 1.11. Ideal features of an analytical leaching system

Speed No doubt, rapidity is central to current technological processes. Laboratory extractions are known to be sluggish and tedious, and to have a marked adverse effect on the overall throughput of the analytical process. It is therefore essential to minimize the extraction time, with no detriment to efficiency, for both practical and economic reasons.

Efficiency The efficiency of the extraction process should be close to 100% and reached in a fairly short time. A yield below 90% should seldom be accepted, even though this should always be dictated by the detection and quantification limit of the analytical determinative technique to be subsequently applied and the concentration of the analytes in the solid samples concerned.

Reproducibility Repeatability and reproducibility are central to solid–liquid extraction and much more difficult to achieve than in other analytical separation techniques. The extraction process is affected by a myriad of factors such as the sample moisture content and degree of dispersion, interactions between the analytes and the sample matrix, etc., which make it rather variable.

Precision Precision is one other weakness of solid–liquid extraction that calls for generous efforts in order to minimize the differences arising from treating several aliquots of the same sample separately.

Prevention of interference Leaching should isolate the analytes from the sample while preventing other, potentially interfering substances from reaching the extracting phase, whether by dissolution or, simply, by sweeping (transport). The type and composition of the extracting phase used should therefore be carefully chosen on the basis of the nature of the analytes, matrix and interferents.

Leaching should never alter the identity of the analytes so they can be determined reliably. Analytes can essentially be degraded as a result of (a) a chemical reaction with the extracting phase or some of its additives, or (b) the effect (e.g. thermal decomposition) of energy applied to the sample during the process. **Prevention of degradation**

The ideal leaching system would provide an adequate volume of extracting phase containing the analytes at concentrations allowing direct application of the intended analytical determinative technique. This is fairly easy to accomplish in the determination of micronutrients, even though these are very often quantified by using classical batch techniques, which usually involve large volumes of extracting phase (1–5 g of solid is treated with 0.5–2 l of solvent). On the other hand, such large volumes preclude direct application of an instrumental technique to trace components as the analytes end up at concentrations below detectable levels. This calls for a preconcentration operation, usually involving evaporation of the solvent (if organic) or liquid–solid extraction with a sorbent material packed in a minicolumn (a clean-up cartridge). This additional step further complicates and slows down the analytical process. **Avoidance of large volumes**

One other significant practical factor is the cost of leaching. If the process is too slow, it redounds to increased expenses in laboratory staff salaries. As a rule, the solvents used are highly pure and proportionally expensive – particularly when large volumes are needed. Unsurprisingly, a high percentage of the overall analytical costs are usually allocated to this preliminary operation. **Cost**

Finally, analytical leaching is subject to major hazards. In fact, if the process is scarcely automated, the operator may be exposed to serious safety threats. In addition to the dangers of using energy sources (thermal, electric, ultrasonic), handling large volumes of toxic and/or inflammable solvents poses serious risks to personnel directly involved in the process or simply present in the laboratory. As a side effect, such large solvent volumes can cause environmental damage if the laboratory waste is not disposed of in compliance with safety regulations, which are currently enforced in many countries. **Hazards**

1.8 Supercritical Fluids and Analytical Chemistry

Supercritical fluids are by now commonplace in many industrial areas, even though their earliest analytical applications were developed fairly recently. Figure 1.12 shows their two analytical variants: extraction (SFE) and chromatography (SFC).

Supercritical fluid chromatography [36–38] opened up brilliant prospects when it was claimed that it would solve problems unaffordable by

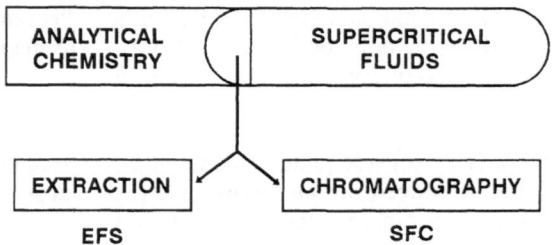

Fig. 1.12. Aspects of the analytical chemistry/supercritical fluid binomial

the other column chromatographies (HPLC, GC) by virtue of the special physical properties of supercritical fluids, amidst those of gases and liquids (see Chap. 2). However, the vast body of applications of SF gathered in over 15 years has not yet materialized in a marketing policy of manufacturers in accordance with the theoretical potential of these fluids – in fact, some SF chromatographs have been deleted from manufacturers' catalogues. Even though SFC surpasses its chromatographic counterparts in some respects (e.g. the ease with which some properties can be altered during the chromatographic process; the ability to use better detectors than those typically employed in HPLC or lower temperatures than those used in GC; the ease of coupling to mass spectrometers and Fourier transform IR spectrophotometers), it is also subject to some serious short-comings such as the limited polarity of affordable supercritical fluids and the need to use thinner capillary columns than those employed in GC, among others. Figure 1.13 shows the comparative scheme proposed by Engelhardt and Gross [39] for column chromatographies in terms of three key parameters: selectivity, efficiency and sensitivity in the detection. Gas chromatography clearly surpasses the other column chromatographic techniques in efficiency and sensitivity. On the other hand, liquid chromatography is usually more selective on account of the ease with which the mobile phase can be altered, and the wide variety of stationary phases and, also occasionally, detection techniques, available. Supercritical fluid chromatography provides good sensitivity and acceptable efficiency; its selectivity, though, is still poorly defined and requires more thorough investigation.

Supercritical fluid extraction (SFE) represents a crucial contribution to facilitating separations, though it can equally be applied to liquid or gaseous samples provided they are appropriately supported on a solid matrix. The definition of this type of extraction can be drawn from its placement among innovative alternatives to preliminary operations (Fig. 1.3) and in the different classifications of separation techniques in general (Fig. 1.5) and extraction techniques in particular (Fig. 1.6).

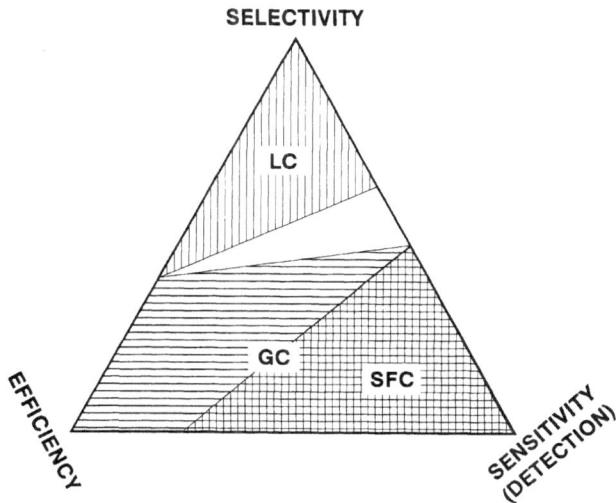

Fig. 1.13. Advantages of each column chromatographic technique [liquid (LC), gas (GC) and supercritical fluid chromatography (SFC)] in terms of the three capital parameters located at the triangle vertices. (Reproduced from [39], with permission of Elsevier Science Publishers)

The SFE technique is somewhat atypical judging by its chronological evolution. On the one hand, its chromatographic counterpart (SFC) was developed at an earlier stage, which is in clear contrast with other extraction or separation techniques whose batch mode was developed much earlier than their continuous mode (e.g. liquid–liquid extraction). On the other, the fairly extensive commercial availability of SFC is inconsistent with the relative scarcity of associated publications and developments compared to most analytical techniques. The manufacture of commercially available apparatuses and instruments, which has no doubt fostered the development of new applications and their customary use, has traditionally followed scientific and technical advances and feasibility tests. These seemingly contradictory facts are easy to explain, though: SFE aids efficient development of the first stage of the analytical process, which, as shown in Fig. 1.1, is also the one calling for the greatest innovations in the context of current analytical chemistry.

The reasons for the effervescent emergence of SFE in analytical chemistry are related to the significant enhancements it has contributed to the resolution of such a traditionally thorny problem as the analysis of solid samples. Many of the ideal properties of a leaching system (Fig. 1.11) are inherent in the SFE technique, as clearly shown in Chap. 5. In

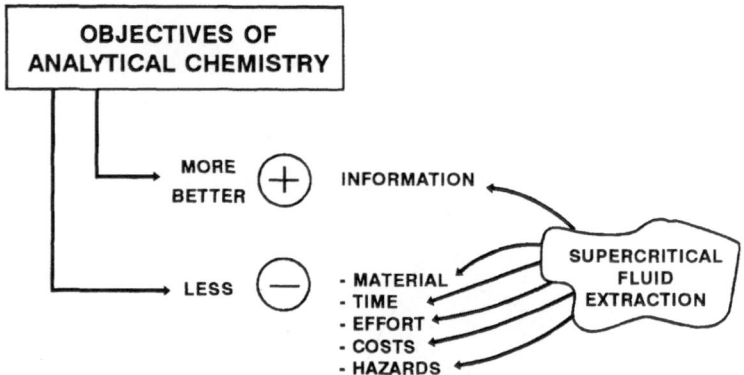

Fig. 1.14. Relationship of supercritical fluid extraction with the general objectives of current analytical chemistry

short, SFE effectively aids fulfillment of the generic analytical objectives (Fig. 1.14), particularly in terms of material, time and economic resource expenditure, and avoidance of hazards.

References

1. Valcarcel M (1990) Quím. Anal. 9:225
2. Valcarcel M (1992) Fresenius J. Anal. Chem. 343:225
3. Valcarcel M, Luque de Castro MD (1988) Automatic methods of analysis. Elsevier, Amsterdam
4. Blanco M, Cerda V (eds) (1988) Quimiometría. Publicaciones de la Universitat Autònoma de Barcelona
5. Massart DL, Vandeginste BGM, Deming SN, Michotte Y, Kaufman L (1988) Chemometrics: a textbook. Elsevier, Amsterdam
6. Valcarcel M, Luque de Castro MD, Tena M (1993) Anal. Proc. 30:276
7. Valcarcel M, Luque de Castro MD (1987) Flow injection analysis. Principles and application. Ellis Horwood, Chichester
8. Valcarcel M, Rios A (eds) (1992) La Calidad en los Laboratorios Analíticos. Reverté, Barcelona
9. Camara C (1992) Calidad en la toma y tratamiento de la muestra, Chapter II of *La Calidad en los Laboratorios Analítico*, Valcárcel M, Ríos A (eds). Reverté, Barcelona
10. Barescu GE, Dimitrscu P, Zugravescy PGh (1991) Sampling. Ellis Horwood, Chichester
11. Smith R, James GV (1981) The sampling of bulk materials. Royal Society of Chemistry, Cambridge
12. Valcarcel M, Gomez-Hens A (1988) Técnicas Analíticas de Separación. Reverté, Barcelona
13. Smyth MR (1992) Chemical analysis of complex matrices. Ellis Horwood, Chichester

14. Anderson R (1987) Sample pretreatment and separation. Anal. Chem. by Open Learning. Wiley, New York
15. Hurst WJ, Mortimer JW (1987) Laboratory robotics. VCH, New York
16. Hawk GL, Strimaitis J. Advances in laboratory automation robotics, Vol. I (1984), Vol. II (1985), Vol. III (1986), Vol. IV (1988), Vol. V (1989), Vol. VI (1990). Zymark, Hopkinton, USA
17. Valcarcel M, Luque de Castro MD (1991) J. Autom. Robot 3:199
18. Valcarcel M, Luque de Castro MD (1993) Analyst 118:593
19. Janata J (1989) Principles of chemical sensors. Plenum Press, New York
20. Nichols GD (1988) On-line process analyzers. Wiley, New York
21. Scheeline T, Coleman M (1987) Anal. Chem. 59:1185A
22. Linares P, Lazaro F, Luque de Castro MD, Valcarcel M (1988) J. Autom. Chem. 20:88
23. Kingston HM, Jassie LB (1988) Introduction to microwave sample preparation. American Chemical Society, Washington
24. Izquierdo A, Luque de Castro MD (1990) J. Autom. Chem. 12:267
25. Giddings JC (1991) Unified separation science. Wiley, New York
26. Valcarcel M, Luque de Castro MD (1991) Non-chromatographic continuous separation techniques. Royal Society of Chemistry, Cambridge
27. Valcarcel M, Luque de Castro MD (1991) Trends Anal. Chem. 10:114
28. Valcarcel M, Luque de Castro MD (1994) Flow-through (bio)chemical sensors. Elsevier, Amsterdam
29. Valcarcel M, Silva M (1984) Teoría y Práctica de la Extracción Líquido–Líquido. Alhambra, Madrid
30. Braun T, Ghersini G (1975) Extraction chromatography. Elsevier, Amsterdam
31. Mizvike A (1983) Enrichment techniques for inorganic trace analysis. Springer-Verlag, Berlin
32. Bergamin H, Krug FJ, Zagatto EAG, Arruda EC, Coutinho CA (1986) Anal. Chim. Acta 190:177
33. Bergamin G, Krug FJ, Reis BF, Nobrega JA, Mesquita M, Souza IG (1988) Anal. Chim. Acta 214:397
34. Chen D, Lazaro F, Luque de Castro MD, Valcarcel M (1989) Anal. Chim. Acta 226:221
35. Lazaro F, Luque de Castro MD, Valcarcel M (1991) Anal. Chim. Acta 242:283
36. Smith RM (1988) Supercritical fluid chromatography. Royal Society of Chemistry, Cambridge
37. Wenclawiak B (1992) Analysis with supercritical fluids: extraction and chromatography. Springer-Verlag, Heidelberg
38. Jinno K (1992) Hyphenated techniques in supercritical fluid chromatography and extraction. Elsevier, Amsterdam
39. Engelhardt H, Gross A (1991) Trends Anal. Chem. 10:64

2 Physico – Chemical Properties of Supercritical Fluids

2.1 Definition of Supercritical Fluid

A supercritical fluid (SF) is a state where matter is compressible and behaves as a gas (i.e. it fills and takes the shape of its container), which is not the case when it is in a liquid state (an incompressible fluid that occupies the bottom of its container). However, a supercritical fluid has the typical density of a liquid (between 0.1 and 1.0 g/ml) and hence its characteristic dissolving power. Finally, an SF can also be defined as a heavy gas with a controllable dissolving power or as a form of matter in which the liquid and gaseous state are indistinguishable.

Phase diagram   A typical phase diagram for a pure substance (e.g. that in Fig. 2.1) shows the temperature and pressure regions where the substance occurs as a single phase [viz. solid (s), liquid (l) or gaseous (g)]. Such regions are bounded by curves indicating the coexistence of two phases (s–g, s–l and l–g, which are involved in sublimation, melting and vaporization equilibria, respectively). The three curves intersect at the so-called **Triple point** *triple point* (TP), where the solid, liquid and gaseous phases coexist in equilibrium.

At a constant pressure, a phase transition takes place at a *transition temperature* that is a function of the pressure. Phase transformations involve enthalpy changes. Thus, in the absence of external influences, two phases can coexist indefinitely at a transition temperature for each pressure. However, only one phase will be stable at such a pressure and a temperature above or below the transition value. For example, the line that represents the pressure and temperature at which the liquid and gaseous phase coexist in equilibrium, known as the "vapour pressure curve", divides the (P,T) plane into two regions: one where the liquid is the stable phase and the other when the gas is.

Coexistence curve   The coexistence curve representing the equilibrium between two phases with a different internal symmetry (e.g. s–l or s–g) tends to infinity or eventually intercepts another coexistence curve. This is not the case with the liquid–gas equilibrium since the vapour pressure curve suddenly breaks **Critical point** at a point called the *critical point* (CP), which can thus be defined as a point in the phase diagram designated by a critical temperature (T_c) and a critical pressure (P_c) above which (a) no liquefaction will take place on raising the pressure and (b) no gas will be formed on increasing the

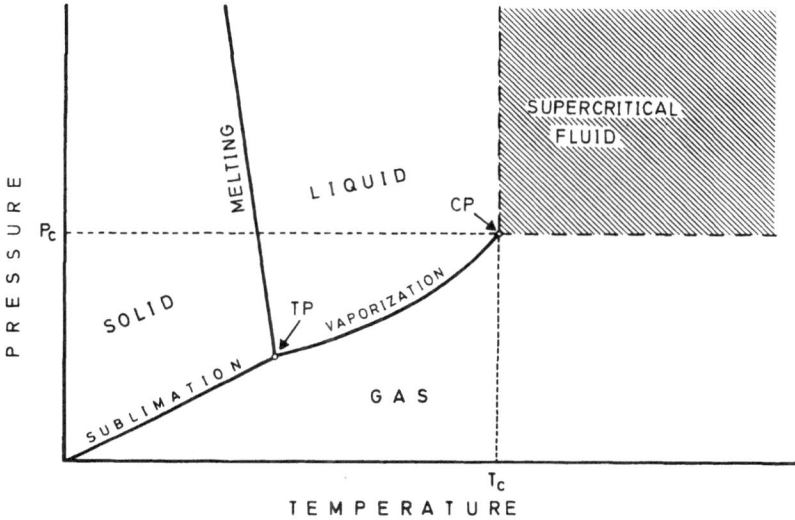

Fig. 2.1. Solid–liquid–gas–supercritical fluid phase diagram. TP = triple point. CP = critical point. P_c = critical pressure. T_c = critical temperature

temperature. This latter property allows for a new definition of supercritical fluids: one that is above its critical pressure and temperature.

Increasing the temperature also increases the pressure at which the liquid and vapour phase coexist on the vapour pressure curve. The increase in the vapour pressure is concomitant with a decrease in the difference between the density of the liquid and gaseous phases. At a given pressure and temperature, the density of the liquid and gas are identical so the two phases are indistinguishable. Above such a temperature or pressure, the liquid and the gas occur as a single phase. This region of pressures and temperatures above P_c and T_c is called the *supercritical region*.

A supercritical fluid is thus a gas which has been heated above its critical temperature and simultaneously compressed above its critical pressure. In fact, this definition is rather arbitrary, as reflected in the dashed lines by which the supercritical region is bounded. Some properties of a liquid (e.g. density, viscosity) subjected to a pressure above its P_c change gradually as the temperature is raised (a horizontal movement in Fig. 2.1, which leads to the formation of a critical fluid); likewise, a gas which has been heated above its T_c becomes a supercritical fluid on gradually increasing the pressure (a vertical movement in Fig. 2.1). As can be seen in Fig. 2.2, there are two different ways of reaching a given final supercritical state. One involves starting at point A in the liquid region, increasing the pressure above P_c and then raising the temperature above T_c until the

Supercritical region

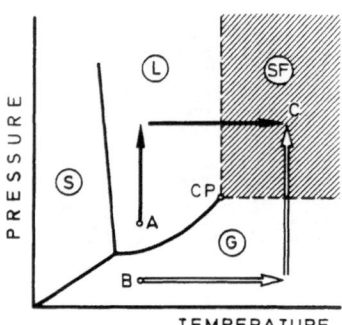

Fig. 2.2. Technical ways of reaching a supercritical state (point C) from a liquid (point A) and a gas (point B). CP = critical point

state denoted by C is reached. The other way entails starting in the vapour region (point B in Fig. 2.2), heating the substance above its T_c and then raising the pressure above its critical value.

The critical point is characteristic for each substance. Table 2.1 lists the critical pressure and temperature for various solvents classified according to their chemical nature, as well as the fluid density at the critical point, **Critical density** which is called the *critical density* (ρ_c). Figure 2.3 shows the situation of several solvents on a three-dimensional graph based on their critical parameters, and Fig. 2.4 shows their critical pressure–temperature (A), critical density–temperature (B) and critical density–pressure projections (C). These two figures allow some interesting conclusions to be drawn, namely:

a) Many solvents lie in a specific region of the three-dimensional diagram. Notable exceptions include hydrogen (no. 8), helium (no. 7), xenon (no. 9), $CClF_3$ (no. 32) and water (no. 5), as confirmed by the two-dimensional diagrams.

b) The critical values for inorganic solvents increase from hydrogen (no. 8) and helium (no. 7) to xenon (no. 9) and water (no. 5).

c) The critical density of hydrocarbons is very similar, as is their critical pressure. On the other hand, their critical temperature increases markedly with increase in the number of carbon atoms.

d) Fluorocarbon solvents have a similar critical density but a highly variable critical pressure and a generally high critical temperature – that of CHF_3 and $CClF_3$ excluded.

e) Oxygen-containing organic solvents lie in the same region of the three-dimensional diagram.

f) All aromatic solvents have a high critical temperature and a similar critical density.

Reduced value   The properties of supercritical fluids are frequently expressed in terms of reduced rather than absolute values. A reduced value is defined as the ratio of the actual absolute value to the critical point value. Thus, the

Table 2.1. Features of various solvents at the critical point

Solvents	Critical temperature (°C)	Critical pressure (bar)	Critical density (g/ml)
Inorganic			
1 CO_2	31.1	72	0.47
2 N_2O	36.5	70.6	0.45
3 NO_2	158	98.7	0.27
4 Ammonia	132.5	109.8	0.23
5 Water	374.2	214.8	0.32
6 Sulphur hexafluoride	45.5	38.0	
7 Helium	−268	2.2	0.07
8 Hydrogen	−240	12.6	0.03
9 Xenon	17	56.9	1.11
10 Hydrogen chloride	51	83.3	0.45
11 Sulphur dioxide	157	76.8	0.52
Hydrocarbons			
12 Methane	−82	46.0	0.169
13 Ethane	32.3	47.6	0.2
14 Propane	96.7	42.4	0.22
15 n-Butane	152	70.6	0.228
16 n-Pentane	196	32.9	0.23
17 n-Hexane	234.2	28.9	0.23
18 2,3-Dimethylbutane	226.8	42.4	0.241
19 Ethylene	11	50.5	0.2
20 Propylene	92	45.4	0.22
21 Benzene	288.9	98.7	0.302
22 Toluene	319	41.1	0.292
Alcohols			
23 Methanol	239	78.9	0.27
24 Ethanol	243.4	72	0.276
25 Isopropyl alcohol	235.3	47.6	0.273
Ethers			
26 Diethyl ether	193.6	63.8	0.267
27 Ethyl methyl ether	164.7	47.6	0.272
28 Tetrahydrofuran	267	50.5	0.32
Halides			
29 Trifluoromethane	26	46.9	0.52
30 Dichlorodifluoromethane	111.7	109.8	0.558
33 Dichlorofluoromethane	178.5	32.9	0.522
32 Chlorotrifluoromethane	28.8	214.8	0.58
33 Trichlorofluoromethane	196.6	28.9	0.554
34 1,2-Dichlorotetrafluoroethane	146.1	78.9	0.582
Miscellaneous			
35 Acetone	235	47.0	0.279
36 Acetonitrile	275	47	0.25
37 Pyridine	347	56.3	0.312

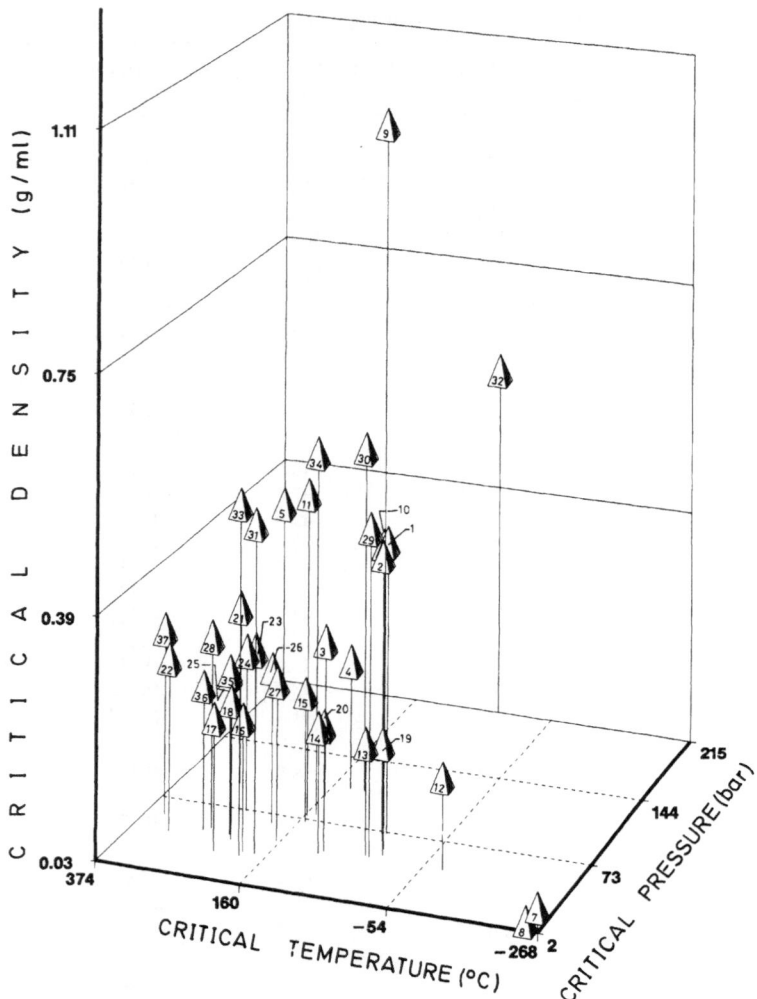

Fig. 2.3. Three-dimensional graph of the critical parameters (density, pressure and temperature) for various substances. Numbers are matched to those used in Table 2.1

reduced pressure (P_r), temperature (T_r) and density (ρ_r) are given by the following expressions:

$$P_r = \frac{P}{P_c}, \quad T_r = \frac{T}{T_c}, \quad \rho_r = \frac{\rho}{\rho_c}$$

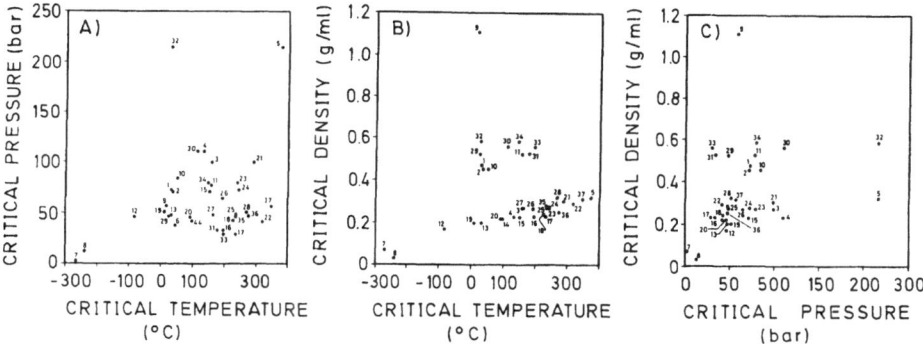

Fig. 2.4A–C. Two-dimensional graphs of the critical parameters for various substances. **A** Pressure–temperature. **B** Density–temperature. **C** Density–pressure. Numbers are matched to those used in Table 2.1

If both P_r and T_r are greater than unity, then the substance in question is in its supercritical state.

2.2   Physical Properties of Supercritical Fluids

Most properties of substances vary widely in the vicinity of the critical point in the phase diagram. Such a variability should be taken into account in studying the behaviour of supercritical fluids and can be exploited for some applications. Often, sharp variations are only observed in the very close vicinity of the critical point but are of no consequence within the supercritical region.

The values of such significant properties as density, diffusivity, viscosity, etc., in the supercritical region lie between those of liquids and gases, which can be used to exploit the advantages of either in some applications (e.g. the separation of substances from solid media).

A distinction will thus henceforward be made between fluid properties at or near the critical point and in the supercritical region.

2.2.1   Properties at or near the Critical Point

Systems near their critical point exhibit some very special properties. In addition, the occurrence of a critical point in the gas–liquid phase transition clearly suggests the absence of essential differences between the liquid and gaseous state. In fact, the fluid can reach the liquid state from

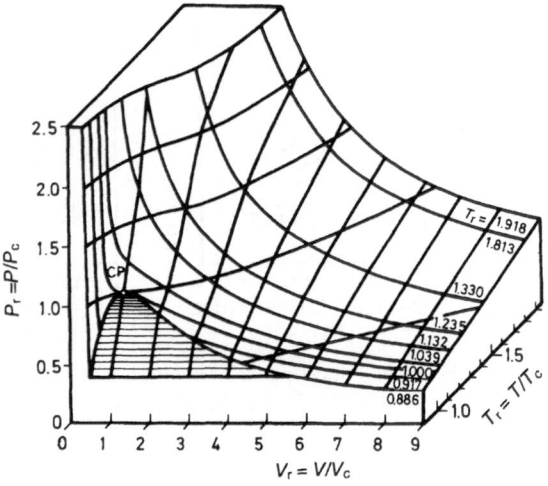

Fig. 2.5. Three-dimensional reduced pressure (P_r)–reduced volume (V_r)–reduced temperature (T_r) graph for argon $(P_c = 48.0 \, atm; \; V_c = 1.884 \, cm^3/g; \; T_c = 150.7 \, K)$. CP = critical point. (Reproduced from [1] with permission of John Wiley)

the gaseous state around the critical point without a break. This observation is based on the classification of the different types of transitions for which a critical point can be expected. Note that a critical point signals the end of a line joining the points where two phases coexist. With a gas and a liquid, the only quantitative differences are those between the properties of the two phases. Such quantitative differences exist because, in addition

Molecular environment

Phase convergence

to the density difference, there is one between the averaged molecular environment of the gas and that of the liquid. Thus, the similarity between the two phases increases as their densities approach each other. If the two densities are identical, both phases converge on a single one with no change in the local symmetry. In the absence of a symmetry difference between the two phases, their coexistence line has a critical point.

If the pressure and temperature are increased simultaneously along the gas–liquid coexistence line, a point defined by P_c and T_c is eventually reached where the gas is so highly compressed and the liquid expanded that $\rho_{liq} = \rho_{gas}$; in the absence of another difference, the two phases are indistinguishable.

Variation of properties

Pressure isotherms

Let us concentrate on the gas–liquid critical point and the variation of some properties in its vicinity by first defining the prevailing conditions at the critical point for a fluid. Figure 2.5 shows part of the (P, V, T) surface for argon by means of constant P, V and T lines. Careful observation of the pressure isotherms as a function of the volume reveals that one of them is horizontal at a point in the (P, V) plane where it changes from a positive to a negative curvature (an inflection point). The first and second derivative of a curve at an inflection point are both zero. Since these are **Inflection point** $P-V$ isotherms, the first and second partial derivative of the pressure with respect to volume at a constant temperature T_c are zero at P_c and V_c, coinciding with the horizontal tangent and inflection point of the curve at the CP, respectively:

$$\left(\frac{\partial P}{\partial V}\right)_{T=T_c} = 0 \tag{1}$$

$$\left(\frac{\partial^2 P}{\partial V^2}\right)_{T=T_c} = 0 \tag{2}$$

These two properties define the critical point of a fluid.

Equations (1) and (2) have three major implications, namely:

a) The isothermal compressibility becomes infinitely positive at the critical **Isothermal** point. Since $(\partial V/\partial P)_T < 0$ for any substance in equilibrium, **compressibility**

$$\lim_{\substack{T \to T_c \\ V \to V_c}} \left[-\frac{1}{V}\left(\frac{\partial V}{\partial P}\right)_T \right] = \infty$$

b) Based on Eq. (1) and other thermodynamic relations, the thermal **Thermal** expansion coefficient at the critical point can be shown to be positive and **expansion** infinite. **coefficient**

c) The enthalpy of vaporization at the critical point is zero. If the **Enthalpy of** Clapeyron equation is written as **vaporization**

$$\Delta H_{vap} = T(v_G - v_L)\left(\frac{dP}{dT}\right)_\sigma,$$

at the critical point, where $T = T_c$, $v_G = v_L$ and $(dP/dT)_\sigma$ is finite, one has

$$\lim_{T \to T_c} \Delta H_{vap} = 0$$

where σ denotes conditions along the liquid–vapour coexistence curve, ΔH_{vap} the enthalpy of vaporization, and v_G and v_L the molar volume of the gas and liquid, respectively.

The fact that many thermodynamic properties of a system become infinitely large or zero at the critical point provides strong enough evidence that the region around the critical point of a fluid possesses some very special properties. The relationship between the behaviour of fluids in the critical region and the underlying molecular interactions has not yet been clearly established.

The critical point is also associated with a great thermal anomaly. In fact, at a constant density equal to ρ_c, the constant-volume heat capacity

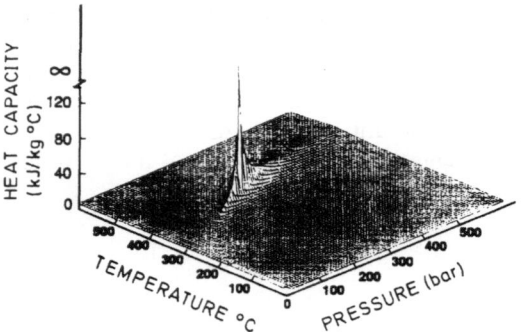

Fig. 2.6. Variation of the heat capacity of water with pressure and temperature. (Reproduced with permission of the American Chemical Society)

Heat capacity

at $v = v_c$ tends to infinity $[c_v(T) \to \infty]$ as $T \to T_c$, both above and below the critical temperature. The surface depicted in Fig. 2.6 shows the variation of the heat capacity of water as a function of temperature and pressure. As can be seen, the heat capacity tends to infinity at the critical point (374 °C and 221 bar) [2]. Even 25 °C above T_c at a pressure of 300 bar, the heat capacity is still higher than its asymptotic value of $4 \, \mathrm{J g^{-1} \, K^{-1}}$ by at least one order of magnitude at very high and very low pressures. The heat capacity is one of the key properties that vary over wide ranges of temperature and pressure around the critical point.

2.2.2 Properties of the Supercritical Region

Figure 2.7 summarizes the basic properties of supercritical fluids. First, the density of a supercritical fluid depends on the pressure and temperature to which it is subjected, even though it is always close to the typical values for liquids (see Table 2.2). This is the origin of the good dissolving properties of supercritical fluids, where interactions between the fluid and solute molecules are quite strong.

Supercritical viscosity

Supercritical viscosity values lie between those of liquids and gases, which endows SFs with more favourable hydrodynamic properties than those of liquids. On the other hand, their very low surface tension allows them to penetrate readily porous solids and packed beds. On constant column dimensions, the pressure drop along a supercritical fluid chromatographic (SFC) column is typically ten times smaller than it is in liquid chromatography (LC), but also, obviously, ten times greater than in gas chromatography (GC) [3].

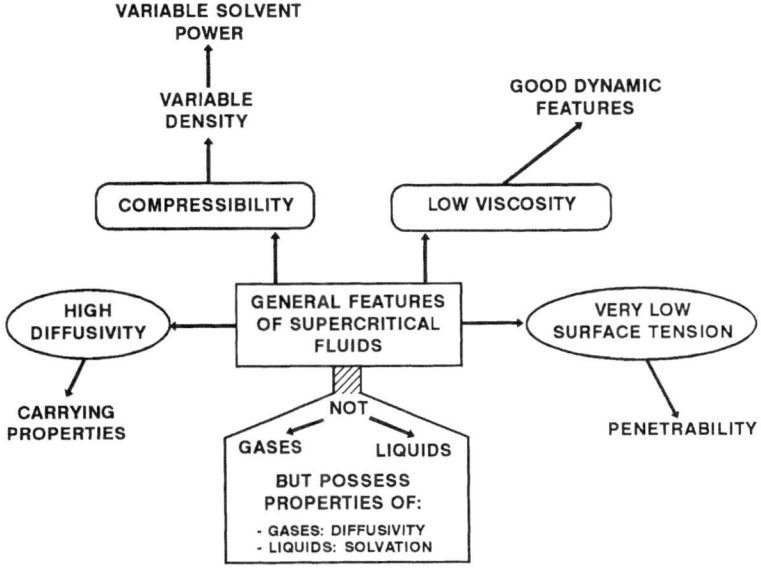

Fig. 2.7. Most salient features of supercritical fluids

Table 2.2. Comparison of the properties of supercritical CO_2 and those of ordinary gases and liquids

		Density (g/cm^3)	Viscosity (g/cm s)	Diffusion coefficient (cm^2/s)
Gases		$(0.1-2)\ 10^{-3}$	$(1-3)\ 10^{-4}$	$0.1-0.4$
Supercritical CO_2	T_c, P_c	0.47	3×10^{-4}	7×10^{-4}
	$T_c, 6P_c$	1.0	1×10^{-3}	2×10^{-4}
Liquids		$0.6-1.6$	$(0.2-3)\ 10^{-2}$	$(0.2-2)\ 10^{-5}$

Diffusion coefficients

The diffusion coefficients of solutes in supercritical fluids are between those they display in liquids and gases. Because diffusion coefficients in supercritical fluids are higher than those in liquids, mass transfer is usually more favourable in SFs.

Two other properties of SFs, viz. a high diffusivity and a low viscosity, are quite significant for leaching purposes as they lead to more expeditious and efficient extractions (see scheme in Fig. 2.8). Supercritical fluids are occasionally regarded as "supersolvents", which is not factual judging by their dissolving power relative to that of liquids. In fact, the dissolving power of a supercritical fluid only approaches that of a liquid solvent at a

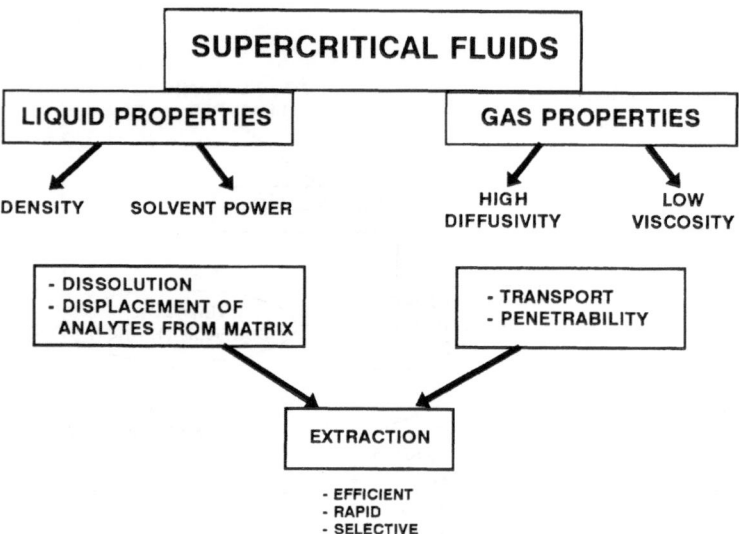

Fig. 2.8. Relationship between the basic properties of supercritical fluids and the features of SFE

high enough density; also, the maximum solubility in most liquids surpasses that in SFs. Even though supercritical fluids offer no advantage over liquid solvents in terms of dissolving power, some other properties of SFs make supercritical fluid extraction the ideal alternative to analytical leaching. In fact, the rate at which extraction can be accomplished is determined by mass transfer constraints. Because solute diffusivities in supercritical fluids are typically higher than those of liquid solvents by one order of magnitude (10^{-4} vs 10^{-5} cm^2/s) and their viscosity is lower by one order of magnitude (10^{-4} vs 10^{-3} N s/m^2), their mass transfer properties are much more favourable than those of liquids. As a result, SFE is much faster than liquid extraction. Thus, a quantitative SF extraction can be finished in 10–60 min, whereas a liquid extraction usually takes from a few hours to several days.

Mass transfer properties

The following sections discuss the influence of pressure and temperature on several essential properties of supercritical fluids, namely: density, diffusivity, viscosity and dielectric constant.

2.2.2.1 Density

The density of a supercritical fluid is markedly dependent on its pressure and temperature. The variation of the density of an SF with the pressure

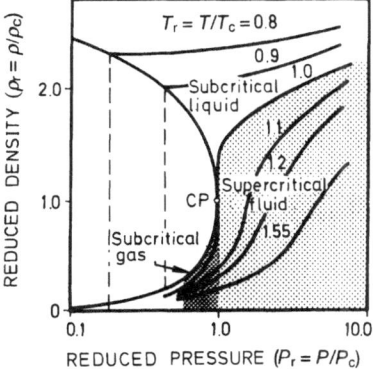

Fig. 2.9. Two-dimensional reduced density–reduced pressure graph including several reduced temperature isotherms. CP = critical point. (Reproduced with permission of Butterworth)

at a constant temperature (an isotherm) is typically non-linear. Figure 2.9 depicts a phase diagram showing both the supercritical (shaded area) and subcritical region [4]. The variables in the diagram are all reduced variables given as the ratios between their actual values and those at the critical point. The critical point results from a triple intersection where the three reduced variables are unity.

Triple intersection

Density in the supercritical region increases sharply with increasing pressure at a constant temperature; also, it decreases with increasing temperature at a constant pressure.

It is worth emphasizing the steep slopes of the curves in the vicinity of the critical point. A small pressure rise results in a sharp increase in the solvent density. Also, the slopes of the curves decrease sharply with increasing distance from the critical point. The zone immediately above the critical point provides the greatest density changes and is thus the most effective for changing this variable by introducing minimal temperature and/or pressure variations.

The darker shaded area at the bottom of the diagram, just left of the supercritical region is also useful for extraction purposes. The slopes of the curves in this region are not so steep as in the supercritical region, yet the density varies significantly enough with the pressure for useful purposes.

The dissolving power of a given fluid depends on its density. Even though the density of a fluid is known to be highly influential on solute–solvent interactions in the bulk of a supercritical solvent, fairly little is known about the nature of the interactions themselves or how they change with density [5]. The dissolving power of an ordinary organic liquid solvent is essentially independent of its pressure since the liquid density is not affected by this variable. On the other hand, the density of a supercritical fluid can be altered over a wide range by changing the pressure, the temperature or both. The close relationship between solubility and density

Solute–solvent interaction

is discussed in detail in Chap. 3. At a given, constant temperature, dissolution of non-polar and scarcely polar analytes will be favoured by low pressures, whereas that of polar and high-molecular weight solutes will be easier at high pressures. This allows the sequential extraction of groups or families of similar compounds, which is dealt with in detail in Chap. 5.

The density of a supercritical fluid at a given pressure and temperature can be of interest in designing applications where the dissolving power is bound to play a central role. There are few available experimental density **Pure fluids** values, whether for pure CO_2, N_2O, $CHClF_2$ or for binary fluids (e.g. **Binary fluids** $CO_2/MeOH$), which encompass relatively narrow ranges of operational pressures and temperatures.

The ideal–gas law can be used to relate the pressure and density of a supercritical fluid at a given temperature [6,7]. Thus, the density of an SF under given pressure and temperature conditions can be obtained from the following expression:

$$PV = zRT, \quad \rho = \frac{M}{V}, \quad \rho = \frac{MP}{zRT},$$

where V is the molar volume, z the compressibility factor, R the gas constant, ρ the density and M the molecular weight. The compressibility factor is given by

$$z = z^{(0)} + \omega z^{(1)}$$

where ω is the acentric factor ($\omega = -\log P_r - 1.0$), and $z^{(0)}$ and $z^{(1)}$ can be obtained from Pizter's tables as a function of P_r and T_r. After the com- **Compres-** pressibility factor, z, has been calculated, ρ can readily be determined as a **sibility factor** function of P and T.

A straightforward, expeditious, reliable procedure was recently developed for determining the density of pure and modified supercritical fluids based on the weight of fluid held in an extraction cell of known volume at various pressures and temperatures [8].

2.2.2.2 Diffusivity

Figure 2.10 illustrates the diffusivity in supercritical CO_2 [4]. The graph is a plot of diffusivity against temperature at different pressure values. It also shows the typical diffusivity range for ordinary liquids as a band running along the temperature axis. The graph allows three immediate, interesting conclusions to be drawn, namely:

(a) The diffusivity of a solute in a supercritical fluid always exceeds that in an ordinary liquid solvent.

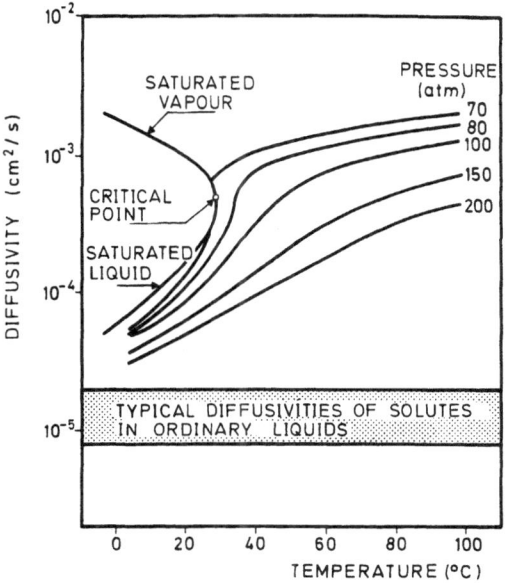

Fig. 2.10. Variation of the diffusivity in CO_2 as a function of temperature at several pressures. (Reproduced with permission of Butterworth)

(b) Diffusivity in an SF decreases with increase in the pressure.

(c) Diffusivity increases with increasing temperature, especially in the vicinity of the critical point. The slopes of diffusivity–temperature isobars increase with decreasing temperature and pressure.

As noted earlier, the significance of this property lies in the fact that the higher the diffusion coefficient of a solute in a fluid is, the faster is mass transfer. Therefore, supercritical fluids excel over ordinary liquids in diffusivity and hence in leaching and chromatographic separation performance.

2.2.2.3 Viscosity

The viscosity of a fluid (be it is a gas, liquid or supercritical fluid) is temperature-dependent. On the other hand, while pressure has little effect on the viscosity of liquids, it markedly influences that of supercritical fluids.

Figure 2.11 shows the variation of the CO_2 viscosity with pressure at several temperatures [4]. At a given constant temperature, viscosity increases with pressure. Also, the change decreases with increasing pressure, as can be seen in the figure despite the non-linear scale of the graph.

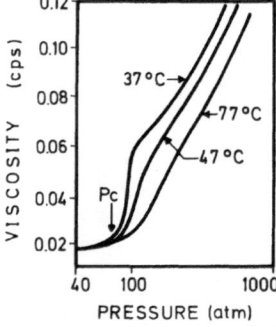

Fig. 2.11. Variation of the viscosity of CO_2 with pressure at three different temperatures. P_c = critical pressure. (Reproduced with permission of Butterworth)

It is worth noting the abrupt jump near the critical pressure, particularly at low temperatures close to T_c.

Increased pressures also result in increased SF viscosity and hence in diminished solute diffusivity and transport phenomena, but also – most often – in increased solubility through decreased density.

2.2.2.4 Dielectric Constant

The dielectric constant is one of the most relevant physico-chemical properties for defining the solubility in fluids.

Water has a relatively high dielectric constant (78.5 at 25 °C and 1 atm). As a result, it effectively masks ionic charges at room temperature, thereby facilitating dissolution of ionic compounds. The dielectric constant of water decreases with increasing temperature and decreasing pressure; thus, it is approximately 12 at 1000 °C and a density of around 1 g/ml; approximately 90 at 0 °C and the same density; and around 6 at the critical point [2].

Ion-pairs   Because of its low dielectric constant at high temperatures, water shields the electrostatic potential between ions weakly, so dissolved ions can freely form ion-pairs. Under these conditions, supercritical water behaves as a non-polar rather than a polar solvent. These properties partly account for its ability to dissolve non-polar organic compounds.

As can be seen from Fig. 2.12, the dielectric constant of CO_2 increases with pressure as does its density [9].

The dielectric constant of CO_2 in a very dense state (200 bar and 40 °C) is ca. 1.5, so it can be assimilated to a highly non-polar solvent which is appropriate for dissolving nonpolar substances [10].

By way of summary, the scheme in Fig. 2.13 shows the basic properties of fluids in the supercritical region, classified according to whether they

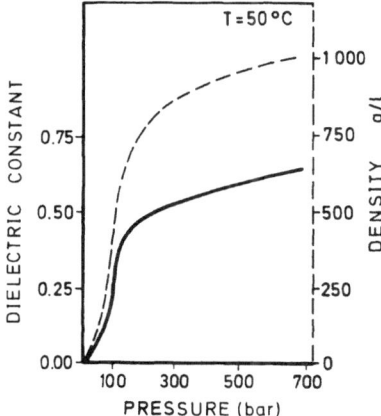

Fig. 2.12. Influence of pressure on the dielectric constant and density of SC CO_2 at a constant temperature. (Reproduced with permission of the American Chemical Society)

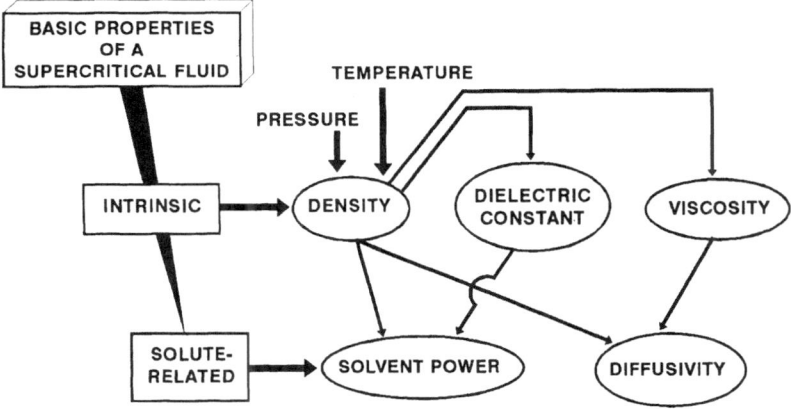

Fig. 2.13. Relationships between the basic (intrinsic and solute-related) properties of a supercritical fluid

are inherent in the SF itself (intrinsic) or related to the dissolved substances (solutes). The key physico-chemical property of a supercritical fluid is no doubt its density, which is determined by the experimental pressure $(P > P_c)$ and temperature conditions $(T > T_c)$. This property is directly influential on the dielectric constant and viscosity. On the other hand, the intrinsic properties of SFs have a decisive effect on their dissolving power and transport phenomena that take place in the bulk fluid (through diffusivity).

2.3 Binary Systems

Phase equilibria

Understanding phase equilibria involving more than one component is central to correct development of SF processes. It would be highly desirable to be able to predict phase equilibria from the properties of the pure components involved – obtaining reliable data from high-pressure equilibria is rather difficult, costly and time-consuming. It may be useful in this respect to extrapolate the few available experimental data in the meantime. The high complexity of high-pressure mixtures endows quantitative equilibrium models with a purely basic interest.

Even though processes involving mixtures are usually very complex, it is highly enlightening to study phase diagrams of binary mixtures. The ex-

Phase rule

pression of the phase rule for a binary mixture (two components, $c = 2$) is

$$l = 4 - f$$

where f is the number of phases and l that of degrees of freedom (i.e. the number of independent variables that must be specified to describe properly the thermodynamic state of the mixture).

Three-dimensional graph

An exhaustive description of a binary system calls for a three-dimensional graph, the most frequently used of which is the pressure/temperature/mole fraction (PTx) diagram. Figure 2.14 shows one such diagram [11] in which the samples present in a single phase are geometrically represented by volumes and the coexistence of two phases by the surface bound by the two vapour pressure lines of each pure component and that joining the two critical points (CP_1 and CP_2), which is far from an immovable "border". For a given composition of the mixture x_a, the outline of the cross-section provides some information on it. At a constant temperature T_a lower than both critical temperatures (T_{c1} and T_{c2}), the cross-section outline is a typical pressure–composition diagram

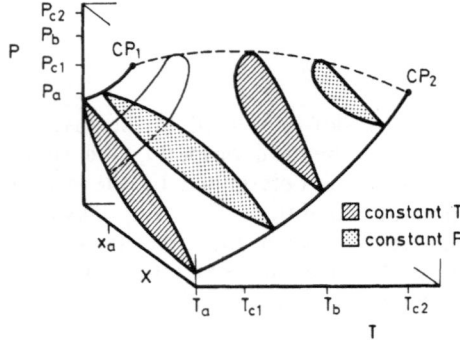

Fig. 2.14. Three-dimensional pressure (P)–temperature (T)–mole fraction (x) graph for a two-component system. For details, see text. (Reproduced with permission of CRC)

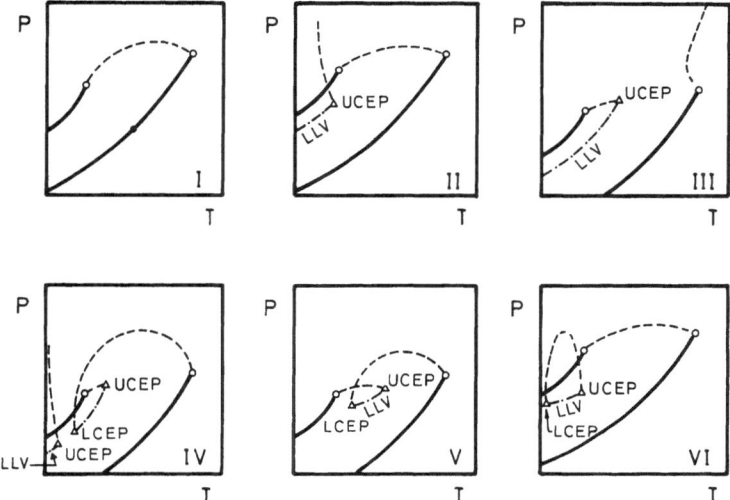

Fig. 2.15. Pressure (P)–temperature (T) projections of three-dimensional graphs also including the mole fraction. For details see text. UCEP = upper critical end point. LCEP = lower critical end point. LLV = liquid–liquid–vapour equilibrium curve. (Reproduced with permission of CRC)

for a binary mixture. A similar profile is also obtained at a constant pressure P_a lower than both critical pressures $(P_{c1}$ and $P_{c2})$. The cross-sections obtained at pressures (P_b) and temperatures (T_b) intermediate between the two critical pressures $(P_{c1}$ and $P_{c2})$ and temperatures $(T_{c1}$ and $T_{c2})$, respectively, are different from the previous ones.

These three-dimensional diagrams can be simplified by projecting on the *PT* plane the surface obtained from the different mixture compositions. The *PT* diagrams in Fig. 2.15 show the vapour pressure lines for the pure components, the two critical points $(CP_1$ and $CP_2)$, the location of the critical points for the binary mixture (dashed line) and the lines reflecting the coexistence of three phases (dashed–dotted line). The intercepts of the three-phase lines with the curves for the critical mixture make the so-called "upper" and "lower critical end point" (UCEP and LCEP, respectively). At a critical end point, two phases merge critically into a single one in the presence of another. Notwithstanding their high variability, van Konynenburg and Scott [12] have classified *PT* projections into six basic types as shown in Fig. 2.15.

Type I equilibria are distinguished by liquid–liquid miscibility under every possible condition, as well as by an unbroken gas–liquid critical curve between the critical points of the pure components. As can be seen

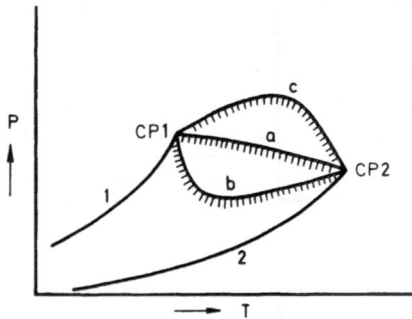

Fig. 2.16. Phase diagram (pressure–temperature projection) for a system of two completely miscible components 1 and 2 (a type I system). CP_1 and CP_2 = critical point of component 1 and 2, respectively. a, b and c = critical curves. (Reproduced with permission of Elsevier)

in Fig. 2.16, the critical curve can join the critical points of the pure components (CP_1 and CP_2) directly (curve a) and via a minimum (curve b) or maximum (curve c) [3]. Under each curve, the system is heterogenous (liquid + vapour). At a pressure and temperature above each critical curve, the two components are completely miscible throughout the composition range and make up a binary supercritical fluid. It is interesting to note that the mixtures represented by curves b and c can give rise to seemingly contradictory situations. Thus, in the case represented by curve c, the mixture can be heterogeneous rather than a homogeneous supercritical fluid at a pressure above both critical pressures; in the case represented by curve b, the opposite seemingly holds true: a homogeneous binary supercritical state is reached at a pressure below the two critical pressures. These facts show that the individual critical conditions of mixture components cannot be directly extrapolated to the mixture in question, particularly in the vicinity of the two critical points.

In type II equilibria (Fig. 2.15), the gas–liquid critical curve is also continuous, but reflects a liquid–liquid phase separation at low temperatures. There is a liquid–liquid–vapour (LLV) equilibrium curve that intercepts the liquid–liquid critical curve at an UCEP, where both liquid phases become identical in the presence of a vapour phase.

Two-segment curve   Type III equilibria (Fig. 2.15) are represented by a two-segment curve. The lower temperature segment, which starts at the critical point of the more volatile component, intercepts a three-phase (LLV) line at an UCEP. The other segment of the critical curve starts at the critical point of the less volatile component and reaches very high pressures much more sharply. This curve can thus take a variety of shapes and include a pressure maximum (or minimum) and/or a temperature minimum.

The critical curve of type IV equilibria (Fig. 2.15) also has two segments, which, however, intercept a three-phase line at critical end points. Like type II equilibria, they include a liquid–liquid immiscibility region at low temperatures.

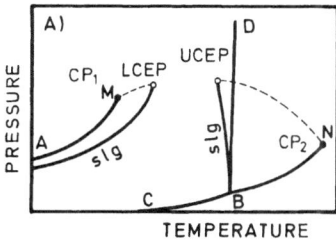 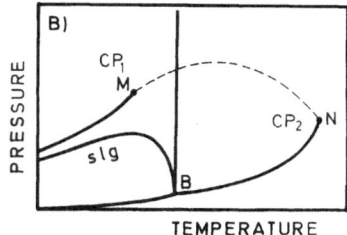

Fig. 2.17A,B. Phase diagram for a mixture consisting of a light component (1) and a heavy non-volatile solid (2). (**A**) Immiscible and (**B**) miscible molten solid and supercritical fluid. For details see text

Type V equilibria (Fig. 2.15) are similar to type IV equilibria except for the absence of an immiscibility region at temperatures below that of the LCEP. Finally, type VI equilibria (Fig. 2.15), the only ones which do not conform to van der Waals state equations, are characterized by an unbroken gas–liquid critical curve and a liquid–liquid critical curve that intercepts an LLV curve at both UCEPs and LCEPs. This type of equili- **Hydrogen** brium typically occurs in systems involving hydrogen bonding. **bonding**

Schneider [13] has constructed experimental phase diagrams for a host of systems; also, he has shown that the boundaries between the different types of equilibria are all ill-defined and continuously involved in transitions between one another. For example, on examination of binary diagrams for CO_2 and the *n*-alkane series it is seen that the CO_2–ethane system is of type I, while the CO_2–octane system if of type II and the CO_2–hexadecane system is of type III.

The behaviour of a binary mixture consisting of a heavy non-volatile solid (component 2) and a light supercritical fluid (component 1) such that the melting point of the heavy solid T_{m2} is higher than the critical temperature of the light component T_{c1}, and the molecular size, form, structure and critical conditions of the two components are considerably different, is very interesting [14]. Figure 2.17A shows a basic *PT* diagram for this type of binary mixture including the vapour pressure curves for the pure components (AM and BN) and the melting (BD) and sublimation curve (CB) for component 2. The critical points for the pure components are denoted by CP_1 and CP_2. The critical mixing curve for this type of system, which represents the critical conditions for mixtures of different compositions, has two branches. One starts at the critical point of the heavy component (CP_2) and intercepts the freezing point depression curve (s–l–g three-phase curve) at an upper critical end point (UCEP). The other starts at the critical point of the light component (CP_1) and intercepts the s–l–g curve at a lower critical end point (LCEP). The depression of the freezing point of the solid arises from the solubility of the light component in the heavy liquid. If such a solubility is high enough, the s–l–g curve starts at the melting point of the heavy component and extends undisruptedly to the lower temperatures. Also, the critical mixing curve joins the critical points of both components without a break (Fig. 2.17B) [15]. If, on the other hand, component 1 is only slightly soluble in the heavy liquid phase, then the depression of the melting point of component 2 is quite small. The s–l–g curve in Fig. 2.17A rises steeply and intercepts the critical mixing curve at both the LCEP and the UCEP. At these critical end points, the liquid and the gaseous phase in the s–l–g line converge on a single fluid phase in the presence of excess solid. The extraction of solids with solvents in a supercritical state may occur in the gas–solid region bound by the two branches of the s–l–g curve.

Even though *PT* projections are useful for classifying phase diagrams, complete understanding of the behaviour of a binary system requires studying three-dimensional *PTx* diagrams, some of which have been discussed at length by Bruno and Ely [11].

Computer application An accurate knowledge of the position of the critical point for a given mixture composition is of great significance with a view to developing procedures involving binary mixtures. Some computer software [16] allows one to calculate the critical temperature and pressure of an ordinary supercritical fluid (e.g. CO_2) modified by an organic solvent. The computations involved are based on the modified Handinson–Brobst–Thomson equation [17]. The critical parameters for a binary mixture, P_{cb} and T_{cb}, are given by the following equations:

$$P_{cb} = \frac{(0.291 - 0.08\,\omega_b)RT_{cb}}{V_b}$$

$$T_{cb} = x_1^2 V_1 T_{c1} + 2x_1 x_2 \sqrt{V_1 V_2 T_{c1} T_{c2}} + x_2^2 V_2 T_{c2}$$

where x denotes the mole fraction, V the characteristic volume, ω the acentric factor, and subscripts 1 and 2 the mixture components, ω_b being obtained from the following expression:

$$\omega_b = x_1 \omega_1 + x_2 \omega_2.$$

V_b is calculated from M_b, which in turn is given by

$$M_b = x_1 M_1 + x_2 M_2$$

where M_1 and M_2 denote the molecular weights of the two components.

2.4 Polarity

Polarity is one of the most influential properties on solubility and also one **Extraction**
that can be altered in order to modify the selectivity of an extraction **selectivity**
process.

A given molecule is polar when the centre of its negative charge does
not coincide with that of its positive charge. One such molecule acts as a
dipole, viz. two charges of the same magnitude and opposite sign that are
separated in space. Such molecules as those of CO_2, which have a zero
dipole moment, are said to be non-polar.

Table 2.3 lists the dipole moments of various substances used as supercritical fluids. According to this parameter, SFs can be classified into
polar, scarcely polar and non-polar.

Non-polar and scarcely polar solvents with moderate critical temperatures (e.g. N_2O, CO_2, ethane, propane, pentane, xenon, SF_6 and some
freons) have a limited dissolving power for solutes of a highly polarity or
molecular weight. On the other hand, these solutes are highly soluble in
polar fluids such as NH_3, use of which may be limited by other factors
such as a high toxicity or reactivity.

The effect of pressure on the dissolving power of several supercritical
fluids has been investigated spectroscopically (e.g. with the solvatochromic
method). Comparisons in this respect rely on the π^* polarity and polarizability scale developed by Taft et al. [18].

2.4.1 The π* Polarizability/Polarity Scale

The dissolving power of liquid solvents can be quantified on the basis of a
number of solvent scales. One of them, the π^* scale, was developed by

Table 2.3. Permanent dipole moment of some supercritical fluids

Fluid	Dipole moment (Debye)
CO_2	0.0
SF_6	0.0
Xe	0.0
Ethane	0.0
n-Butane	0.0
N_2O	0.2
Freon-12	0.2
Freon-11	0.5
Freon-22	1.4
NH_3	1.5
CHF_3	1.6
MeOH	1.7

Solvato-chromic effect

Kamlet and Taft to correlate different solute–solvent interactions by the solvatochromic effect of the solvent on the $\pi \to \pi^*$ transitions of some probe solutes.

The basic relation on which the π^* scale relies includes terms accounting for the polarity (bipolarity) and polarizability, acidity (hydrogen bond donor character), basicity (hydrogen bond acceptor character) and other properties of the solvent [19]:

$$v = v_0 + s\pi^* + a\alpha + b\beta + \cdots$$

The wavelength of maximum absorbance of the solute (expressed as a frequency, v) depends on the polarity/polarizability of the solvent (π^*), the specific solute (s), the ability to form hydrogen bonds as donor (α) or acceptor (β), and several other properties of the solvent concerned. Based on these parameters, Kamlet and Taft successfully predicted the solubility of various solutes and their chromatographic behaviour, as well as a wide variety of solvent properties [20].

Electron absorption spectra

The π^* solvent scale of Kamlet and Taft was developed from a group of probe solutes with electron absorption spectra that were known to be sensitive to the polarity/polarizability of solvents but formed no hydrogen bonds. Cyclohexane (a non-polar solvent) was arbitrarily assigned $\pi^* = 0.0$, and dimethyl sulphoxide (a highly polar solvent), $\pi^* = 1.0$. In this way, the more positive the π^* value for a solvent is the more polar(izable) it will be. Solute–solvent interactions with the solvent in vapour phase are minimal, so π^* is close to -1 [i.e. a "gaseous solvent" behaves like a non-polar(izable) liquid solvent].

π^* varies widely between supercritical fluids of the same density. Any comparison in this respect is more enlightening if it is based on reduced densities ($\rho_r = \rho/\rho_c$, where ρ_c is the critical density) [5]. Figure 2.18 shows the variation of π^* with the reduced density for seven fluids. It also shows

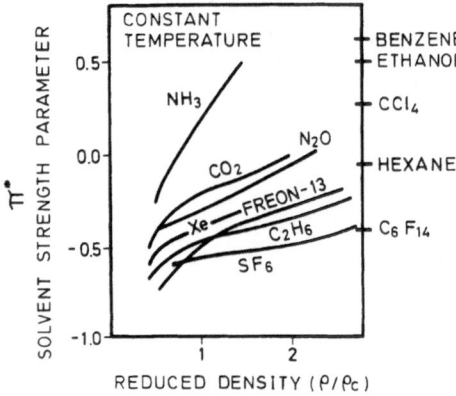

Fig. 2.18. Variation of the dissolving power as a function of the reduced density for various supercritical fluids. The parameter values for several common liquid solvents are also included for comparison. (Reproduced with permission of the American Chemical Society)

the position of some liquid solvents on the scale. As can be seen, fluids of the same reduced density have rather different π^* values, which suggests large differences in their effective polarity/polarizability. Ammonia has the highest π^*, which is consistent with its highest polarity, whereas SF_6 has the lowest. The π^* values for all the other solvents (CO_2, N_2O, Xe, Freon-13 and C_2H_6) lie in-between.

However, the high π^* of NH_3 may be incorrect; in fact, it reflects a contribution from hydrogen bonds which should be assessed by contrast with other probe molecules.

The data shown in Fig. 2.18 are self-illustrative of the different effective solvent polarities of the supercritical fluids studied. Thus, NH_3 is clearly the most polar of all these fluids – on the assumption that no hydrogen bonds are formed with the solute concerned. At the highest density studied, the π^* value of NH_3 on the Kamlet–Taft scale is comparable to that of methanol (0.60). The π^* values for SF_6 are markedly negative, so sulphur hexafluoride is an extremely non-polar solvent. In fact, its polarity/polarizability (π^*) at high densities is similar to those of liquid perfluoroalkanes; this is hardly surprising in view of the high symmetry of SF_6 and the high ionic character of S–F bonds. The data for Xe are interesting because they show it to be a "better" solvent than SF_6 (its π^* values are slightly greater than those of C_2H_6). Also, the polarizability of Xe is substantially higher than that of CO_2 and NH_3. However, the dissolving power of this noble gas can only be ascribed to dispersion forces and the polarizability of its electron cloud. **Dispersion forces**

The effective polarity/polarizability of a solvent for a given solute depends on the density of the supercritical fluid, which is a function of both pressure and temperature. Increasing the fluid density increases π^* for all the solvents studied, which is consistent with the experimental fact that the dissolving properties of an SF can be adjusted by changing the pressure and temperature. It is interesting to note that the rate of change of π^* with density is not constant. At low densities (below ρ_c), π^* is strongly dependent on ρ, whereas at reduced densities above $0.70 \pm 0.15\,g/cm^3$, the dissolving properties change more gradually. Polar solvents such as NH_3 exhibit more marked changes in their dissolving power with increase in the density than do less polar solvents (e.g. SF_6, C_2H_6).

The polarizability of CO_2 is lower than those of all hydrocarbons except methane. For its polarizability per unit volume to be comparable to that of liquid cyclohexane at 45 °C, carbon dioxide must be subjected to a pressure of 2700 bar [21].

Because unary polar solvents such as NH_3 pose major practical problems, small proportions of other solvents (modifiers) are frequently added to scarcely polar solvents such as CO_2 in order to enhance their dissolving power [22]. The ability to adjust the nature and degree of interaction of binary fluids provides separations and extractions of better **Modifiers**

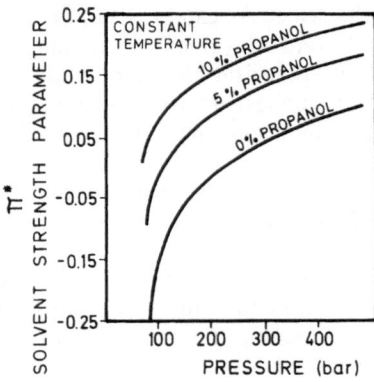

Fig. 2.19. Influence of the proportion of polar modifier on the dissolving power of supercritical CO_2 at a constant pressure and temperature. (Reproduced with permission of the American Chemical Society)

dissolving features and modified selectivities. By way of example, Fig. 2.19 shows the variation with pressure of solvatochromic measurements of the dissolving power for a binary system consisting of pure CO_2 and CO_2 plus different concentrations of 2-propanol as modifier. As can be seen, π^* increases with increasing 2-propanol content. This is consistent with the predicted increase in the dissolving power with increase in the pressure and modifier concentration.

Dipole moment   A pressure rise or the presence of a modifier does not increase the polarity of a fluid as this is determined by the dipole moment of the fluid molecules. The fact that π^* increases with increase in the pressure and the presence of a modifier reflects a stronger interaction between the molecules of the solute and those of the supercritical fluid and modifier, respectively; in addition, the solvatochromic shifts – on which calculation of π^* relies – produced by such modifiers as alcohols are not the sole result of an increased polarity since alcohols clearly behave as both hydrogen donors and acceptors.

Several fluids including CO_2, N_2O, $CHClF_2$ and $95:5$ $CO_2/MeOH$ have been used to compare the rates and recoveries achieved in supercritical fluid extractions. Based on the initial results, the fluid polarity is the most influential parameter on recovery in the extraction of polycyclic aromatic hydrocarbons (PAHs), polychlorobiphenyls (PCBs) and aromatic nitro compounds from environmental solid samples [24].

Supercritical N_2O [24] and Freon-22 [25] have been assayed as substitutes for CO_2 in the extraction of polar compounds such as amines and steroids, respectively, with improved efficiency and reduced extraction times.

2.5 Reactions in or with Supercritical Fluids

The use of a supercritical fluid as reaction medium can be solely aimed at exploiting its peculiar solvent properties, i.e. for reactions *in* supercritical fluids. However, some fluids are not chemically inert, so they can function as reactants in some processes, i.e. in reactions *with* supercritical fluids. Such a clear distinction is difficult to establish in most cases.

2.5.1 Reactions in Supercritical Fluids

The high diffusivity of solutes in supercritical fluids relative to ordinary **Miscibility**
solvents and their virtually unlimited miscibility with other gases are two
major assets that allow high reactant concentrations to be accommodated
in the bulk fluid, thereby facilitating development of both homogeneous
and heterogeneous chemical reactions.

The rate and/or selectivity at/with which a reaction takes place in a **Reaction rate**
supercritical fluid can be altered through pressure thanks to the peculiar
properties of SFs – a catalytic effect can also occasionally arise from the **Catalytic effect**
formation of solute–solvent molecular aggregates. In addition, super-
critical fluids make promising media for development of enzyme-catalysed
reactions.

2.5.1.1 Influence of Pressure on the Reaction Rate

The strong dependence of the properties of supercritical fluids on pressure
makes it necessary to consider the potential effect of this experimental
variable on the rate of reactions taking place in their bulk.

The overall effect of the pressure to which a solvent is subject in a **Effect on the**
supercritical state on the rate of a reaction that takes place in it is the **rate constant**
result of two individual effects: one on the reaction rate constant, k, and
the other on the reactant concentrations (availability) [11].

For a chemical reaction such as

$$A + B \rightleftharpoons M^* \rightarrow \text{products}$$

the derivative of the rate constant with respect to pressure at a given,
constant temperature is the activation volume, ΔV^+ (viz. the volume **Activation**
change undergone by the reactants on reaching the transition state): **volume**

$$\left(\frac{\partial \ln k}{\partial P}\right)_T = -\frac{\Delta V^+}{RT}$$

Most liquid solutions call for pressure changes in the kbar range in order to produce large enough changes (ΔV^+ is usually about $\pm 30\,cm^3/mol$). However, small pressure changes in supercritical solutions can result in substantial variations in the rate constant because the absolute value of ΔV^+ is much larger. On the other hand, the activation volume can re-

Intrinsic volume   asonably be divided into two terms: the intrinsic volume, ΔV_I^+, which reflects intrinsic molecular size differences between the reactants in their original state and the transition state, and the effect of the number of bonds formed or cleaved in the establishment of the transition state, and a second term, ΔV_{II}^+, which is related to solvent–transition state interactions – this latter term is much more significant than ΔV_I^+ for supercritical fluids [2].

The effect of pressure on the rate constant is related to that on the density of the supercritical fluid as well as on its diffusivity, viscosity, dielectric constant and dissolving power. These effects on solvent pro-perties allow k to be increased or decreased in order to adjust the overall rate or selectivity of reaction by adjusting the pressure and hence the density, dielectric constant and solubility parameter – or any other measure of dissolving power. Experiments with guaiacol [26] and dibenzyl and

Hydrolysis rate   phenethyl phenyl ether have shown the hydrolysis rate constant to increase with the solution polarity. The increase in k could be ascribed to a displacement to M^+ in the transition state equilibrium, where the stability of the transition state increases more sharply with the polarity than does the stability of A and B.

Effect on reactant availability   Consider a two-phase system consisting of a pure reactant A in equili-brium with a second phase of supercritical water containing A. Component A will only be hydrolysed in the second phase, at a rate proportional to its concentration in that phase, i.e. $v = k[A]$. Reactant availability is in-fluenced by the solution polarity. The maximum solubility for a reactant is achieved when its polarity equals that of a dilute solution. If the solution polarity is higher than that of the reactant, this will be less readily available in the supercritical phase, so any potential increase in the rate constant will be offset and the overall reaction rate diminished as a result. There-fore, a pressure increase can result in a concomitant increase in k and a decrease in [A]; this suggests that the maximum overall rate will be achieved at an optimal supercritical fluid polarity. These phenomena are illustrated qualitatively in Fig. 2.20, which shows the variation with pres-sure of the rate constant and reactant availability (concentration) (Fig. 2.20A), as well as the overall effect (Fig. 2.20B).

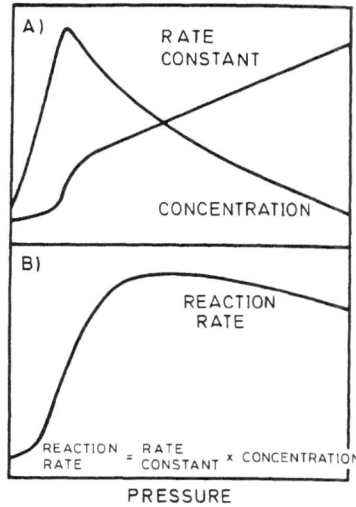

Fig. 2.20A,B. Influence of pressure on reactions in supercritical fluids. **A** Effect on the rate constant and reactant concentration (availability). **B** Influence on the reaction rate (overall effect). (Reproduced with permission of CRC)

2.5.1.2 Catalytic Effects

The catalytic effects of supercritical fluids on the reaction rate originate in their uncommon solvent properties and interactions with reactants and the transition state. These systems are usually highly asymmetric. The molecular size of the solvent and those of the solutes are usually rather different, as are intermolecular forces between solute and solvent molecules and those between the solvent molecules themselves. These solutions behave very differently from the liquid solutions chemists are used to handling. For this reason, classical equations cannot be reliably applied to these asymmetric solutions as they were developed on the assumption of similar molecular sizes and intermolecular forces.

Local properties in the vicinity of a solute molecule dissolved in a supercritical fluid are considerably different from the average bulk solution properties. While the word "cluster" was formerly used to describe an aspect of this phenomenon (viz. the increased local density), there may not be a cluster proper. Instead, the region around a solute molecule may be highly dynamic, with solvent molecules continuously entering and leaving it while still increasing the local density.   **Clustering**

Clustering, which is facilitated by supercritical fluids, may alter a reaction mechanism by lowering its activation energy (i.e. by introducing a catalytic effect).

Supercritical water may provide new reaction pathways by forming special structures with reactant molecules, bond cleavage and formation in which would lower the activation energy [2].   **Supercritical water**

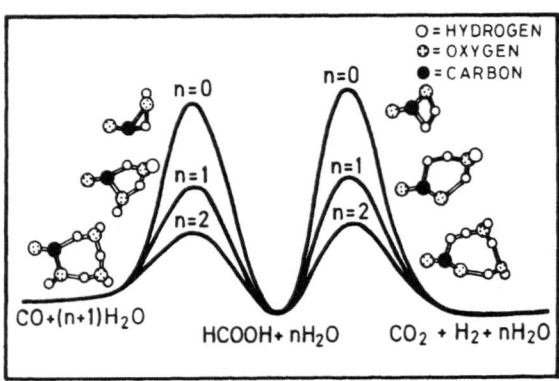

Fig. 2.21. Activation energies of reactions taking place in the supercritical water. For details see text. (Reproduced with permission of the American Chemical Society)

Quantum chemical calculations on the reaction

$$CO + (n + 1)H_2O \rightarrow HCOOH + nH_2O \rightarrow CO_2 + H_2 + nH_2O$$

indicate that, provided no further molecules of H_2O take part in the reaction (i.e. $n = 0$), the activation energy for the first and second step is 61.7 and 64.9 kcal/mol, respectively. If one additional water molecule takes part in the first step ($n = 1$), then a cyclic transition state can be formed with carbon monoxide molecules which requires roughly half the activation energy (35.6 kcal/mol) needed with $n = 0$. If the additional water molecule also takes part in the second step, then the activation energy for such a step is lowered by around 40% (to 37.3 kcal/mol).

The activation energies involved when two molecules rather than a single molecule take part in the reaction ($n = 2$) are even lower (19.3 and 21.5 kcal/mol for the first and second step, respectively). Consequently, the activation energies for $n > 2$ should presumably be even lower. Figure 2.21 shows the transition structures involved and their respective activation energy peaks.

The water acts as a catalyst by lowering the activation energy. Involvement of the additional H_2O molecules in the reaction is more likely if this takes place in supercritical water, the high compressibility of which favours the formation of solute–solvent clusters.

Catalytic phenomenon   Under SFE conditions, the fluid may give rise to a different catalytic phenomenon. This is probably why the mevinoline to hydroxy acid (L-154,819) ratio in the extract obtained with a CO_2/co-solvent system is much higher than that in the starting material [27]. Acidic conditions favour lactonization of the hydroxy acid, whereas the reverse reaction can only be accomplished by refluxing with a base. Residual moisture in the extracor may convert some CO_2 to carbonic acid and a pH decrease in the reactor may favour extraction of mevinoline and/or the extraction and subsequent conversion of L-154,819 to mevinoline.

2.5.1.3   Supercritical Water as an Exceptional Reaction Medium

The extremely high solvating power and compressibility of supercritical water, in addition to its favourable mass transport properties, make it a superior medium for reaction development. In addition to its ability to dissolve many non-polar organic compounds (e.g. PCBs), properly compressed supercritical water is miscible in virtually any ratio with organic substances and oxygen; in this way, such substances can readily be oxidized in supercritical water, which thus makes an excellent medium for decomposing hazardous organic materials in a wide variety of waste waters.

However, the use of supercritical water goes beyond that of a special reaction medium; in fact, SC water can often act as a reactant or catalyst itself.

Supercritical water might be used in some types of selective synthesis as each of the pathways along which a given reaction can take place is favoured by definite pressure and temperature conditions. However, the most promising use of SC H_2O as a reaction medium is not for synthetic purposes, but for destroying potentially hazardous organic molecules.

The decomposition of various heterocyclic organic substances in supercritical water has been found to take place via parallel hydrolytic and pyrolytic pathways [11,28]. The mechanism for these reactions has the following salient features: **Decomposition of hazardous organic molecules**

- The reaction always involves a C atom bearing a leaving group that includes a heteroatom. For example, aromatic amines undergo substitution of an NH_2 by an OH group to yield a phenol compound plus ammonia.
- The process has an alternative parallel pyrolytic pathway via free radicals.
- The hydrolytic selectivity increases with increasing water density, with a break at the critical point.
- The rate constant frequently increases with increase in the H_2O density.
- The rate constant of hydrolysis increases on addition of salts. Hence acid catalysts can be used to boost the reactivity of some reactants.

2.5.1.4   Enzymatic Reactions in Supercritical Fluids

The use of solvents other than water is currently being evaluated in relation to enzyme catalysis. Many enzymes are more active in mixtures of

Table 2.4. Effect of supercritical CO_2 treatment on enzyme activity

Enzyme[a]	% Residual activity			
	CO_2	CO_2 + ethanol (3% w/w)	CO_2 + ethanol (6% w/w)	CO_2 + water (0.1% w/w)
α-Amylase	94	96	93	105
Glucoamylase	102	96		105
β-Galactosidase	98	98		93
Glucose oxidase	97	93	101	99
Glucose isomerase	102	95		97
Lipase	96	88	82	106
Thermolysin	101	96		102
Alcohol dehydrogenase	97	87		94
Catalase	90	96		92

[a] Water content of the enzyme preparations: 5–7% (w/w).
Treatment conditions: 35 °C, 200 atm, 1 h.

water and organic solvents than they are in pure water [29]; some are active and stable even in pure alcoholic media [30].

Non-aqueous solvents   Non-aqueous solvents are quite appealing as hosts for enzymatic reactions for several reasons. In their natural cellular microenvironment, enzymes do not occur in pure water, but in a mixed medium consisting of lipids, proteins and ionic species in addition to water. An enzyme in a non-aqueous solvent may interact with the solvent similarly as in its natural environment and thus be more active than in pure water. Also, substrates may be more readily soluble and hence more reactive in a non-aqueous solvent than they are in water. Other potential advantages of non-aqueous solvents include easier separation and enhanced enzyme specificity. Supercritical fluids make interesting alternative solvents on account of their advantages over conventional liquid solvents. Supercritical CO_2 in particular is of especial interest to the pharmaceutical and biochemical industry for several reasons. Thus, its critical temperature (31.1 °C) enables processing of thermolabile organic substances without a risk of denaturation or thermal decomposition. Also, CO_2 is highly inert, especially towards proteins, which it does not dissolve.

Enzyme activity and stability   The two most significant aspects of supercritical fluids in relation to enzymatic reactions are the activity and stability of the enzyme concerned in the SF used.

Table 2.4 shows the effect of SC CO_2, both pure and mixed with different proportions of ethanol and water, on nine commercially available enzyme formulations. As can be seen, these treatments hardly decrease enzyme activities [31].

Randolph et al. [32] found the enzyme alkaline phosphatase (EC **Alkaline**
3.1.3.1) to be active in a system consisting of supercritical CO_2 as solvent. **phosphatase**
They used a batch reactor loaded with disodium p-nitrophenylphosphate,
de-ionized water (0.1% of the reactor volume) and the enzyme encapsulated in a glass tube. After air was flushed from the reactor by passing
several volumes of CO_2, further carbon dioxide was pumped up to a
pressure of 100 atm at 35 °C. The reactor was shaken in order to break the
glass vial containing the enzyme so as to release it and start the reaction.
In the presence of alkaline phosphatase in the reactor, p-nitrophenol was
formed with yields of up to 71%; on the other hand, no reaction between
water and disodium p-nitrophenylphosphate was detected after 24 h in a
control experiment (the same procedure with no enzyme in the reactor).
The enzyme stability was then studied by exposing it to supercritical CO_2
at 100 atm and 35 °C for various lengths of time and measuring its activity
in the aqueous solution. The enzymes thus treated still proved to be active
in water after as long as 24 h.

Also, the enzyme polyphenol oxidase (EC 1.14.18.1) was found to be **Polyphenol**
catalytically active in supercritical CO_2 and fluoroform [33], where it **oxidase**
rapidly oxidized p-cresol and p-chlorophenol to their corresponding o-
benzoquinones. The solute conversions obtained by using static and flow
systems in the experiments were quite similar (70–80%). Exposure to
supercritical CO_2 or fluoroform containing O_2 but no phenolic substrate,
resulted in little or no deactivation of the enzyme; however, the enzyme **Enzyme**
was deactivated by the presence of phenolic substrate during the oxidation **deactivation**
process.

2.5.2 Reactions with Supercritical Fluids

Using reactants in a supercritical state can make it easier to purify special
products (e.g. polymers from supercritical monomers) or enable decomposition of hazardous organic substances and processing of wastes with
supercritical water. However, reactivity in a supercritical fluid is undesirable for such applications as extraction and chromatographic separation.

One example of the use of supercritical reactants is the production of **Low-density**
low-density polyethylene from supercritical ethylene, where compressed **polyethylene**
ethylene acts both as monomer and as reaction medium (solvent). Notwithstanding the capital efforts using the high-pressure technology that
some supercritical fluids demand, reactions with SFs are of great potential
interest for manufacturing other special polymers on account of the high **High purity**
purity with which they could be obtained [34]. **polymer**

Occasionally, the reactivity of a supercritical fluid can be a limiting
factor for use in supercritical fluid chromatography (SFC) or extraction
(SFE). Thus, the high reactivity of NH_3 restricts its use as extractant

despite the ease with which it can dissolve polar compounds. Also, the corrosiveness of supercritical H_2O, derived from the high pressure and temperature it requires, entails the use of reactors made of special alloys.

A supercritical fluid may react with an analyte to yield some derivative, which should also be identified and quantified in order to avoid errors. Supercritical methanol has been used for the extraction of soil- and plant-**Pesticide** bound pesticide residues that often go undetected in routine analyses. **residues** Under the typical conditions required to produce supercritical methanol (250 °C and 150 bar), the fluid can react with the pesticide and/or its metabolites. Thus, atrazine and its monodealkylated derivative, which are present in some soils and plant materials, react with methanol to yield their corresponding methoxy derivatives [35].

One of the greatest – though fairly unpublicized – assets of N_2O as solvent for both SFE and SFC is its low chemical reactivity. Because nitrous oxide is not an acid like CO_2, it could be advantageously used for the extraction or elution of organic bases, some of which tend to react with CO_2 to form insoluble carbonates [24]. However, some authors have warned that, if N_2O is to be used for processing organic materials, it should be in the knowledge that, under given conditions, it decomposes into a strong oxidant [34].

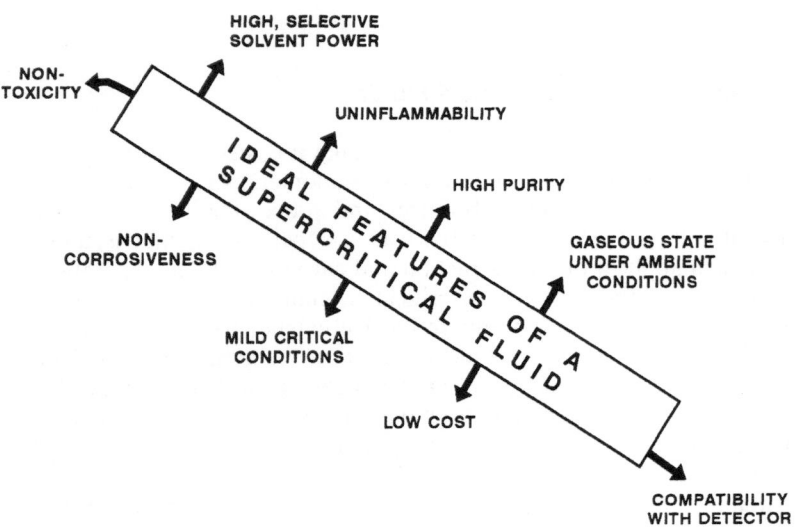

Fig. 2.22. Ideal features of a supercritical fluid for use as solvent

Table 2.5. Advantages and disadvantages of the most commonly used supercritical fluids

Properties	Inorganic				Organic		
	CO_2	NH_3	H_2O^a	N_2O	CFCs	HC	MeOH
Toxicity	+	−	+		+	+	−
Inflammability	+	−	+		+	−	−
Cost	+	−	+		+	+	−
Reactivity	+	−	−	−	+	+	−
Ease of reaching supercritical conditions	+	−	−	+	+	+	−
Environmental aggressiveness	+	+			−		
Gaseous under ambient conditions	+	−		+	+	+	−
Compatibility with detector	+				−		
Polarity	−	+		+	+	−	+

^{a} Not used as extracting fluid.
CFCs = chlorofluorocarbons; HC = hydrocarbons; MeOH = methanol.

2.6   Other Properties of Supercritical Fluids

The ideal or desirable properties of a supercritical fluid are summarized in Fig. 2.22; some of them are common to all solvents, while others only apply when an SF is used as solvent (the choice in this case should also be dictated not only by these properties but also by others such as polarity, dealt with in the previous sections). Table 2.5 summarizes the advantages and disadvantages of the most common supercritical fluids.

One of the basic properties of SFs is their low or even nil toxicity and inflammability. Both are of a great interest owing to the hazards posed by leakage of toxic or inflammable gases – many supercritical fluids are gaseous under ambient conditions. Their non-toxic character is also of paramount significance in relation to some industrial uses on account of the nature or ultimate purpose of the processed substances (e.g. in the food and pharmaceutical industries). Supercritical methanol is toxic; also, low-molecular weight hydrocarbons used as SFs are highly inflammable.

The choice of supercritical fluids for analytical extraction purposes is severely limited by the need to use a sensible critical temperature and pressure, and a fluid that is gaseous under normal conditions. The thermal stability of the extracted compound at the working temperature is another factor determining the choice. Consequently, a low critical temperature is usually desirable – particularly with thermolabile analytes.

Occasionally, use of a supercritical fluid is impeded by instrumental limitations derived from a high critical pressure or corrosiveness. This is **Instrumental limitations** why solvents with moderate critical pressures are usually to be preferred. A very high critical pressure poses instrumental constraints resulting from the high working pressures involved. In addition, the corrosiveness of some critical fluids can be further increased by high working pressures and temperatures. Corrosive SFs entail the use of apparatuses made of special alloys, which obviously leads to increased costs. Supercritical fluids that can react with the analytes or processed materials may pose unsurmountable problems and should thus be avoided as a rule – chemically inert fluids under the working conditions to be used are therefore to be preferred. Polar fluids such as NH_3 have some very useful properties above their critical points and continue to attract growing attention; however, the complications derived from their reactivity impose severe restrictions on their use. Finally, N_2O is a powerful oxidant at high temperatures, so it should be avoided in treating large amounts of organic materials.

Disposal   For a supercritical fluid to be preferred to an ordinary liquid solvent it must be a gas at atmospheric pressure and ambient temperature so it can readily be disposed of leaving no waste.

The availability and affordability of an SF in a high degree of purity are two other major factors to be seriously considered. If a supercritical fluid is to be used in large amounts, it obviously should be affordable enough not to raise analytical costs unduly.

Polluting hazards   Due attention should also be paid to the potential polluting hazards involved in using some SFs. In fact, supercritical fluids are intended to replace liquid organic solvents in order to avoid the problems posed by disposal of their wastes. Some Freons [25] have been proposed as substitutes for supercritical CO_2 as they not only share some of its advantages (e.g. reasonably low P_c and T_c, chemical inertness, a low toxicity and gaseous state under ambient conditions), but also have favourable dipole moments. However, chlorofluorocarbon (CFC) Freons have been associated with ozone depletion and the so-called "greenhouse effect", which is bound to constrain their future analytical use. The effects are somewhat less adverse for hydrogen-containing Freons (HCFCs) such as Freon-22 ($CHClF_2$), which has a much higher dipole moment than CO_2 and other Freons.

Compatibility with detector   Compatibility between the supercritical solvent and the detector to be used is decisive for some analytical applications. One of the assets of SFC using a CO_2 mobile phase is its compatibility with both liquid and gaseous phase detectors in general, and the flame ionization detector (FID) in particular. The FID can only be used with scarcely ionizable fluids (e.g. CO_2). While an N_2O SFC mobile phase is compatible with UV detection, it is unsuitable for FTIR spectrometry. Xenon, which has a favourable critical temperature ($T_c = 16.65\,°C$) and pressure ($P_c = 58.0\,atm$), and is monoatomic – so it does not specifically interact with solutes – provides an optimal spectroscopic window for applications involving detection or analysis [5].

Of the large variety of supercritical fluids available, which encompass wide ranges of critical temperatures and pressures, molecular weights and polarity, CO_2 is unarguably the most commonly used. This SF has a moderately low critical pressure (P_c = 72.85 atm) and quite a low critical temperature (T_c = 31 °C), so it is highly suitable for the extraction of many thermolabile compounds. In addition, it can readily be separated from solutes, poses no environmental problems thanks to its non-toxic character, and is uninflammable and inexpensive. However, it has some limitations, particularly for the extraction of polar compounds, as discussed in the following chapters.

2.7 General Applications of Supercritical Fluids

Table 2.6 summarizes the most salient practical uses of supercritical fluids. Though far from exhaustive, it reveals the large number and wide variety of possible applications, which take advantage of one or more features of SFs and, except for critical-point drying, all also rely on the ability to

Table 2.6. General applications of supercritical fluids

Industrial processes	– Foodstuffs	Decaffeination of coffee beans
		Extraction and refining of oils and fats
		Extraction of flavours, scents, drugs, etc. from plants
	– Polymers	Polymer synthesis
		Impurity removal
		Molecular weight fractionation
		Precipitation–polymerization media
	– Petroleum and heavy hydrocarbons	Petroleum deasphalting
		Coal distillation
		Hydrocarbon fractionation
	– Pharmaceuticals	
Analytical	– SFC	
	– SFE	
	– Others	
Waste detoxification		
Miscellaneous	– Crystallization under SC conditions	
	– Adsorbent regeneration	
	– Deposition of materials on microporous solids	
	– Critical-point drying	
	– Others	

control the properties of a solvent by compressing it or allowing it to expand [36].

2.7.1 Industrial Processes

Supercritical fluids offer chemical industries the ability to accomplish separations with reduced energy expenses; they are also being assessed by the food industry as substitutes for traditional solvents (hydrocarbons and organochlorine compounds). Lately, they have also been used to purify and fractionate polymers. Other industries, including those of petrochemical and pharmaceutical products, are currently evaluating the potential of supercritical fluids for development of their typical processes.

While most of the industrial applications of SFs involve their use as extraction solvents, some industrial systems are also taking advantage of the peculiar properties of these fluids.

Typical industrial extraction   In a typical industrial extraction process, one or more components are separated from a mixture. The product of interest can be either the extract, the raffinate or both. Thus, in the decaffeination of coffee with supercritical CO_2, the starting material, the refined material and the extract are the initial coffee beans, decaffeinated coffee beans and caffeine, respectively.

Figure 2.23 depicts a simplified supercritical fluid extraction process. In the loading step, the supercritical fluid and the starting mixture are brought into close contact in order to extract the soluble components from the mixture. The extraction conditions can be adjusted in order to isolate some components selectively. After extraction, the extract-loaded solvent is separated from the refined phase. At a later stage, the extract is isolated from the solvent by introducing small temperature and/or pressure changes in order to lower the dissolving power.

The extracted components condense as liquid drops or solid particles, depending on the conditions under which the extract is separated from the solvent. After separation, the solvent is readjusted to the extraction conditions for recycling. Any solvent loss is offset by adding fresh solvent to the recycled solvent.

The solvent can be fully separated from the extract by taking the mixture to ambient pressure, where the solvent is a gas; this, however, results in extra re-compression costs. Because the solvent is recycled, strictly complete separation is unnecessary in many cases, so a small pressure drop is usually sufficient. In fact, the whole process can be carried out at a constant pressure by adjusting the temperature for the separation step or using a different operational (adsorption, absorption) unit to separate the extract from the supercritical solvent.

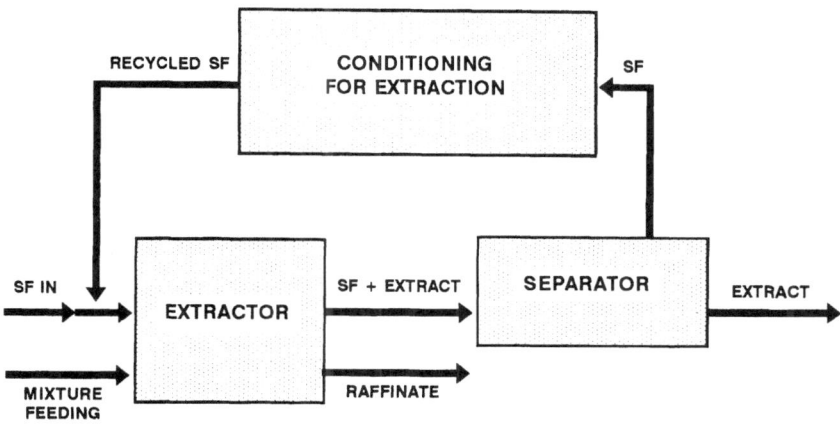

Fig. 2.23. Supercritical fluid extraction of an industrial mixture including recycling of the SF. (Reproduced with permission of the American Chemical Society)

One other industrial use of supercritical solvents is the fractionation of mixtures of similar compounds with widely variable molecular weights (e.g. synthetic polymers and hydrocarbon mixtures) [37]. In the process, illustrated in Fig. 2.24, the material to be fractionated is completely dissolved in the supercritical fluid. A first column is used under such pressure and temperature conditions that the solvent density is similar to that of a liquid and the whole material of interest is soluble in this phase. Insoluble impurities precipitate as a result and build up in the bottom of the column, from which they can readily be removed. The extract-loaded solvent is then driven to the next column. So far, the process is similar to the above-described extraction. In the subsequent steps, the purified extract is fractionated. The extract–solvent mixture is introduced into a second column under such pressure and temperature conditions (e.g. a temperature 10 °C higher than in the purification step) such that the solvent density is low enough to allow condensation as a heavy phase and hence separation of the compounds with the higher molecular weights. This phase is collected as the first fraction and the remaining mixture is **Continued** driven to the next column for further fractionation. In each successive **fractionation** step, the pressure and temperature are adjusted in order to lower the solvent density so as to allow condensation of the next heavier fraction. In principle, the material can be fractionated by successive pressure and temperature adjustments until the density becomes too low for any component to be dissolved.

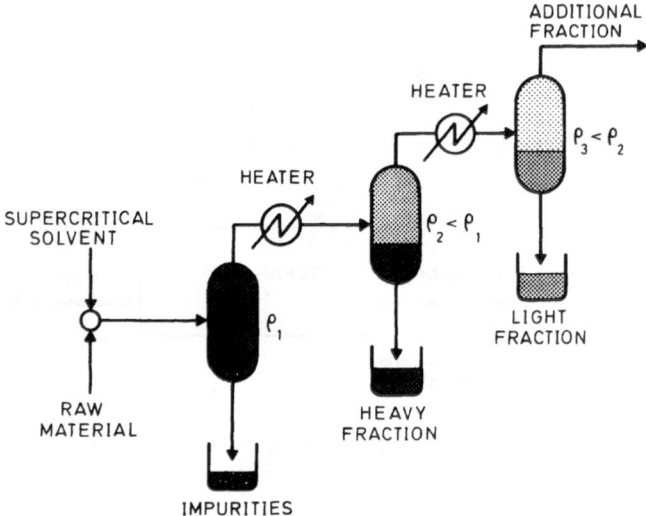

Fig. 2.24. Industrial fractionation procedure based on the use of supercritical fluids. (Reproduced with permission of Pergamon)

2.7.1.1   Supercritical Fluids in the Food Industry

Because supercritical CO_2 is neither toxic nor inflammable, it can safely be used to process natural products and foods, which require mild treatments in order to preserve their nutritional and sensory properties. The non-toxic character and mild working conditions under which SC CO_2 can be used make it the ideal solvent for the agricultural food industry. Below a few uses of supercritical carbon dioxide as extractant in industrial food processing are commented on [38].

Food industrial applications

Decaffeination, which removes caffeine but none of the aroma substances of coffee, calls for an acceptably selective solvent. The presence of moisture increases the selectivity of caffeine extraction, so the coffee beans are suspended in water prior to extraction. Caffeine is extracted from green coffee beans by using wet supercritical CO_2 at pressures between 160 and 220 bar and temperatures in the range 70–90 °C. It is not very readily soluble in CO_2 under these conditions, but the extraction is highly selective. The caffeine content in coffee, which typically ranges between 0.7 and 3%, is thereby reduced to 0.02% with no effect on aroma components.

One major application of heavy gas technology is the extraction and refining of edible oils. The use of dense CO_2 allows several complex **Edible oils** extraction steps of the classical refining process to be combined in a single one. One advantage of extractions with supercritical CO_2 is that they provide a high-quality extract that can be directly used for human feeding with no subsequent refining (removal of lecithins, free fatty acids, mucins, bad odours, bitter contaminants, etc.). Coconut, peanut, soya bean and sunflower seed oil can all be extracted with CO_2 at a pressure of 280–350 bar and temperatures close to the T_c of the fluid (45–50 °C). Some authors have shown oils extracted from soya beans and rice bran with supercritical CO_2 to have some advantages over those obtained with an organic solvent such as n-hexane. Thus, CO_2-extracted oils have lower iron and phosphorus contents, basically as a result of the low solubility of phospholipids relative to triglycerides in SC CO_2. Also, the oils are more lightly coloured by virtue of the low solubility of carotenes and other pigments in supercritical CO_2, so they need no refining. Oils containing large amounts of phospholipids are inedible as they deteriorate very rapidly. However, oils processed with SC CO_2 contain non-polar lipids only since phospholipids (polar) are very sparsely soluble in this fluid. Removal of oil from raw lecithin is one other major application of supercritical CO_2 in food refining.

One further use of supercritical carbon dioxide by the oil industry is the recycling of frying oils. These are typically dark-coloured and contain large amounts of free fatty acids and peroxides.

The SC CO_2 extraction of animal fats from fish and eggs has also received much attention. The interest of these types of food lies partly in the appreciable amounts of glycerides (including polyunsaturated fatty acids) that some fish species contain, which have important pharmaceutical uses. Also, the development of new techniques for extracting cholesterol **Cholesterol** from fat is warranted by the central role this substance plays in the **extracting** pathogenesis of some cardiovascular diseases (e.g. artherosclerosis) and the growing social concern with the matter.

Supercritical CO_2 has also been used to obtain low-cholesterol foods. In their extraction from egg yolk, total lipids – particularly cholesterol – are largely reduced, especially at high pressures and temperatures. Milk fat can be fractionated with CO_2 to obtain fractions of different triglyceride composition and melting point.

Production of food colourings, flavours and aromas is another potentially fertile field for application of supercritical CO_2. For example, in the context of extraction from spices, herbs and drugs with SC CO_2, the extraction of bitter components from hops is of great significance to beer **Hops** manufacturers. The extraction of beer hops was one of the earliest uses of SFE with supercritical CO_2. Hops used to be extracted with dichloromethane, which must be evaporated from the extract down to a 2.2%

Spices

content after extraction. Hops are composed chiefly of humulones and lupulones, which give beer its bitter taste. Humulones in hops can be extracted with a 99% yield – which is higher than the required minimum, 95% – by using supercritical CO_2. Supercritical fluid extracts containing the principal constituents of spices have the advantages of being sterile and usable for longer periods of time. In most cases, extraction from spices is performed in two steps. In the first, dry SC CO_2 is used to extract essential oils. In the second, water-saturated CO_2 is employed to extract flavour components. This second step, however, is not always necessary (e.g. in the extraction of pepper, where the main pungent component has already been extracted in the first step).

2.7.1.2 Polymer Processing with Supercritical Fluids

Low-density polyethylene

Even though polymer processing with supercritical fluids is merely anecdotic to many, manufacturers have already developed commercially available technologies for this purpose. Thus, supercritical ethylene has been used since the 1950s for massive manufacturing of low-density polyethylene. Ethylene acts both as monomer and solvent in the process. In view of the results, this technology could reasonably be extended to industrial processing of other polymers. The high pressures required by supercritical fluids call for substantial investments. In some applications, though, the added cost of the process may be offset by the advantages derived from the high purity of the polymers obtained. Below are described a few other potential uses of SFs in relation to polymers that exploit both the adjustable dissolving power of the fluids and their ability to penetrate the substrate.

Many procedures for purification and fractionation of synthetic polymers involving supercritical fluids have proved to be very efficient and offer valuable advantages over other alternatives [34].

Impurity removal from the end polymer

The presence of impurities (e.g. solvent residues, unreacted monomer, catalysts and products of side reactions) has an adverse effect on the properties of a polymer material and can also increase its odour or toxicity. α-Olefin–maleic anhydride copolymers, which usually contain large amounts of unreacted monomer, are purified with supercritical fluids. Because of its low vapour pressure, the monomer is very difficult to remove by conventional methods such as vacuum distillation or thin layer evaporation. On the other hand, supercritical propane at 130 °C and 1100–4000 psi effectively provides oligomer fractions containing no unreacted monomer.

Some end uses of polymers call for a narrow molecular weight distribution (MWD), which is rather difficult or even impossible to achieve by using such methods as free-radical or step-growth polymerization. This entails the use of a fractionation technique following polymerization. Supercritical fluid extraction offers substantial advantages in this respect over conventional alternatives such as distillation, precipitation and filtration, all of which have proved to be scarcely effective for this purpose. Thus, N_2O has been used to fractionate poly(n-butylacrylate) of a wide MWD. Also, a light hydrocarbon has been employed to fractionate styrene/acrylic resins. The calorimetric and viscosity data for the fractions obtained suggest that supercritical fluid fractionation is useful for producing materials with variable viscosity profiles, which is particularly interesting with a view to obtaining resins for use as binders in coatings and paints with a high solid content.

Fractionation of polymers according to molecular weight

The above examples testify to the effectiveness of supercritical fluids for purifying and fractionating mixtures of polymers, which in turn facilitates the obtaining of highly pure materials with precisely controlled physicochemical properties. However, it should also be noted that supercritical fluids have a solid potential for use in other polymer processes such as the production of finely divided polymer powder by rapid depressurization of polymer solutions in a supercritical fluid. Also, SFs have been used as precipitation–polymerization media in which the degree of polymerization can be adjusted by altering the density of the supercritical fluid. This technique is of a great potential for manufacturing polymers with a narrow MWD.

2.7.2 Processing of Heavy Hydrocarbons

Such processes as coal distillation and related operations require high temperatures in order that coal particles can be broken into smaller units, as well as high pressures ensuring preservation of a high dissolving power, i.e. supercritical conditions. A supercritical solvent can penetrate coal very readily and selectively extract the low-molecular weight fractions of interest from its decomposition products. Other advantages of using supercritical fluids in coal distillation include the separation of solid-free extracts and a porous residue (coke) that is fairly easy to gasify [39].

Coal

Deasphalting of petroleum fractions relies on the low density and viscosity of a phase enriched with a supercritical component. In this regard, a supercritical fluid offers a low solvent–petroleum ratio, easy solvent recovery and operational flexibility. Other related applications are based on the ability of SFs to dealt with scarcely volatile compounds and fractionate similar substances.

Petroleum

The fractionation of petroleum tar with supercritical toluene as an intermediate step in the production of raw materials for manufacturing carbon fibres of a higher quality than those provided by other tar processing methods has been investigated [37].

2.7.3   Analytical Applications of Supercritical Fluids

As stated in Chap. 1, the two most prominent applications of supercritical fluids in analytical chemistry are *supercritical fluid chromatography* (SFC) [3,22] and *supercritical fluid extraction* (SFE) [40].

Even though supercritical fluid chromatography was devised before HPLC, it was not significantly developed until the last decade. The SFC technique combines the features of both column chromatographies. The low viscosity of supercritical fluids and the high diffusivity of solutes in the bulk fluid result in faster separations than with liquid chromatography (LC); also, the high dissolving power of SFs allows processing of less thermally stable and volatile compounds than those afforded by gas chromatography (GC). SFC separations can be based on chemical (sorption and solubility, mainly) and/or melting point differences. The instrumentation required for implementation of SFC is quite similar to that used in LC and GC, with some slight modifications. Future developments in SFC will probably be determined by application ranges and detector sensitivities.

Supercritical fluid extraction, the subject matter of this book, makes the potentially ideal alternative to leaching of solid samples.

Supercritical fluids are used in supersonic jet spectroscopy as substitutes for ordinary gas carriers when non-volatile or thermolabile analytes are involved. As the carrier fluid expands freely in vacuo, the sample molecules within are cooled, which allows one to obtain highly resolved spectra (fingerprints) [41].

Supercritical carbon dioxide has also been used as sample carrier in atomic absorption spectrometry. By transferring organic solutions of metal complexes to a heated restrictor with the aid of supercritical CO_2, aerosols or vapours for introduction into flame atomic absorption spectroscopy can readily be obtained. This sample introduction procedure has been compared to conventional pneumatic nebulization of continuously aspirated solutions and injections into a water carrier flow. The determination of copper pyrrolydinedithiocarbamate in 4-methyl-pentan-2-one is 1.2 times more sensitive when the solution is injected into a CO_2 carrier at 1000 psi than if conventional nebulization with a water carrier stream is used; in any case, the high sensitivity obtained by continuous nebulization of the sample solution can hardly be approached [42].

2.7.4 Waste Detoxification with Supercritical Fluids

Oxidation in supercritical water can destroy over 99.9% of hazardous **Sewage**
organic materials in a few minutes, which allows processing of a wide **processing**
variety of sewage including sewage sludge, paper pulp waste, etc. [2].
Since the main reaction products are water and carbon dioxide – plus
ordinary acids for halogen-containing organic substances – the final
aqueous mixture can be harmless enough for direct disposal with no
further treatment. Otherwise the end product can be neutralized or
removed from the aqueous mixture by distillation or evaporation.

Unlike an incinerator, a supercritical H_2O reactor is a closed system
that emits no polluting gases. In addition, it can work at 500–600 °C, i.e.
well below the typical temperatures used in incinerators (2000–3000 °C).
These fairly low working temperatures probably prevent the formation of
nitrogen oxides and save fuel. If the carbon content of the waste to be
processed is at least 10%, the supercritical treatment may require no
additional energy; this is not the case with incinerators, which call for
external energy unless the carbon content in the waste exceeds 30%.

The chemistry of supercritical water involves some practical shortcomings. Thus, the high pressures and temperatures required make SC
H_2O highly corrosive, particularly when processing halogen-containing
compounds, which calls for reactors made of special alloys. Extensive
implementation of this process has been delayed by the high cost of
reactors and the drastic working conditions involved. However, the high
corrosiveness of supercritical water solutions, which hinders their use,
makes them specially suitable for decomposing some substances that are
rather difficult to process otherwise. The proton concentration in SC H_2O
can be altered by changing its density, via which the ionic character of its
solution can be adjusted. This property, together with the high miscibility
of supercritical water with organic compounds, results in dramatic destruction efficiencies (around 99.99%) for most chlorinated organic compounds.

Supercritical water reactor technology is seemingly a very promising
alternative for processing of hazardous wastes, which vary widely from
place to place; however, much fundamental research is still needed in
order to make it profitable for use in the environmental field. This opens
up fascinating research prospects for the development of major applications such as waste detoxification as part of environmental pollution
control.

2.7.5 Other Applications of Supercritical Fluids

In addition to extraction, the pressurization and subsequent depressuriza- **Insect**
tion of plants and other materials with CO_2 can be used for such purposes **destruction**

as destroying cell structures in order to disperse lipophilic solutes in small portions or exterminating insects, their eggs and larvae [3].

Deposition on microporous solids

Deposition of materials on microporous solids by means of supercritical fluids is based on the nil surface tension of the latter. Solutes dissolved in an SF are driven by this into the solid and subsequently deposited as the solvent expands.

Pressure release via a valve can be achieved at intervals of a few microseconds or even shorter. Therefore, nucleation rates in supercritical fluids can be much larger than those provided by traditional crystallization and precipitation procedures, thereby allowing production of extremely fine solid particles. The high crystallinity and small particle size (10–50 μm) of pure materials dissolved in supercritical CO_2 (with or without a co-solvent) on subsequent precipitation suggests that supercritical crystallization could be used as an advantageous alternative to conventional grinding [27]. The main asset of supercritical "grinding" – in fact, there is no grinding, but a particle nucleation phenomenon that results in substantially reduced particle sizes – is its non-destructive character, which is very important for a size reduction technique. Many compounds are extremely unstable under ordinary grinding conditions. Supercritical grinding could be an advantageous alternative to cryogenic grinding for them; in addition, it would be less likely to raise the problem posed by residual solvent in the previous crystallization step.

Grinding

Micrographs of crystals obtained by conventional grinding and SC CO_2 dissolution followed by precipitation show them to be very similar in size. If 5% methanol is added to the CO_2, then the size of the particles obtained is significantly larger. However, this may be the result of condensation of methanol in the expansion valve, which may cause the precipitate particles to grow. Therefore, the assembly for product recovery must be carefully designed if particle size is to be precisely controlled.

Recycling such materials as active carbon, adsorbents, filters and catalysts relies on the ease with which the solvent can be first removed from the adsorbent material and then recovered, its ability to penetrate solid materials and the use of mild treatment conditions to avoid adsorbent losses as far as possible.

Finally, critical-point drying entails taking a solvent to its critical conditions before it is removed. In this way, formation of a liquid–vapour interface and any capillary forces that might break the structure are avoided.

References

1. Berry RS, Rice SA, Ross J (1980) Phase equilibria in one-component systems. In: Physical chemistry. John Wiley
2. Shaw RW, Brill TB, Clifford AA, Eckert CA, Franck EU (1991) Supercritical water. A medium for chemistry. C&EN Special Report, pp. 23–26
3. Engelhardt H, Gross A (1991) Trends Anal. Chem. 10:64
4. McHugh M, Krukonis V (1986) Supercritical fluid extraction. Principles and practice. Butterworth, Stoneham, MA
5. Smith RD, Frye SL, Yonker CR, Gale RW (1987) J. Phys. Chem. 91:3059
6. Pitzer KS (1955) J. Am. Chem. Soc. 77:3427
7. Pitzer KS, Lippman DZ, Curl RF, Huggins CM Jr, Petersen DE (1955) J. Am. Chem. Soc. 77:3433
8. Langenfeld JJ, Hawthorne SB, Miller DJ, Tehrani J (1992) Anal. Chem. 64:2263
9. Supercritical fluid extraction and chromatography. Techniques and applications, Charpentier BA, Sevenants MR (eds). ACS Symposium Series 366, p. 130
10. Schneider GM, Stahl E, Wilke G (eds) Extraction with supercritical gases. Verlag Chemie, Weinheim, p. 73
11. Bruno TJ, Ely JF (1991) Supercritical fluid technology. Reviews in modern theory and applications. CRC Press
12. van Konynenburg PH, Scott RL (1980) Critical lines and phase equilibria in binary van der waals mixtures. Philos. Trans. Royal Soc. London 298:495
13. Schneider GM (1978) High-pressure phase diagrams and critical properties of fluid mixtures. In: Chemical thermodynamics, Vol. 2, Specialist periodical reports. The Chemical Society, London, 105
14. McHugh MA, Seckner AJ, Yogan TJ (1984) Ind. Eng. Chem. Fundam. 23:493
15. Rowlinson JS, Richardson MJ (1959) Adv. Chem. Phys. 2:85
16. Tehrani J, Myer L (1992) Supercritical fluid extraction method optimization using novel computer software. Pittcon'92, New Orleans
17. Reid RC, Prausnitz JM, Poling BE (1987) The properties of gases and liquids, 4th ed. McGraw-Hill Book Co. New York
18. Kamlet MJ, Abboud JL, Taft RW (1977) J. Am. Chem. Soc. 99:6027
19. Kamlet MJ, Abboud JL, Abraham MH, Taft RW (1983) J. Org. Chem. 48:2877
20. Sadek PC, Carr PW, Doherty RM, Kamlet MJ, Taft RW, Abraham MH (1985) Anal. Chem. 57:2971
21. Kim S, Johnston KP (1986) Effects of supercritical solvents on the rates of homogeneous chemical reactions. ACS Symp. Ser. 329:42
22. Smith RD, Wright BW, Yonker CR (1988) Anal. Chem. 60:1323A
23. Langenfeld JJ, Hawthorne SB, Miller DJ, Paschke T (1992) Comparison of fluids in supercritical fluid extraction (SFE). Pittcon'92, New Orleans
24. Ashraf-Khorassani M, Taylor LT, Zimmermann P (1990) Anal. Chem. 62:1177
25. Li SFY, Ong CP, Lee ML, Lee HK (1990) J. Chromatogr. 515:515
26. Huppert GL, Wu BC, Townsend SH, Klein MT, Paspek SC (1988) Hydrolysis in supercritical water identification and implications of a polar transition state. Ing. Eng. Chem. Res. 27:143
27. Larson KA, King ML (1986) Biotechnol. Prog. 2:73
28. Houser TJ, Tiffany DM, Li Z, McCarville ME, Houghton ME (1986) Fuel 65:827
29. Butler LG (1979) Enzyme Microb. Technol. 1:253
30. Zaks A, Klibanov AM (1984) Science 224:1249
31. Taniguchi M, Kamihira M, Kobayashi T (1987) Agric. Biol. Chem. 51:593
32. Randolph TW, Blanch HW, Prausnitz JM, Wilke CR (1985) Biotechnol. Lett. 7:325

33. Hammond DA, Karel M, Klibanov AM, Krukonis VJ (1985) Appl. Biochem. Biotechnol. 11:393
34. Scholsky KM (1987) Chemtech, 750
35. Capriel P, Haisch A, Khan SU (1986) J. Agric. Food Chem. 34:70
36. Hoyer GG (1985) Chemtech, 440
37. Hutchinson KW, Roebers JR, Thies MC (1991) Carbon 29:215
38. Hierro MTG, Santamaria G (1991) Cromatografi*1a y Te*1cnicas Afines 12(2):62
39. Gangoli N, Thodos G (1977) Ind. Eng. Chem. Prod. Res. Dev. 16:208
40. Hawthorne SB (1990) Anal. Chem. 62:633A
41. Sin CH, Linford MR, Goates SR (1992) Anal. Chem. 64:233
42. Bysouth SR, Tyson JF (1992) Anal. Chim. Acta. 258:55

3 Theoretical and Practical Aspects of Supercritical Fluid Extraction

3.1 Introduction

In Chap. 1, leaching was defined as a solid–liquid separation process whereby the analytes of interest in a solid sample are extracted by means of a condensed liquid or supercritical fluid, whether organic or inorganic. The most salient properties and functions of supercritical fluids were described in Chap. 2. This Chapter is devoted to describing in a broad sense the theoretical and practical aspects of supercritical fluid (SF) leaching (extraction). In this context, the purity and affordability of SFs typically used in analytical applications are discussed first. Next, two essential elements that are central to the extraction process are examined, viz. analyte dissolution and transport phenomena, in terms of both the way they are influenced by various experimental factors and the physico-chemical parameters usually employed to characterize them. Finally, factors affecting supercritical leaching are reviewed in groups according to whether they influence the supercritical fluid, the solid sample, the solute–analyte, the working conditions or modifier systems (with or without derivatization). A more detailed description of the development and applications of supercritical fluids for analytical purposes is provided in Chap. 4, devoted to extractors and operational modes, and Chap. 5 is concerned with the features and applications of SFE.

3.2 Foundation of Leaching

Selective solid–liquid extractions involving the separation of major and/or minor components from a solid for purification purposes and isolation of the separated components and/or their determination are chemical processes of great industrial, synthetic and analytical interest. The use of liquids in this context poses a number of problems that can be minimized by employing supercritical extraction phases.

Because leaching encompasses a variety of steps that depend on the particular fluid, the nature of the sample and analytes and the working conditions used, it is far from simple. Broadly speaking, leaching involves

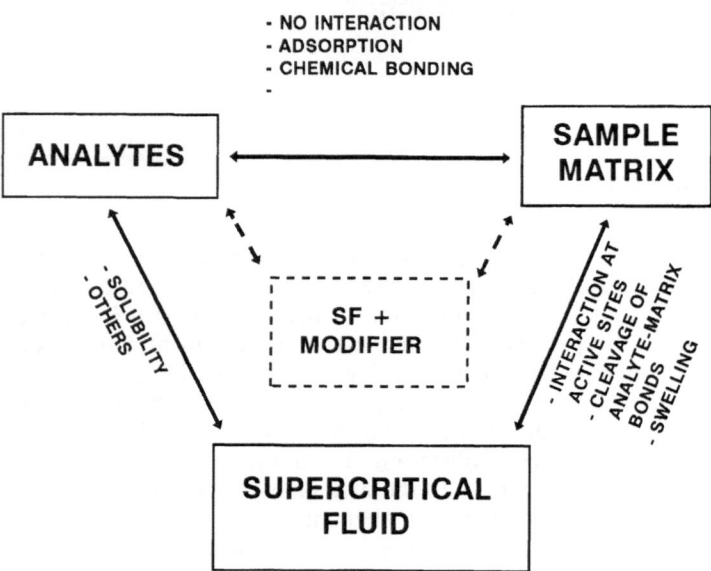

Fig. 3.1. Binary relationships between the components of a supercritical fluid extraction system

two essential phenomena: dissolution of the components of interest in the supercritical fluid and transfer from within the solid (or its surface) to an area beyond the point where the extraction takes place. However, this is rather an oversimplistic view of the process, which often involves additional phenomena that can be as significant as, or even more so than, these two.

In fact, one should always consider potential binary interactions between the three elementary ingredients of a leaching process, namely: the components to be extracted (analytes), the starting solid (sample) and the supercritical fluid (extractant). Figure 3.1 illustrates the different possible binary interactions.

First, one should take into account the potential interaction between the sample matrix and the analytes, which can simply be deposited, adsorbed or chemically bonded to it – the ease of separation decreased in this order. In addition, one should consider the analyte location – at the solid surface, within a rigid or expansible, porous or non-porous material, etc.

Second, one should consider the way in which the SF is bound to interact with the analytes. The most immediate interaction is through solubility, yet a variety of other possible actions ranging from physical sweeping to chemical reaction should also be taken into account.

Finally, one should bear in mind the relationship between the SF extractant and the sample matrix. Such a relationship, which is frequently overlooked, plays a central role in many leaching processes. The supercritical fluid may alter the matrix structure by causing it to expand or contract, which will obviously influence the extraction speed and efficiency. Other SF effects influence analyte-related aspects: on the one hand, the SF molecules can displace those of the analytes from the matrix active sites; on the other, they can have a chemical action (e.g. cleavage of analyte–matrix bonds).

Modifiers

The presence of an SF modifier has major effects on the diagram that reflects the relationships between the ingredients of an analytical leaching system. It essentially affects SF–analyte and SF–matrix interactions by altering solubility, swelling and interactions of the analytes with the matrix active sites. Adding small amounts of suitable modifiers to SFs opens up previously unexpected prospects and broadens the scope of application of supercritical leaching.

The advantages of supercritical fluids over ordinary liquid leaching solvents are not directly related to the solvent power, which is higher for liquids, but with the peculiar physical properties of SFs, which result in increased efficiency and speed through decreased viscosity and diffusivity; these lead to boosted transport phenomena and hence more effective leaching. The ability to remove SF excesses by simply depressurizing and collecting the concentrated analytes is another major asset in this context. It is also worth emphasizing the high selectivity provided by SFE relative to ordinary liquid leaching. As shown in Fig. 3.2, the selectivity of SFE arises from three main sources, namely: (a) the SF itself, whose solvent power (density) can be adjusted by changing the pressure and temperature above their critical values and whose physico-chemical properties can be altered by addition of a modifier; (b) the temperature, which affects the solubility, volatility and diffusion of solutes in the supercritical fluid as well as its density – the fact that the working temperature is usually lower than in conventional leaching allows application to thermolabile analytes; and (c) the collection (and elution/rinsing) system, which can be used to remove interferences as described in Chaps. 4 and 5.

High selectivity

3.3 Purity of Supercritical Fluids

One of the most salient and advantageous features of supercritical fluids used as mobile phases in chromatographic (SFC) or extraction processes (SFE) is the ability to adjust their solvent power by regulating their pressure and temperature. Impurities in supercritical fluids give rise to baseline rises or isolated peaks when the fluid is used as the mobile phase.

Solvent power adjustment

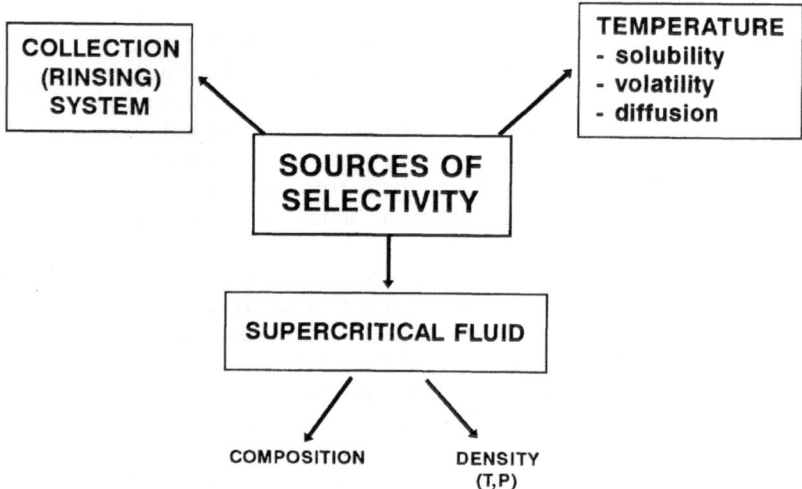

Fig. 3.2. Origin of selectivity in SFE

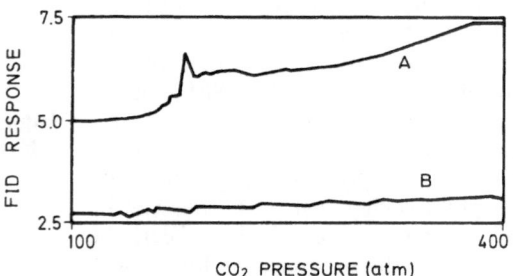

Fig. 3.3. Differences between the behaviour of ultrapure (B) and less pure CO_2 (A) on passage through a FID detector under a pressure gradient. For details, see text. (Reproduced with permission of Air Products & Chemicals)

Figure 3.3 shows the baseline drift and background signal of an SFC/FID chromatogram obtained by adding CO_2 to a detector gaseous stream at a programmed pressure (100–400 atm) and a constant temperature of 70 °C using SFC/SFE-grade CO_2 (B) or contaminated CO_2 (A). As can be seen, purity becomes decisive as the SF goes through the detector. However, the role played by supercritical fluid impurities in SFE is much more prominent than it is in SFC. Small amounts of analytes (a few nanograms or picograms) in sample matrices can frequently be extracted with as little as 10–20 g of SF. Therefore, impurity concentrations as low as a few nanograms or even picograms per gram in the fluid can be significant relative to the analyte mass involved if every non-volatile impurity in the SF is collected with the analytes. Consequently, extraction SFs should be impurity-free – to even greater extent than chromatographic SFs owing to the high preconcentration of impurities that take place during extraction and collection.

Need for SF purity

There are two basic types of commercially available supercritical fluids according to purity (Scott Specialty Gases Inc): SFE-grade and SFC-grade. Some manufacturers (e.g. Air Products) supply their fluids as SFC/SFE-grade. Table 3.1 lists commercially available SFE- and SFC-grade fluids; their maximum acceptable concentrations of extraneous components are given in Table 3.2.

Impurities in the extracting fluid must be minimized or, alternatively, impurity-blind reactors must be used, if quantitative results are to be obtained in the subsequent determination. Because impurities can be of a widely variable nature, the latter is not always feasible. In any case, one should determine the blank value (by passing the SF through an extraction cell containing no sample) in order to ensure that impurities in the SF to be used will introduce no significant errors in the analyses if they are sensed by the same procedure as the sample extracts [1]. Before SFE-grade fluids were commercially available, the detection limit for a typical analysis was more markedly determined by the fluid purity than by the detector used. However, some impurities are occasionally retained with analytes extracted into CO_2, even if the fluid is SFE-grade. In this situation, the problem does not arise from CO_2 purity, but from the fluid dispensing system used.

SFC- and SFE-grade fluids are carefully manufactured under precise conditions in order to exclude impurities that may influence SFC resolution or raise SFE detection limits. For this purpose, the fluid is subjected

Table 3.1. Supercritical fluids most frequently used in SFE and SFC

SFE grade fluids
– Carbon dioxide (CO_2)
– Nitrous oxide (N_2O)

SFC grade fluids
– Carbon dioxide (CO_2)
– Nitrous oxide (N_2O)
– Ammonia (NH_3)
– Sulphur hexafluoride (SF_6)
– Xenon (Xe)
– Sulphur dioxide (SO_2)

Table 3.2. Specifications of SFE and SFC fluids

SFC grade fluids	Maximum concentration (ppm)					
	CO_2	NH_3	N_2O	SF_6	Xe	SO_2
Argon (v/v)	5	–	–	–	–	–
Hydrogen (v/v)	5	–	–	–	4	10
Nitrogen (v/v)	50	–	–	–	20	100
Oxygen (v/v)	2	2	2	10	4	4
Carbon monoxide (v/v)	5	–	–	–	–	10
Methane (v/v)	2	3	3	5	–	4
Water (v/v)	3	1	3	3	1	6
Non-volatile organic compounds (w/w)[a]	0.1	0.1	0.1	0.1	1	–
Particulates (w/w)[a]	1	1	1	1	1	1
UV cut @ 200 nm (au)	0.1	0.1	0.1	0.1	0.1	0.1

SFE grade fluids		Maximum concentration (ppm)		
		Research grade CO_2	Ultrapure CO_2	N_2O
Hydrogen	<	2	2	2
Nitrogen	<	20	20	20
Oxygen	<	2	2	2
Carbon monoxide	<	2	2	2
Water	<	0.1	0.5	1

		Maximum extractable concentration of residue		
		Research grade CO_2	Ultrapure CO_2	N_2O
ECD response (ppt)[b]	<	10	100	50
FID response (ppb)[c]	<	0.1	10	1

[a] Only applicable to returnable aluminium cylinders.
[b] ECD = Electron capture detector.
[c] FID = Flame ionization detector.

Fluid purification   to an effective stepwise purification process and subsequently transferred to a packing system that ensures cleanliness and appropriate preparation within the dispensing cylinders and their valves. Carbon dioxide is an excellent solvent for such substances as organic solvent wastes, polymers, polymer additives and lubricants. These contaminants, particularly valve lubricants, tend to build up in CO_2 during routine handling and treatment and must be removed in the purification process. The purity and quality of **Highly sensitive analytical techniques**   the end product are assured by using highly sensitive analytical techniques. The overall purity of an SF can exceed 99.9999%. Maximum organic contaminant levels appear on the label of each SFC- and SFE-grade cylinder, in addition to the resolution, as flame ionization detection (FID)

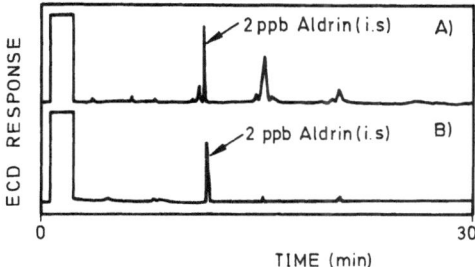

Fig. 3.4A,B. Effect of contaminants on the chromatogram obtained with an electron capture detector using aldrin as internal standard and (**A**) impure CO_2, and (**B**) high-purity CO_2 in a tenfold amount. (Reproduced with permission of Air Products & Chemicals)

and electron capture detection (ECD) background chromatograms. Residues that can be sensed by an ECD are present at concentrations below 10 ppt (parts per trillion) in SFE-grade CO_2.

Trace contaminants in CO_2 should not be allowed to concentrate with the analyte in order to avoid detection interferences and have CO_2 contribute minimally to peak area in the final chromatogram. Figure 3.4 shows the GC/ECD chromatograms obtained for two blank extracts to which 2 ppb of the pesticide aldrin was added as internal standard (i.s.) by using 200 g of SFE-grade CO_2 (B) and 20 g of contaminated CO_2 (A). This pesticide was chosen as non-specific internal standard in order to relate its concentration to the overall background peak area because the ECD response varies between halocarbons, as does its linear relationship to the concentration. In this example, the amount of impurities to which the ECD responded was equivalent to less than 0.1 ppb aldrin. Identified impurities are usually sensed as cleavage products of the lubricant chloro-fluoropolyether by GC/MS and FTIR spectroscopy.

In on-line coupled SFE/GC, supercritical fluid impurities are preconcentrated at the cryogenic trap and subsequently driven to the gas chromatograph, which poses serious constraints on trace analyses. The small overall amounts of water and hydrocarbons (a few parts per million) present in research-grade CO_2 (purity > 99.999%) detract from recoveries. Water limits the cryogenic trap temperature to values above 0 °C, while hydrocarbons give rise to parasitic signals at the flame ionization detector. Too high temperatures in the cryogenic trap result in losses of the more volatiles solutes. Also, organic residues can form films acting as stationary phases in the trap, thereby reducing its desorption efficiency and giving rise to a memory effect [2].

Occasionally, supercritical fluids are further purified in the laboratory. The fluid purity is substantially improved by taking the gas from the **Purification in laboratory**

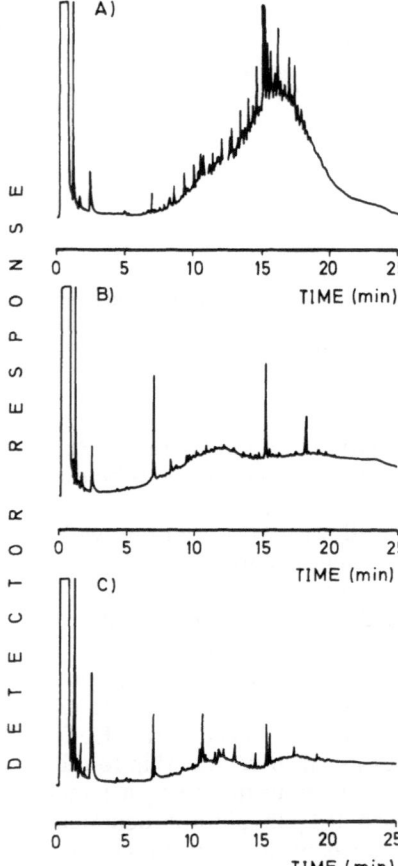

Fig. 3.5A–C. Influence of the SF delivery mode on chromatographic sharpness. **A** Direct insertion of liquid CO_2. **B** Delivery from the cylinder head. **C** Delivery via an active carbon clean-up trap. For details, see text. (Reproduced with permission of Huethig)

cylinder head and condensing it in the pump rather than dispensing the liquid with a probe [3] since impurities, which are less volatile than CO_2, are more concentrated in the liquid phase than they are in the cylinder gas. Figures 3.5A and B show the gas chromatograms for the impurities contained in the same mass of CO_2 taken from the liquid and gaseous phase, respectively.

Traps   Alternatively, SFs can be further purified by means of traps. Figure 3.5C shows the gas chromatogram of CO_2 from the liquid phase on passage through an active carbon adsorption trap placed between the CO_2 reservoir and the pump. As can be seen, impurities are substantially reduced in relation to Fig. 3.5A.

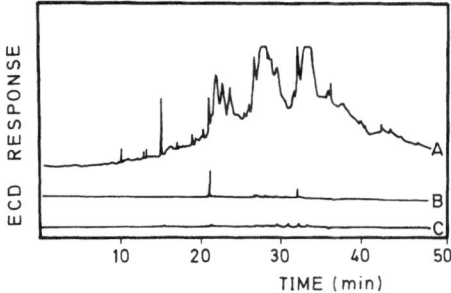

Fig. 3.6. Effect of internal extraction of impurities on the performance of an ultrapure CO_2 extraction system. Chromatograms were obtained by (*A*) passing the SF initially through the system, (*B*) passing the SF directly through the detector, and (*C*) passing the SF after the internal extraction. (Reproduced with permission of the American Chemical Society)

The extraction system itself may be another source of contamination (e.g. at the pump or valve sealings). The extracting fluid passes through all these elements during normal functioning, so any contaminant or degradation product will be dissolved in the fluid and carried through the extractor. Sealing systems are usually based on high-density polymers such as Teflon, polyethylene or poly(ether-ether-ketone) (PEEK), depending on the particular manufacturer. The polymer used to manufacture the sealing systems can contain a mixture of oligomers of variable molecular weight in addition to any contaminants or side products formed in the polymerization process. During SFE, these components may be extracted "internally" since SFE is an excellent technique for extraction of polymers. Figure 3.6 shows a typical example of polymer contamination [4].

The role of modifiers in SFE is one of the most interesting subjects in analytical research on supercritical fluids. As in SFC, the properties of SFE media can be adjusted chemically by addition of organic solvents. In fact, modifiers make a powerful, ingenious tool for increasing the efficiency and selectivity of an extraction or chromatographic separation. The solvent power of CO_2, a non-polar SF, is inadequate for extracting polar analytes from a sample matrix, so addition of a polar modifier is required in order to achieve quantitative extraction. Commercially available supercritical fluid mixtures are useful in those cases where the instrumentation needed to prepare the fluid from its individual components is unavailable, as well as in routine extractions with mixtures of constant momposition. While there is a wide range of commercially available CO_2-compatible modifiers, the choice for other SFC- and SFE-grade fluids is more limited (Table 3.3). Nevertheless, mixtures containing unusual modifiers

Commercially available fluid mixtures

Table 3.3. Common mixtures of supercritical fluids

Modifiers	CO_2	SF_6	N_2O	NH_3	Xe	SO_2
Oxygen-containing						
Methanol	A	O	U	O	A	U
Ethanol	A	O	U	O	A	U
Isopropyl alcohol	A	O	U	O	A	U
Acetone	A	O	U	O	A	O
Tetrahydrofuran	A	O	U	O	A	O
Nitrogen-containing						
Ammonia	U	O	U	–	A	U
Acetonitrile	A	A	U	O	A	O
n-Propylamine	U	O	U	A	A	U
Hydrocarbons						
n-Hexane	A	A	U	A	A	O
Toluene	A	A	U	A	A	O
Halocarbons						
Methylene chloride	A	A	O	A	A	O
Chloroform	A	A	O	A	A	O
Carbon tetrachloride	A	A	A	A	A	A
Trichlorofluoromethane	A	A	A	A	A	A
Acids						
Formic acid	A	A	U	U	A	O
Formic acid	A	A	U	U	A	O
Sulphur-containing						
Carbon disulphide	A	A	U	U	A	O
Sulphur dioxide	A	A	U	U	A	–
Sulphur hexafluoride	A	–	O	O	A	A

A = Commercially available.
O = To order.
U = Commercially unavailable.

or modifier concentrations can also be purchased to order from some manufacturers.

3.4 Solubility in Supercritical Fluids

The process by which a molecule becomes part of a supercritical phase can be described as vaporization inasmuch as molecules shift from a condensed phase to an expanded phase, or as dissolution since it involves solute–solvent interactions. This combined vaporization–dissolution reflects the intermediate nature of the supercritical state.

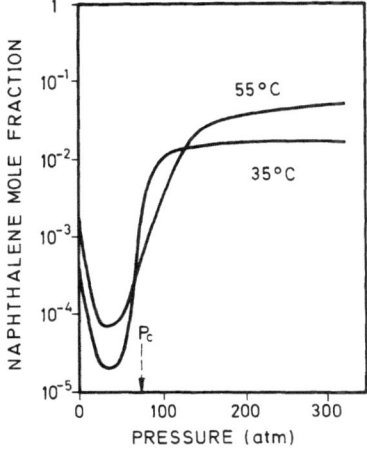

Fig. 3.7. Variation of the solubility of naphthalene in supercritical CO_2 as a function of pressure at two different temperatures. P_c = critical pressure. (Reproduced with permission of the American Chemical Society)

Solute–solvent interactions in dense CO_2 increase the solubility of naphthalene by a factor of up to 10^4 relative to that one would expect from an ideal gas at 200 atm and 35–55 °C. In fact, solubility enhancement factors can be as high as 10^{10}.

By way of example, consider the solubility of a solid such as naphthalene in an SF such as CO_2. Figure 3.7 shows the variation of the solubility of naphthalene in CO_2 with pressure at two different temperatures, both of which lie above T_c for carbon dioxide (31 °C). At low pressures, the solubility is determined by the vapour pressure of naphthalene, which is consistent with the expectations since CO_2 is a relatively ideal gas with no special solvent properties under these conditions. The mole fraction of naphthalene initially decreases with increase in the overall pressure. If this is raised further, the CO_2 density also increases up to a point where the solubility of naphthalene also starts to increase. As the CO_2 density increases, the mean intermolecular distance decreases and specific interactions between the solvent and solute increase proportionally. Near the critical pressure of CO_2 (72.8 atm), the solubility increases sharply as a result of the marked increase in density with pressure. At higher pressures, further pressure rises give rise to small solubility elevations because the density increases more gradually with increasing pressure. At extremely high pressures, the solubility reaches a maximum (Fig. 3.7). A pressure increase near the critical point results in a sharp increase in the solubility of naphthalene at both temperatures.

Intermolecular distance

The effect of temperature on solubility is somewhat more complex. Below 60 atm and above 120 atm, the solubility of naphthalene increases with increase in the temperature. On the other hand, the solubility de-

Retrograde vaporization

creases with increasing temperature at pressures between 60 and 120 atm, contrary to expectations. This phenomenon is commonly known as retrograde vaporization.

As the temperature rises, two competing effects come into play: on the one hand, the vapour pressure of naphthalene increases, and so does solubility; on the other hand, the density and solvent power of CO_2 decrease, so the solubility tends to decrease as well. Below 60 atm and above 120 atm, the CO_2 density does not depend so markedly on the temperature, so the increase in the vapour pressure of naphthalene prevails and the solubility rises. At intermediate pressures (60–120 atm), the CO_2 density does depend on the temperature: it decreases with increasing temperature, so the solubility decreases as a result. At 60 and 120 atm, the two effects cancel out, so the solubility is virtually independent of the temperature.

As can be seen from Fig. 3.8A, a plot of naphthalene solubility vs CO_2 density reflects no retrograde vaporization. In fact, the solubility of naphthalene increases as a result of both increasing the CO_2 density at a constant temperature and raising the temperature at a constant density. The isotherms in the solubility–pressure plot (Fig. 3.8B) intercept at a given pressure (about 120 atm), which suggests that the solubility decreases with increasing temperature below that point.

The above observations allow one to draw the following conclusions:

a) The solubility increases with increasing pressure. The increase is very sharp near the critical point as a result of marked changes in solvent density.
b) The solubility increases, remains constant or decreases with increase in the temperature at a constant pressure (Fig. 3.9), depending on the predominant factor, viz. the solute vapour pressure or the solvent

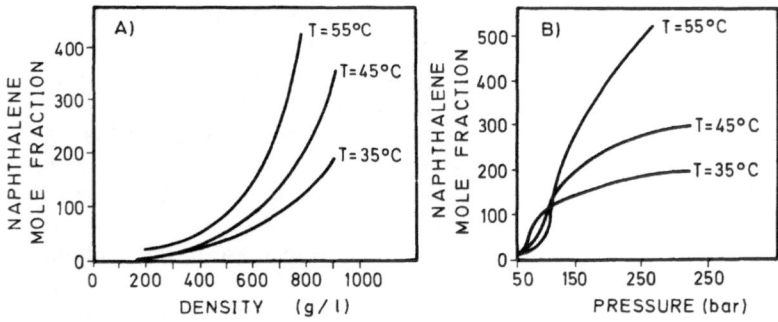

Fig. 3.8A,B. Influence of (**A**) density and (**B**) pressure on the solubility of naphthalene in CO_2 at three different temperatures

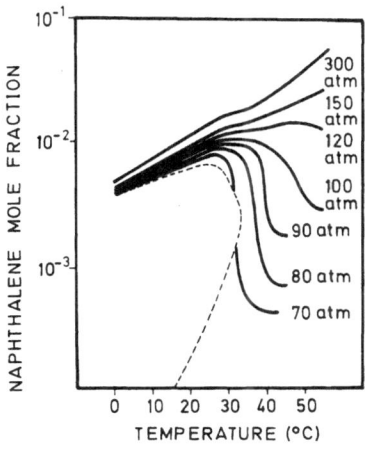

Fig. 3.9. Influence of temperature on the solubility of naphthalene at different pressures. (Reproduced with permission of the American Chemical Society)

density. At low pressures, the solubility decreases somewhat with temperature; at high pressures, it increases markedly with temperature. The maximum solubility increases with increasing temperature at high pressures. The former effect arises from the decrease in density with temperature, whereas the latter, which involves the dense, less readily compressible region, results from the vapour pressure of the solid. At a given solvent density, its solubility increases with increasing temperature.

c) The solubility increases with increasing solvent density (Fig. 3.8A).

However, the solvent power of an SF cannot be exclusively ascribed to an increased density as it is subject to the effect of a state component and a chemical component. Density is the principal variable of the physical state component, whereas the chemical effect defines the interaction between the SF and the solute, which varies between solute–SF systems as it depends on the relative polarity and acid–base and hydrogen-bonding properties.

Figure 3.10 shows three different plots of phenanthrene solubility in SC ethylene at a constant temperature (25, 45 or 70 °C), viz solute mole percent concentration against pressure and density, and decimal logarithm of the solubility against density. As can be seen from the solubility vs density plot at a constant temperature, a potential dependence of the solvent density on pressure can be ruled out. Also, the logarithm of the solubility at a constant temperature varies virtually linearly with the solvent density [5].

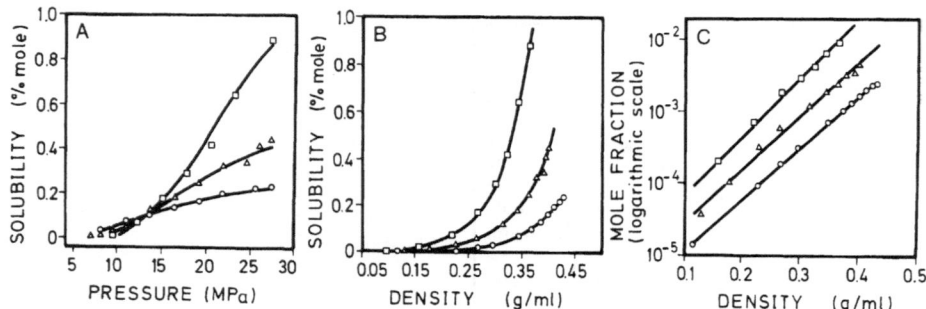

Fig. 3.10A–C. Variation of the solubility of phenanthrene in supercritical ethylene as a function of (**A**) pressure and (**B**) density, and variation of the logarithm of solubility as a function of density (**C**) at three different temperatures (○ 25 °C, △ 45 °C, □ 70 °C). (Reproduced with permission of the American Institute of Chemical Engineers)

The concentration of a solute in a supercritical fluid can be calculated by using a equation derived from the law of mass action and/or the entropies of the components [6]:

$$c = \rho^k - \exp\left(\frac{a}{T} + b\right) \tag{1}$$

where c is the solute concentration in the fluid (in g/l) and k, a and b are three constants that can be determined experimentally from $\log c$ vs $\log \rho$ plots. According to this equation, a plot of the logarithm of solubility against that of density will be linear. In fact, experimental solubility data obtained at a constant temperature for various types of solute over wide pressure and concentration ranges conform accurately to the equation.

3.4.1 Solubility Measurements in Supercritical Fluids

Static, flow-through and recirculation systems

The equilibrium solubilities of various compounds in supercritical fluids have been determined by using static [7,8], flow-through (dynamic) [9,10] and recirculation systems [11,12]. In flow systems, the supercritical fluid is passed through the sample at different flow-rates on the assumption that the fluid and solid reach equilibrium immediately. On the other hand, static solubility determinations are performed in cells filled with a preset amount of solute by varying the pressure or temperature until the whole sample is dissolved in the supercritical phase [8]. Finally, recirculation systems use a fixed fluid volume that is continuously recirculated through them until equilibrium is reached. All recirculation units usually include an on-line coupled UV-visible detector, an injection and a recirculation pump.

Solubility measurements in flow systems usually involve the following steps: the extraction tube is packed with alternate layers of the solid whose solubility is to be determined and glass wool in order to avoid the formation of "preferential channels". Pressure drop in this type of system is negligible. The supercritical fluid is passed through the extraction tube to dissolve the solid and then allowed to expand by means of a thermostated valve. Solids precipitate in a U-tube after the valve. A second tube is normally used to ensure that the solid is fully retained (99% of it is usually retained in the first tube). The mass of precipitated solid is obtained by weighing the previously tared tubes after each experiment. From the mass and the overall solvent volume used the solid concentration in the supercritical fluid is calculated. It is important to check that the calculated solubility is not affected by such factors as the experiment duration, the amount of solid and its location in the extraction tube or the solvent flowrate in order to ensure that the extractor is isothermal and equilibrium is reached very rapidly.

Most solubility data are obtained gravimetrically. Other than the solid weight collected in U-tubes, the extractor weight loss after the experiment can also be used to determine solubility. This can be a more accurate procedure than measuring the separator weight gain since solids occasionally precipitate in the lines connecting the vent valve and separator, thereby resulting in underestimated values. On the other hand, the weight loss procedure may introduce positive errors in solubility measurements as a result of some solid being lost through depressurization. The chief disadvantage of using gravimetric methods for determining solubility is the difficulty involved in measuring small weight changes. The problem is worsened by the very low solubilities of some compounds in SC CO_2 (e.g. <0.3% by weight), which call for very long equilibration times in order to obtain measurable weights and accurate solubility values.

Disadvantage of gravimetric methods

Table 3.4 lists the solubility of various solutes in supercritical fluids, both alone and mixed with cosolvents. It also includes the conditions under which measurements (type of system and pressure and temperature range, if applicable) were made.

3.4.2 Solubility and Chemical Structure

Comparisons of the solubility of a wide variety of substances has allowed some correlations between structure and solubility in dense CO_2 to be established [22,23].

The liquid CO_2 solubility data for over 250 compounds compiled by Francis [23] have been used for three decades as qualitative guidelines to predict the solubility of substances in supercritical carbon dioxide. Dandge et al. [22] established relationships between the solute structure and its

Table 3.4. Experimental conditions used for solubility measurements in solute (solid)–SC solvent systems

Solute(s)	SC solvent	System type	Pressure range	Temperature range (°C)	Reference
Anthracene, o-anisic acid	CO_2	Recirculation	12.0–33.8 MPa	40	12
Naphthalene, 2,3-dimethylnaphthalene, 2,6-dimethylnaphthalene, phenanthrene, benzoic acid	CO_2, C_2H_4	Flow	80–280 bar	35–65	9
Imipenem, Nevinoline, Efrotomicyn (pharmaceuticals)	CO_2	Flow	5000 psi	40	13
Naphthalene, dibenzothiophene (mixtures)	CO_2	Static–flow	76.85–276.80 bar	36–65	14
Pure fatty acids (lauric, myristic, palmitic), pure triglycerides (trilaurin, trimyristin and tripalmitin), mixtures of triglycerides	CO_2	Flow	8–30 MPa	40	15
Phenol, p-chlorophenol, 2,4-dichlorophenol	CO_2	Flow	80–250 atm	36, 60	16
Naphthalene, biphenyl	CO_2	Flow	80–500 atm	35–65	17
Stearic, oleic and behenic acid, tributyrin, tripalmitin, tristearin, triolein, trilinolein, palmityl and behenyl behenate, α-tocopherol	CO_2	Static	80–250 atm	40–80	6
Naphthalene, anthracene, phenanthrene	Ethylene	Flow	5.6–41.5 MPa	25–85	5
Phenanthrene, naphthalene, 2,3- and 2,6-dimethylnaphthalene, benzoic acid (binary mixtures of solids)	CO_2, ethylene	Flow	120–280 bar	45	10

Table 3.4. *Continued*

Solute(s)	SC solvent	System type	Pressure range	Temperature range (°C)	Reference
Pesticides (linuron, methoxychlor, methyldichlofop, dichlofop, 2,4-D), 3,4-dichloroaniline	CO_2	Recirculation	200 bar	40	11
Octacosane	CO_2	Flow Static	80–325 atm	34.7–52.0	8
Naphthalene[1], Anthracene [1,3], Phenanthrene[1,3,2], Triphenylmethane[1,3], Hexamethylbenzene[2,3], Fluorene[2,3], Pyrene[2,3]	Ethane[1], Ethylene[2], CO_2[3]	Flow	4.23–48.35 MPa[1], 6.3–48.35 MPa[2], 6.99–48.35 MPa[3]	[1] 20–70, [2] 25–70, [3] 30–70	18

Solid solute(s)	SC solvent	Cosolvent	System type	Temperature (°C)	Pressure range	Reference
Benzoic acid, 2-aminobenzoic acid, phthalic anhydride, 2-naphthol, acridine	CO_2	Methanol, acetone, n-octane, SO_2	Flow	35	90–350 bar	19
Anthracene, 2-aminobenzoic acid, benzoic acid, hexamethylbenzene, 2-naphthol	CO_2	Methanol	Flow	35, 45, 55	120–350 bar	20
Phenol Blue	CO_2	Acetone	Flow	35	100–300 bar	21

2,4-D = 2,4-dichlorophenoxyacetic acid.
The superscript numbers in brackets indicate the analyte-working conditions relationship.

solubility in CO_2, which can be used for the qualitative estimation of solubilities in CO_2.

Below is discussed how the solubility in CO_2 (at 2500 psi and 25 or 32 °C) is affected by structural changes in each of various groups of substances established according to chemical nature and/or functional groups.

3.4.2.1 Hydrocarbons

The structural features most markedly influencing the solubility of hydrocarbons in SC CO_2 are chain length, the presence of unsaturations, branching and the number of rings and type and position of their substituents in cyclic compounds.

Alkanes   Normal alkanes of less than 13 carbon atoms are fully miscible with SC CO_2. Solubility in this fluid decreases sharply above 13 carbon atoms; thus, while *n*-dodecane is still miscible with CO_2 in any proportion, the solubility of *n*-tetradecane, *n*-hexadecane and *n*-octadecane in this fluid is only 16, 8 and 3%, respectively.

Alkenes   Alkenes behave similarly in this respect, even though unsaturations have a favourable effect on solubility. At a given number of carbon atoms, the solubility of an alkene is much higher than that of its corresponding *n*-alkane; thus, the CO_2 solubility of 1-octadecene is 10%, whereas that of *n*-octadecane is only 3%.

Isoalkanes and branched alkanes   Isoalkanes and branched alkanes in general are comparatively more soluble than their straight-chain counterparts. Their maximum limit (full miscibility) lies between 19 and 30 carbon atoms against only 12 in *n*-alkanes. While 2,6,10,14-tetramethylpentadecane (the C_{19} isoalkane) and other lower isoalkanes are miscible with CO_2, squalene (the C_{30} isoalkane) is only partially miscible with this fluid. Increased methyl branching reduces the number of methylene groups in the hydrocarbon chain, which in turn decreases intermolecular interactions in isoalkanes and hence leads to lower solubility parameter values.

Bicyclic hydrocarbons   As a rule, bicyclic hydrocarbons, whether alicyclic or aromatic, are not fully miscible with CO_2. Their solubility increases with the extent of hydrogenation [e.g. naphthalene (2%) < tetralin (12%) < decalin (22%), or, cyclohexylbenzene (8%) > biphenyl (2%)]. Substitution of protons by methyl groups also increases the solubility of bicyclic aromatic compounds; thus, α-methylnaphthalene (6%) and β-methylnaphthalene (9%) are both more soluble than naphthalene (2%).

The fact that SC CO_2 is fully miscible with aliphatic and monocyclic hydrocarbons but not with naphthenic and aromatic bicyclic hydrocarbons with similar boiling points is suggestive of the greater significance of chemical structure relative to physical properties.

Side branching in aromatic compounds also increases solubility. While *p*-di-*tert*-butylbenzene (C_{14}) and 1,3,5-tri-*tert*-butylbenzene (C_{18}) are fully miscible with CO_2, 1-phenyldecane (C_{16}) is only partially miscible with it.

The effect of substituting an α-proton by a halogen atom in toluene is somewhat different. While dichloromethylbenzene is fully CO_2-miscible, the solubility of trichloromethylbenzene is only 2%. Substituting aromatic protons by chlorine atoms has no effect on solubility in either direction (both chlorobenzene and dichlorobenzenes are soluble in SC CO_2).

3.4.2.2 Hydroxyl Compounds

Chain length, branching and type (primary, secondary or tertiary) all influence the solubility of alcohols. **Alcohols**

The solubility of primary alcohols decreases sharply with increase in the length of the carbon chain above 6 atoms. Thus, while *n*-hexyl alcohol and lower homologues are fully miscible with SC CO_2, the solubility of *n*-heptyl an *n*-decyl alcohol in this fluid is only 6% and 1%, respectively. As with hydrocarbons, branching in primary alcohols increases their solubility in dense CO_2. Thus, 2-ethyl-1-hexanol (C_8) is 17% miscible and 2,4,4-trimethyl-1-pentanol (C_8) is fully miscible. The solubility of short-chain diols and triols is markedly altered by partial and complete etherification. Thus, ethyleneglycol and glycerol are only 0.2% and 0.5% soluble, whereas their di- and trimethyl ethers are fully miscible. Also, β-methoxyethanol, β-ethoxyethanol and the monomethyl ether of diethyleneglycol are all fully soluble in SC CO_2. On the other hand, phenyl-substituted primary alcohols tend to be less readily soluble. In fact, benzyl alcohol and 2-phenylethanol are 8% and 3% soluble, respectively. Because branching increases the miscibility of primary alcohols with SC CO_2, secondary and tertiary alcohols of 7 or more carbon atoms should be more readily soluble in this fluid than is *n*-heptanol; in fact, 2-methyl-3-hexanol is fully miscible. Finally, cyclohexanol and 4-methylcyclohexanol are ohly 4% miscible.

As with aliphatic and alicyclic alcohols, etherification of phenolic OH results in substantially increased solubility. Thus, while phenol is only 3% soluble in SC CO_2, anisole is fully miscible with the fluid. **Phenols**

The nature of substituents and their position have a great effect on the solubility of phenol compounds. Thus, introduction of a chloro or nitro substituent in *ortho* with respect to the OH group in phenol makes it fully miscible. This is consistent with the fact that, because of the occurrence of an intramolecular hydrogen bond, *o*-nitrophenol is soluble in non-polar organic solvents. On the other hand, *para* chloro and nitro groups scarcely increase the solubility of phenol (*p*-chlorophenol and *p*-nitrophenol are 8% and 1.8% soluble in SC CO_2, respectively). The presence of a second **Substituents**

Second Substituent

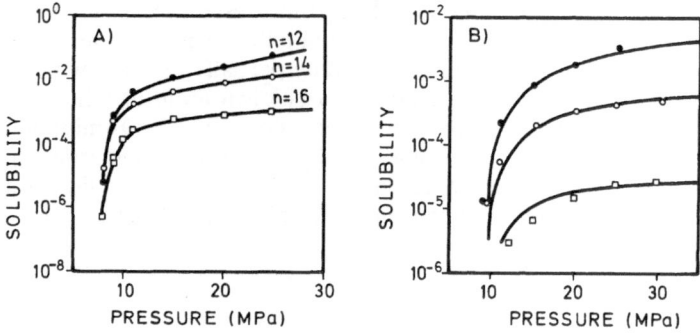

Fig. 3.11A,B. Influence of pressure on the SC CO_2 solubility of (**A**) fatty acids of variable chain length (n) and (**B**) their corresponding triglycerides. ($n = 12$, lauric acid; $n = 14$, myristic acid; $n = 16$, palmitic acid). (Reproduced with permission of the American Chemical Society)

chloro or nitro substituent at any position in the ring has an adverse effect on the solubility of o-substituted phenols. Thus, 2,4-dichlorophenol and 2,6-dinitrophenol are only 0.4% and 0.9% soluble, respectively.

Finally, a methyl substituent at any position in phenol usually increases its solubility. Thus, o-, m- and p-cresol are 30%, 20% and 30% miscible with SC CO_2.

3.4.2.3 Carboxylic Acids

Carboxylic acids of more than 10 C atoms are immiscible with SC CO_2. Thus, decanoic and lower acids are soluble, whereas dodecanoic acid is only 1% soluble. The full miscibility of ethyl dodecanoate suggests that the solubility of dodecanoic and higher acids can be substantially increased by converting them to esters of short-chain alcohols. The influence of **Chain length and solubility** chain length on the solubility of fatty acids and triglycerides is shown in Figs 3.11A and B, respectively. As can be seen, the solubility decreases with increasing chain length [15].

Other groups and solubility The presence of halogens, hydroxyl groups and aromatic rings decreases the solubility of carboxylic acids. Thus, chloroacetic acid and 2-hydroxypropanoic acid are 10% an 0.5% soluble in SC CO_2, respectively, while phenylacetic acid is insoluble in this fluid.

3.4.2.4 Ethers

As noted above in dealing with alcohols, etherification of a hydroxyl group usually increases the solubility of a compound. The chief reason for the increased solubility may be the decreased polarity of the resulting compound and the loss of intermolecular hydrogen bonds. One typical example is 1,4-dimethoxybenzene, which is much more readily soluble than its parent alcohol, 1,4-benzenediol. Etherification does not significantly increase the solubility of polynuclear aromatic alcohols, though. In fact, the solubility of 2-methoxybiphenyl and α-methoxynaphthalene is only 5% and 1%, respectively. The presence of nitro and amino groups in aromatic ethers detracts from their solubility (p-anisidine and o-nitroanisole are less than 2% soluble in SC CO_2).

3.4.2.5 Esters

In the same way as etherification, esterification usually increases the solubility of carboxylic acids, both aliphatic and aromatic. Thus, while 2-hydroxypropanoic acid is only 0.5% soluble, its ethyl ester is fully miscible with SC CO_2; also, ethyl phenylacetate is fully miscible with this fluid whereas its parent acid is utterly insoluble. The fact that, for example, methyl benzoate and methyl 2-hydroxybenzoate are both SC CO_2-soluble whereas their parent acids are not, shows that conversion of polar acid groups into non-polar ester functions has a dramatic effect on the SF solubility of carboxylic acids.

The solubility of trilaurin and palmitic acid is quite similar [15] notwithstanding the large molecular weight difference (the MW of the ester is around 2.5 times greater than that of the acid). The polar carboxyl group of the fatty acid makes it less readily soluble in CO_2, while the triester function of the triglyceride, less polar, facilitates mixing with the fluid.

3.4.2.6 Aldehydes

Lower aliphatic aldehydes such as ethanal, pentanal and heptanal are all SC CO_2-miscible. Unsaturations in aliphatic aldehydes seemingly have no adverse effect on their solubility since such compounds as 2-propenal and trans-2-butenal are fully miscible with SC CO_2. On the other hand, a phenyl substituent decreases the solubility of unsaturated aldehydes. 3-Phenyl-2-propenal is 4% soluble, while 3-phenylpropanal is 12% soluble. Benzaldehyde and 2-hydroxybenzaldehyde are both insoluble in CO_2.

3.4.2.7 Nitrogen-Containing Compounds

N-alkyl substituents have a dramatic effect on the CO_2 miscibility of amides, as shown by the fact that N,N-dialkyl (methyl and ethyl) for-   **Amides** mamides and acetamides are fully miscible while their unsubstituted homologues are virtually insoluble.

Amines Because carbon dioxide is weakly acidic, it has no special affinity for moderately basic amines such as aniline ($K_b = 4.2 \times 10^{-10}$), pyridine ($K = 2.3 \times 10^{-10}$) or picoline ($K_b = 9.38 \times 10^{-9}$). However, it forms salts with stronger bases such as ammonia and aliphatic amines ($K_b = 10^{-4}$).

The CO_2 solubility of amines usually decreases in the order ternary > secondary > primary, as reflected in the following solubility sequence: N,N-dimethylaniline (miscible) > N-methylaniline (20%) > N,N-diethylaniline (17%) > N-ethylaniline (13%) > aniline (3%). Therefore, N-alkyl substituents increase the solubility of amines, whereas aromatic substituents have the opposite effect; thus, N-ethyl, N-benzylaniline is only 4% soluble in SC CO_2.

Nitro compounds As with amino groups, nitro groups decrease solubility, particularly if more than one is present.

The solubility of aromatic nitro compounds varies widely. Thus, while nitrobenzene is fully miscible with SC CO_2, bicyclic nitro compounds are very sparsely soluble in this fluid. The nature and position of the substituent relative to the nitro group in an aromatic compound has a marked effect on its solubility. Thus, while o-nitrotoluene is fully miscible with CO_2, p-nitrotoluene is only 20% soluble in it. Likewise, as noted before, o-nitrophenol is soluble in CO_2, whereas p-nitrophenol is less than 2% soluble. Any other substituent has an adverse effect on the solubility of nitrobenzene.

Figure 3.12 summarizes the effects of chemical structure on SC CO_2 solubility classified according to whether they boost it or diminish it.

3.4.3 The Solubility Parameter

Giddings' theory The solubility of a solute in a supercritical fluid can be quantitatively estimated on the basis of Giddings' theory, which relies on differences between the Hildebrand solubility parameters for the SF and the solute.

Solubility in a supercritical fluid can be understood by examining the Gibbs–Helmholtz equation:

$$\Delta G = \Delta H - T\Delta S \qquad (2)$$

Dissolution will take place when the free energy of mixing, ΔG, is negative. Because the dissolution of some solutes is always associated with a large entropy of mixing increment, ΔS, solubility will ultimately be determined by the heat of mixing, ΔH.

SOLUBILITY IN SC CO_2

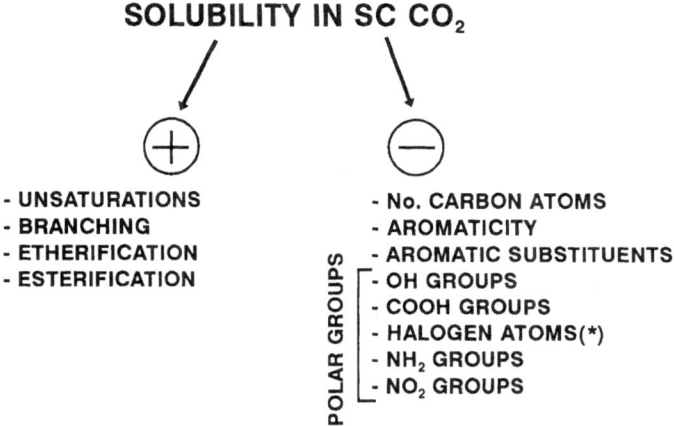

- UNSATURATIONS
- BRANCHING
- ETHERIFICATION
- ESTERIFICATION

POLAR GROUPS
- No. CARBON ATOMS
- AROMATICITY
- AROMATIC SUBSTITUENTS
- OH GROUPS
- COOH GROUPS
- HALOGEN ATOMS(*)
- NH_2 GROUPS
- NO_2 GROUPS

(*) NORMALLY (EXCEPT o-CHLOROPHENOL)

Fig. 3.12. Schematic summary of the beneficial and adverse effects of chemical properties on the solubility of organic compounds in supercritical CO_2

According to Hildebrand [24], the heat of mixing can be defined in mathematical terms as

$$\Delta H = v_1 v_2 (\delta_1 - \delta_2)^2 \tag{3}$$

where ΔH is the energy change arising from the formation and cleavage of intermolecular bonds; v_1 and v_2 are the partial volumes of the solvent and solute, respectively; and δ_1 and δ_2 are the solubility parameters of the solvent and solute, respectively. Solubility is therefore dependent on the differences between the solubility parameters of the fluid and solute: the smaller the difference between such parameters, the higher the solubility. As a rule, in the absence of crystal phases and hydrogen bonds, a solute will be soluble in a fluid if $(\delta_1 - \delta_2) < 1.7-2.0$.

In order to ensure that an analyte is dissolved in a fluid, the cohesion energy of the crystal structure of the substance concerned must be overcome through interaction with the fluid molecules.

The solubility of solutes is defined by the *solubility parameter*, which is given by the following expression:

$$\delta = \sqrt{\frac{\Delta E_v}{v}} = \sqrt{\frac{\rho(\Delta H_V - RT)}{M}} \tag{4}$$

where ΔE_v is the vaporization energy, v the molar volume, ρ the density, ΔH_v the heat of vaporization, and M the molecular weight of the solute, R being the gas constant and T temperature.

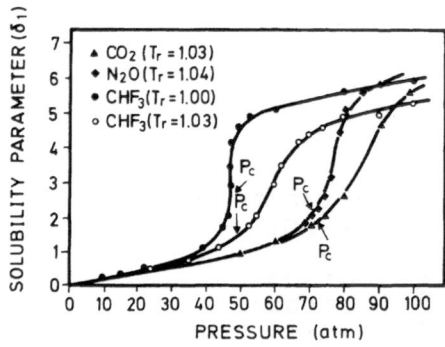

Fig. 3.13. Variation of the solubility parameter with pressure for various supercritical fluids. P_c = critical pressure. (Reproduced with permission of the American Chemical Society)

At a given temperature, the solubility parameter of a gas can be increased by raising the pressure. However, δ is not proportional to ΔH_v for dense gases at a high pressure because vaporization is impossible under such conditions.

The solvent power of a supercritical fluid is more accurately defined by the equation of Giddings et al. [25], in which the solubility parameter of the fluid is given by

$$\delta_1 = 1.25\sqrt{P_c}\frac{\rho_{r.SF}}{\rho_{r.L}} \tag{5}$$

where P_c is the critical pressure, $\rho_{r.SF}$ the reduced density of the supercritical fluid and $\rho_{r.L}$ that of the fluid in a liquid state. This equation reflects the variation of the solvent power of the gas as a function of pressure and temperature, and is related to the fluid properties via P_c.

The solubility parameter of a supercritical fluid, δ_1, increases with increasing density. It is related to pressure via density, as can be inferred from the previous equation and a comparison of the δ vs P and ρ vs P graphs in Figs 3.13 and 2.9, respectively. The solubility parameter is markedly dependent on pressure. The greatest solubility increments occur in the region immediately above P_c. This can clearly be seen in Fig. 3.13, which shows the variation of δ with pressure for several fluids (CO_2, N_2O, CHF_3). The solubility parameter of fluids also depends strongly on temperature, particularly at high pressures, as can be seen in Fig. 3.14, a plot of the solubility parameter of N_2O against the temperature at several pressures. Consequently, the solubility parameter of a fluid can be adjusted

Adjustment through pressure and temperature to the desired value through both pressure and temperature. If the difference between the solubility parameters of analytes and interferents is large enough, one can select appropriate pressure and temperature conditions where the analytes of interest will be soluble in the fluid but interferents will not.

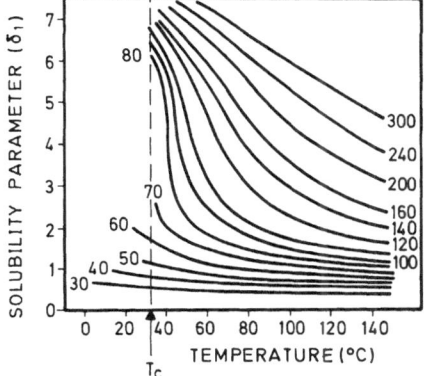

Fig. 3.14. Influence of temperature and pressure on the solubility parameter of SC N_2O. T_c = critical temperature. (Reproduced with permission of the American Chemical Society)

The solvent power of a supercritical fluid can be altered by changing its density to bring it close to those of a wide spectrum of liquid solvents. Figure 3.15 shows a solubility parameter scale including some of the most commonly used liquid solvents and supercritical CO_2 of variable density.

The solubility parameter for a dissolved analyte (δ_2) can be either obtained from the literature [26] or calculated from group contributions provided its structure is known. Not correcting solubility parameters for the effect of temperature can lead to substantial errors in the estimation of the pressure at which the solubility of a given solute in a supercritical fluid will be maximal. In the absence of experimental values at given temperatures, δ_2 values can be adjusted via thermal expansion coefficients. While the variation of δ_2 with temperature may seem of little relevance, it can be used to lower the costs involved in compressing the supercritical fluid by approaching δ_2 to δ_1. In practice, both δ_1 and δ_2 are temperature-dependent, but the decrease in δ_2 with increase in the temperature may offset the δ_1 drop undergone by the supercritical solvent. Phase changes in the solute molecules may lead to significantly decreased δ_2 values. An increase in the extraction temperature may decrease the solute cohesion energy, δ_2^2, and hence its solubility parameter, δ_2. The melting point of solutes is specially relevant to SFE since most solutes are more readily soluble in SFs when they are in a liquid state. Figure 3.16 shows two solubility isotherms for phenol in SC CO_2 run at 36 and 60 °C, respectively. As can be seen, the variation of the solubility of phenol (standard melting point, 41 °C) in supercritical CO_2 with pressure is markedly different at the two temperatures. Thus, the pressure-dependence at 36 °C is typical of a solid–supercritical fluid system. However, phenol is liquid at 60 °C, so its isotherm at this temperature is typical of a liquid–supercritical fluid system, where the solubility continues to increase with pressure, even at high pressure values.

Typical pressure dependence

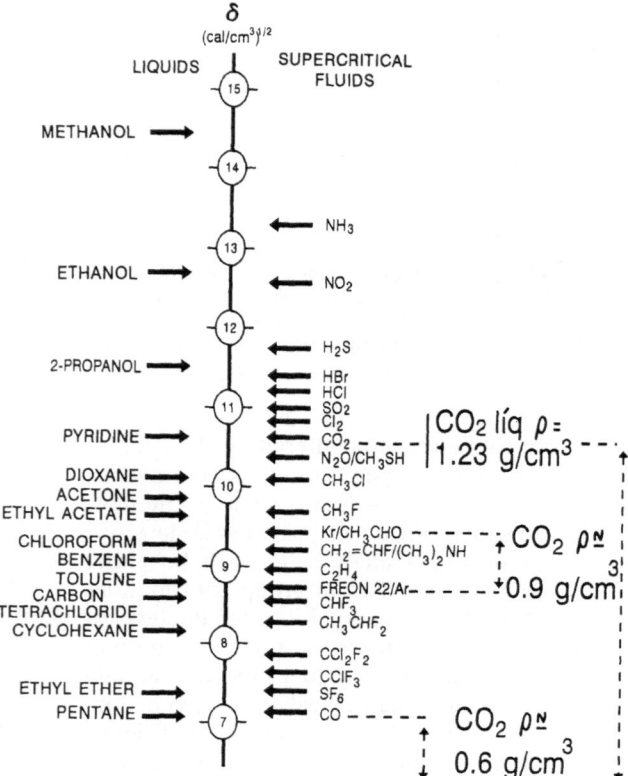

Fig. 3.15. Solubility parameters of various liquids (1 atm) and supercritical fluids. The maximum value is given when the density coincides with that of the fluid in the liquid state

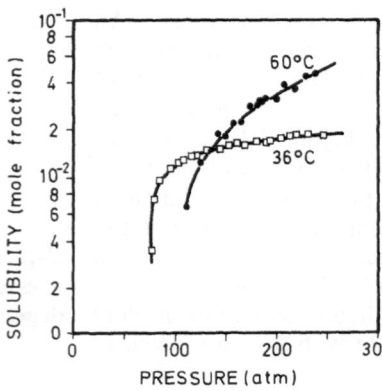

Fig. 3.16. Influence of the state of aggregation of phenol (36 °C solid, 60 °C liquid) on the variation of its solubility in SC CO_2 with pressure. (Reproduced with permission of the American Chemical Society)

Table 3.5. Calculation of the solubility parameter for caffeine

Group		$\Delta E v_i$ (cal/cm^3)	Δv_i (cm^3/mol)
3 CH$_3$		3 375	100.5
2 C=		2 060	−11.0
1 CH=		1 030	13.5
3 N		3 000	−27.0
1 −N=		2 800	5.0
2 C=O		8 300	21.6
2 5/6-Member rings		500	32.0
2 Conjugate double bonds		800	−4.4
$\delta_2 = 12.96$		21 865	130.2

The solubility parameter of a solute can be estimated by using Fedor's group contribution method provided its molecular structure is known. This method allows δ_2 to be calculated for many solutes with complex structures including carotenoids, alkaloids, pesticides, sterols and antibiotics. Table 3.5 illustrates the procedure used to estimate the solubility of a solute of known structure by using Fedor's method. Thus, the solubility parameter for caffeine was calculated by adding up the contributions of the individual groups to the vaporization energy and the molar volume of the overall structure, $\Sigma_i(\Delta E_v)_i$ and $\Sigma_i(\Delta v)_i$, respectively. The solubility parameter for the solute is calculated as the square root of the ratio of the summation of all energy contributions to the summation of all the group volumes:

Fedor's method

$$\delta_2 = \sqrt{\frac{\sum_i (\Delta E_v)_i}{\sum_i (\Delta v)_i}} \tag{6}$$

δ_2 values estimated by Fedor's method are consistent with those obtained from other sources. The method is very useful for estimating the extraction potential of complex molecules with supercritical fluids [27].

Fedor's method also allows one to estimate the solubility of polymers. Table 3.6 shows the relationship between the number of carbon atoms and δ_2 for a series of hydrocarbons of variable chain length. As can be seen, δ_2 increases with increasing chain length, the differences being particularly outstanding at the shorter lengths.

The modified solubility parameter, $\delta_r = \delta/\sqrt{P_c}$, provides a good correlation between experimental solute solubilities in various supercritical fluids [28]. Figure 3.17 shows solubility data for 2,6-diphenylnaphthalene in supercritical CO_2 and ethylene at 318 K over the pressure range 11.6–27.0 MPa. Figure 3.17A, which shows the variation of the solubility in the

Modified solubility parameter

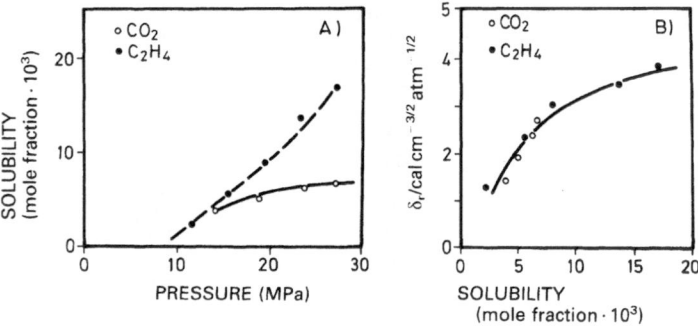

Fig. 3.17A,B. A Variation of the solubility of 2,6-dimethylnaphthalene in SC CO_2 and SC ethylene with pressure. **B** Variation of the modified solubility parameter with solubility. For details, see text. (Reproduced from Nippon Kagakukai)

Table 3.6. Variation of the solubility parameter of the solute (δ_2, 25 °C) as a function of the saturated hydrocarbon chain length

Number of C atoms	δ_2 (cal$^{1/2}$/cm$^{3/2}$)
3	6.2
4	6.618
5	7.021
6	7.242
7	7.423
8	7.554
9	7.648
10	7.722
12	7.841
14	7.927
16	7.993
18	8.040
20	8.080
Polyethylene (MW = 50 000)	8.35

two fluids with pressure, reveals a marked difference between the solvent power of CO_2 and ethylene. The solubility curves for the two supercritical fluids diverge throughout the pressure range studied: that of ethylene rises more sharply with increase in pressure than does that of CO_2. Judging from these two curves (Fig. 3.17A), the behaviour of the two fluids is rather different. Figure 3.17B shows the variation of the modified solubility parameter, δ_r, with pressure. As can be seen, the data for the two solvents converge on a single curve, so there is good correlation between the solubility of 2,6-dimethylnaphthalene and parameter δ_r.

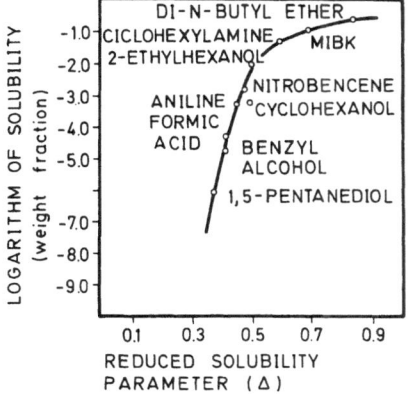

LOGARITHM OF SOLUBILITY (weight fraction)

DI-N-BUTYL ETHER
CICLOHEXYLAMINE   MIBK
2-ETHYLHEXANOL
                    NITROBENCENE
ANILINE   °CYCLOHEXANOL
FORMIC
ACID   BENZYL
         ALCOHOL
         1,5-PENTANEDIOL

-1.0
-2.0
-3.0
-4.0
-5.0
-7.0
-8.0
-9.0

0.1   0.3   0.5   0.7   0.9
REDUCED SOLUBILITY
PARAMETER (Δ)

Fig. 3.18. Variation of the logarithm of the solubility of various solutes in SC ethylene as a function of the reduced solubility parameter. (Reproduced with permission of Elsevier Science)

King and Friedrich [27] used solubility parameter ratios to establish quantitative correlations between the molecular structure of solutes and their solubility in supercritical fluids. For this purpose, they introduced the so-called "reduced solubility parameter", Δ, which is defined as $\Delta = \delta_1/\delta_2$, where δ_1 and δ_2 can be calculated from Eqs (5) and (6), respectively.

Solubility parameter ratios

The ratio of the solubility parameter of the solvent to that of the solute is a measure of the strength of solute–solvent interactions under different pressure and temperature conditions. Differences in the solubility of solutes in supercritical fluids depend not only on differences between the molecular structures of the solutes, but also on the pressure and temperature to which the supercritical fluid is subjected. These factors can alter the solubility of a given solute by a few orders or magnitude, even though solubility in a supercritical fluid is usually lower than in an ordinary liquid solvent. By way of example, Fig. 3.18 shows the variation of the natural logarithm of the solubility (weight fraction) as a function of the reduced solubility parameter, Δ, for various solutes dissolved in SC ethylene. The ethylene solubility data for the compounds shown encompass a wide range of reduced pressures (1.19–3.96 atm) and virtually isothermal conditions (20–26 °C). The y-axis is scaled logarithmically in order better to reflect the large variations between the solubility data. The solubility parameters of the solutes shown range between 7.7 and 14.4 cal$^{1/2}$/cm$^{3/2}$, and those of the supercritical fluids between 3.84 and 6.4 cal$^{1/2}$/cm$^{3/2}$. Increasing reduced solubility parameter values up to 0.5 results in increasing solubility; also the solubility at $\Delta < 0.5$ is extremely sensitive to the molecular structure of the solute. The effect is much less marked at $\Delta > 0.5$ and the maximum solubility is achieved at $\Delta \simeq 1$. The data in Fig. 3.18 seemingly conform to a function that can readily be fitted to a polynomial throughout the experimental data range. This correlation

Reduced solubility parameters

is very useful as it allows one to estimate rapidly the solubility of a solute under extraction conditions similar to those where the graph was run, provided the solubility parameter or structure of the solute is known.

The reduced solubility parameter, Δ, for a solute under given extraction conditions (fluid, P, T) can be calculated from Eqs (5) and (6), and used to obtain the corresponding solubility value by interpolation on a graph such as that shown in Fig. 3.18.

For a mixture consisting of a supercritical fluid (e.g. CO_2) and a modifier solvent, the solubility parameter for the modified fluid, δ, is given by

$$\delta = \phi_{CO_2}\delta_{CO_2} + \phi_m\delta_m \tag{7}$$

where ϕ is the volume fraction and subscript m denotes the modifier.

3.4.4 Theoretical Models

Theoretical treatments for estimation of the solubility of solutes in supercritical phases rely on the assumption that the chemical potential of the solute in the solid and fluid phase are the same.

Fugacity   The solute fugacity in the fluid phase can be described by considering the latter as an "expanded" liquid

$$f_2^L = f_2^{0L}\gamma_2 x_2 \exp\left(\int_{P^0}^{P} \frac{\bar{v}_2}{RT} dP\right) \tag{8}$$

or a far from ideal gas

$$f_2^V = \phi_2 y_2 P \tag{9}$$

where f_2^L and f_2^V are the solute fugacities in an expanded liquid and a non-ideal gas, respectively; f_2^{0L} is the fugacity of component 2 as a pure liquid; x_2 and y_2 are the solute mole fractions in the liquid and gaseous phase, respectively; γ_2 and ϕ_2 are the activity and fugacity coefficient, respectively; and $\bar{v}_2$ is the solute partial molar volume.

The expression for the fugacity of the solid phase is comparatively simpler:

$$f_2^S = f_2^{0S} \exp\left[\frac{v_2^S(P - P_{v2})}{RT}\right] \tag{10}$$

where f_2^S and f_2^{0S} are the fugacities of component 2 in the solid phase and in pure solid form; P_{v2} is the vapour pressure of the solid; and v_2^S is the molar volume of the solid.

If the fluid phase is considered to be a non-ideal gas, then the mole fraction of the solute in the fluid phase (solubility) is given by

$$y_2 = \frac{P_{v2}}{P} \frac{1}{\phi_2} \exp\left[\frac{v_2^S(P - P_{v2})}{RT}\right] \tag{11}$$

where ϕ_2 is the fugacity coefficient of the gaseous phase, which is a function of T, P and y_2, and is defined in thermodynamic terms as

$$RT\ln \phi_2 = \int_0^P \left(\bar{v}_2 - \frac{RT}{P} \right) dP \tag{12}$$

where $\bar{v}_2$ is the partial molar volume of the solute in the fluid phase.

Both alternatives entail solving the integral of $\bar{v}dP$, which is made difficult by the scarcity of critical partial molar volume data at different pressures and temperatures in the critical region.

The hard-sphere van der Waals (HS–VDW) equation of state and its many modified versions can be used in conjunction with a mixing rule in order to calculate $\bar{v}$. Modifications of the HS–VDW equation differ in the expressions they use for attractive (a) and repulsive terms (b) in the parent equation. Constant b is related to the size of the hard spheres considered. Parameter a can be regarded as a measure of intermolecular attractive forces. While b is assumed to be temperature-independent, a is only considered to be constant in the VDW equation.

Van der Waals equation

The Redlich-Kwong (RK) equation and Soave's modification (SRK), in addition to the Peng–Robinson (PR) equation [29], have successfully been used to correlate solubility and pressure. Parameters a and b in these equations are estimated from the critical properties of the components, which are occasionally unavailable or difficult to combine, particularly for mixtures of compounds with rather disparate critical parameters.

Correlation of solubility and pressure

The Carnahan–Starling–van der Waals (CS–VDW) model [5] requires none of the critical solute properties to be known, which is a substantial advantage over the above-mentioned models; rather, it correlates the term reflecting the interaction between solvent and solute, a_{12}, with the enthalpy of vaporization of the solute.

An improved version of the VDW equation, the Augmented van der Waals (AVDW) equation [18], includes an additional second-order term that takes account of the increased attractive forces in a mixture, which is experimentally more outstanding at low densities and temperatures near the critical value. This second-order term represents oscillations in the fluid structure arising from molecular compression in energetically favoured zones resulting from attractive forces. This non-random distribution of the fluid molecules or aggregation (clustering around the solute molecules) can have a particularly significant effect on the fugacity coefficient at a high compressibility.

Table 3.7 compares in terms of percent uncertainty the performance of the RK equation on the basis of critical properties with that of the CS–VDW model in the prediction of phenanthrene and anthracene solubility in supercritical ethylene at relative densities above 1.3. The CS–VDW model can be considered appropriate for this purpose since 10% uncertainty is fairly close to experimental uncertainty. However, as can be seen from the table, the model is less appropriate at relative densities between

Comparison of models

Table 3.7. Comparison of the performance of the CS–VDW and AVDW models, as well as the Redlich–Kwong (RK) equation (expressed as percent uncertainty)

	T (°C)	RK	CS–VDW	CS–VDW	AVDW
Minimum ρ/ρ_c	–	1.3	1.3	1.0	1.0
Phenanthrene–ethylene	25	7.6	8.1	21	12
	45	13	11	25	6.7
	70	30	11	17	14
Anthracene–ethylene	50	38	9.1	9.1	7.5
	70	24	14	17	7.3
	85	17	12	22	12

1.5 and 1.0. Its predicitive value at such densities is surpassed by that of the AVDW model (right-most column in Table 3.7), which is thus applicable over a wider density range.

The success of these theoretical models relies on two essential factors: the applicability of the hard-sphere van der Waals model and selection of the most physically meaningful method for calculating the parameters in the van der Waals equation.

The lattice model was developed to correlate equilibrium data for mixtures of substances with rather disparate molecular sizes [15]. It assumes molecules to occupy the cells of a three-dimensional cubic lattice of coordination number z and cell size v_H where each molecule occupies r sites and some sites are left vacant.

3.4.5 Influence of Cosolvents on Solubility: the Entrainer Effect

The entrainer effect is defined as the increase in the solubility of a solute on addition of a fairly small amount of a second solvent to the primary solvent (a supercritical fluid). Such an increase arises from interaction between the solute and cosolvent in the supercritical fluid through preferential intermolecular forces such as those involved in hydrogen bonding.

The increased solvent power of many systems is a result of the increased density of the solvent mixture, so it has no effect on selectivity. The solubility gain experienced by such systems is comparable to that obtained by adjusting the pressure and temperature of pure solvents. On the other hand, the entrainer effect is characterized by a sharp increase in both the solvent power and selectivity as a result of the presence of some cosolvent. It is a specific effect arising from chemical association between the cosolvent and solute [21,3], so investigating it requires a fair knowledge of the chemical interactions involved.

Altering supercritical fluids by addition of polar cosolvents adds a fur- **Polar**
ther dimension to supercritical fluid extraction as substances can be iso- **cosolvents**
lated on the basis of polar forces such as those involved in the formation
of electron donor–acceptor complexes, as well as on volatility differences,
for example. Polar cosolvents usually cannot be used as pure supercritical
solvents because their critical temperatures are too high for thermolabile
substances. Instead, CO_2 is used as primary solvent and a small amount of
a cosolvent including polar, acidic or basic groups is added in order to
combine the polarity and density effects. In this way, a small amount of
liquid solvent added to SC CO_2 may give rise to a solvent of higher
polarity or polarizability (or both), but still supercritical and widely
adjustable. The solubility of systems forming strong hydrogen bonds can
be increased by a factor of 3–7.2 by adding a 3.5 mol% of a polar
cosolvent.

A fairly high extraction yield can be maintained however much the pressure is reduced by simply adding a small amount of a suitable cosolvent;
thus, the solubility at 120 bar of benzoic acid in SC CO_2 containing
5.8 mol% SO_2 is comparable to that in pure CO_2 at 280 bar [19].

The solubility increase resulting from the use of a cosolvent is a function
of the chemical properties of the pure components. The solvent properties
of a polar or non-polar supercritical multi-component mixture can be
ascribed to the occurrence of dispersion, dipolar and acid–base forces.
Thus, a multi-component supercritical fluid can be highly selective towards
some solutes as a result of polar, hydrogen-bonding or specific chemical
forces.

As can be seen in Fig. 3.19, which shows the variation of the solubility
of benzoic acid in CO_2 containing different concentrations of sulphur

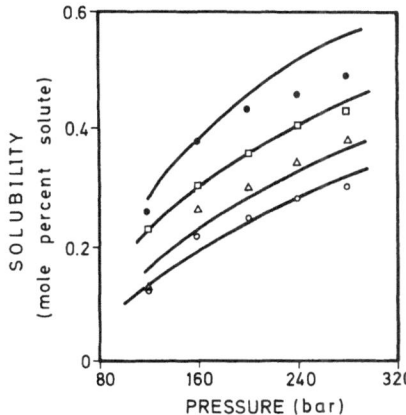

Fig. 3.19. Influence of the presence of different concentrations of sulphur dioxide in
supercritical CO_2 (● 5.85, □ 4.35, △ 1.86
and ○ 0 mol%) on the variation of the
solubility of benzoic acid with pressure at
35 °C. (Reproduced with permission of the
American Chemical Society)

Ternary isotherms dioxide as cosolvent with pressure at 35 °C, the shape of ternary isotherms (solute + CO_2 + cosolvent) is similar to that of binary isotherms (solute + CO_2). The ratio of solubility in the ternary system to that in the binary system is relatively pressure-independent. This simplifies the development of theoretical models for phase equilibria. The effect of density is roughly the same in a mixture of solvents as in pure CO_2; however, the polarity effect of the solvent mixture shifts the solubility isotherm to higher values by a fairly constant factor. The solubility gains produced by cosolvents rarely follow the same trend as their polarity since the more specific interactions of functional groups have a greater effect. Table 3.8 lists the solubility gains obtained with various cosolvents at 35 °C and a constant density of 20.5 g-mol/l (25 MPa for pure CO_2) [19]. Gains can be qualitatively explained on the basis of both attractive and repulsive forces. Thus, even though n-octane is the most polarizable of all the cosolvents shown, it also possesses the largest molecular volume, which leads to the strongest repulsive forces. Therefore, it is more useful to relate solubility gains to a ratio of energy to volume (i.e. a cohesion energy density or solubility parameter).

Hildebrand solubility parameter There is a wide variety of solubility parameters such as the total solubility parameter, δ^T (also called the Hildebrand value, which is calculated from heats of vaporization and undercooled liquid volumes). There are also solubility parameters descriptive of dispersion (δ^D), orientation forces or permanent dipoles (δ^O), inductive forces (δ^I), acidity (δ^A)

Table 3.8. Effect of various cosolvents on the solubility at 35 °C of some solids at a concentration of 20.5 mol/l in CO_2

Solid	Cosolvent	y_2 (ternary)/ y_2 (binary)
Benzoic acid	Methanol	3.7
	Acetone	2.1
	n-Octane	2.3
2-Aminobenzoic acid	Methanol	7.2
	Acetone	3.1
Phthalic anhydride	Acetone	1.7
Hxamethylbenzene	Methanol	1.1
	Acetone	1.2
	n-Octane	2.1
	n-Pentane	1.8
Acridine	Methanol	2.3
	Acetone	1.7
2-Naphthol	Methanol	4.5

and basicity (δ^B). Most solubility gains by effect of a cosolvent can be qualitatively justified on the basis of the dispersion and acid–base solubility parameters of the solute and cosolvent. For example, the increase in the solubility of benzoic acid ($\delta^A = 9.3 \, cal^{1/2}/cm^{3/2}$) in SC CO_2 is much greater with methanol ($\delta^B = 8.3 \, cal^{1/2}/cm^{3/2}$) than with acetone ($\delta^B = 3.0 \, cal^{1/2}/cm^{3/2}$), which is a weak base.

In the dense supercritical region ($\rho_r > 1.3$), the effect of a cosolvent on the solvent density contributes only moderately to increasing the solubility. Solubility gains, which are virtually independent of pressure, can be qualitatively predicted by using solubility parameters for the components to calculate the solute–cosolvent attraction parameters included in the corresponding equations of state. This approach avoids theories of state based on critical properties and takes account of dispersion, orientation and acid–base interactions, in addition to density effects. Predictions are quantitative for non-polar systems, the local composition effects of which are seemingly minimal. The solubility gain increases exponentially with increase in the a_{23}/a_{12} ratio, i.e. the ratio of the solute–cosolvent attraction parameter to the CO_2–solute attraction parameter – a's denote attractive van der Waals parameters and the subscripts CO_2 (1), the solute (2) and cosolvent (3); however, it also decreases markedly with increase in the cosolvent molecular volume (b, van der Waals volume). Thus, cosolvents of small molecular size such as methanol, which form strong hydrogen bonds with the solute, are predicted to raise the solubility by over 300% in some cases. On the other hand, the effects of orientation forces are comparatively negligible for solutes with a dipole moment of less than 5 debye.

Solute–cosolvent attraction

In choosing the solvent and cosolvent to be used in an SFE process, one should bear in mind Lewis acid–base properties [dissociation constants and solvatochromic parameters, viz. α (hydrogen bond donor ability) and β (hydrogen bond acceptor ability)]. Table 3.9 lists the α and β values for some of the liquid solvents most commonly used as cosolvents [30].

Choice of solvent and cosolvent

The entrainer effect is influenced by the chemical nature of the solute. Thus, a comparison of the solvent strength of pure CO_2 and CO_2 + 5% methanol for five solids of similar molecular size and structure but different functional groups (phenanthrene, fluorene, fluorenone, dibenzofuran and acridine) reveals that only the solubility of acridine is significantly raised (by a factor of 4–6) by the mixed solvent (CO_2–methanol) relative to the pure supercritical fluid (CO_2) [31]. Likewise, only those cosolvents that can interact with a given solute will exert an entrainer effect on it. Thus, only acetone among three other solvents (benzene, cyclohexane and methylene chloride) has a significant entrainer effect on benzoic acid dissolved in supercritical ethane [32].

The observed effects of methanol on acridine and acetone on benzoic acid can be ascribed to the formation of Lewis acid–base complexes by

Lewis acid–base complexes

Table 3.9. Solvatochromic parameters for some cosolvents

Compound	α	β
Benzene	0.00	0.10
Cyclohexane	0.00	0.00
Methylene chloride	0.30	0.00
Acetone	0.08	0.48
Chloroform	0.44	0.00
Diethyl ether	0.00	0.47
Dimethyl sulphoxide	0.00	0.76
Ethyl acetate	0.00	0.45
Methanol	0.93	0.62
Triethylamine	0.00	0.71

hydrogen bonding between the solute and cosolvent. Methanol, the first modifier, is a Lewis acid ($pK_a = 16.0$); also, acridine is the sole strong Lewis base ($pK_b = 3.35$) among the four solutes compared. The affinity of methanol as a Lewis acid for acridine (a strong Lewis base) accounts for the marked increase in its solubility. Likewise, acetone, the modifier for benzoic acid, is a Lewis base ($pK_6 = 21.1$) with a high affinity for the acid ($pK_a = 4.2$) in supercritical ethane. In fact, these differential chemical associations are reflected in spectroscopic measurements [30]. Based on solvato-chromic parameters, methanol (cosolvent) is both a strong hydrogen bond donor ($\alpha = 0.93$) and acceptor ($\beta = 0.62$). Phenanthrene should exhibit no hydrogen bond donor or acceptor trend since its mono-cyclic homologue, benzene, has $\alpha = 0.00$ and $\beta = 0.10$, respectively. On the other hand, pyridine (monocyclic) and quinoline (bicyclic), two acridine homologues and good hydrogen bond acceptors, have the same parameter values ($\alpha = 0.00$ and $\beta = 0.64$), so acridine is comparable to methanol as a hydrogen bond acceptor.

Even though the interaction that essentially determines the entrainer effect is that between the solute and cosolvent, the role of the supercritical fluid should not be overlooked. Acetone significantly increases the solubility of benzoic acid in supercritical ethane, but has a much less marked effect on its solubility in SC CO_2. Understanding why these systems are seemingly subject to no entrainer effect requires considering both the functional groups of the substances involved and their relative concentrations. Carbon dioxide contains two carbonyl oxygens that compete for hydrogen bond donors with other dissolved hydrogen bond acceptors. In the fluorenone/CO_2/methanol system, CO_2 competes with fluorenone to form hydrogen bonds with methanol, so, because the concentration of CO_2 is very much higher, no entrainer effect is observed. In the benzoic acid/CO_2/acetone system, CO_2 competes with acetone to form

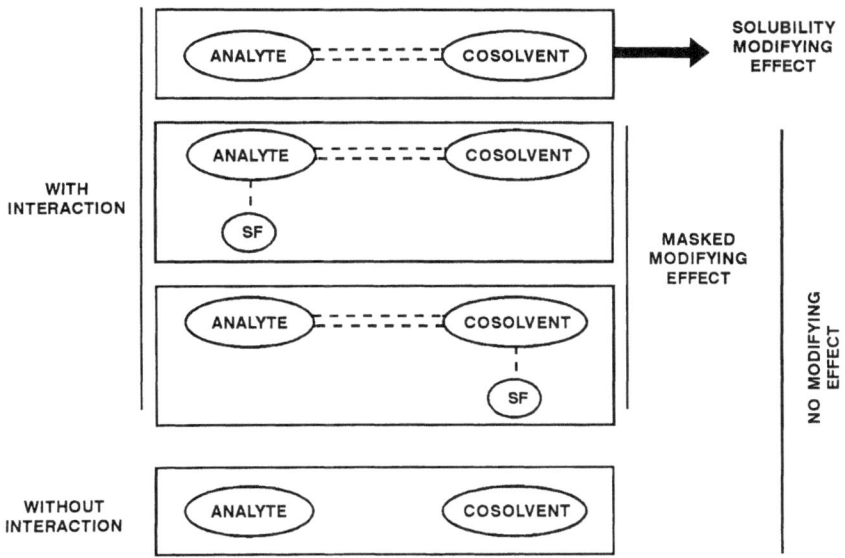

Fig. 3.20. Schematic diagram of interactions between the ingredients of an SF extraction that influence solubility: analyte–cosolvent, analyte–SF and cosolvent–SF

hydrogen bonds with benzoic acid. Again, the combined effect of the weak hydrogen bond acceptor character of CO_2 and its much higher concentration masks the entrainer effect. Figure 3.20 depicts the different possible situations in the presence and absence of analyte–cosolvent, analyte–supercritical fluid and cosolvent–supercritical fluid specific interactions. In the presence of the latter two, the entrainer effect is not apparent, not even if the analyte interacts with the cosolvent, so it is said to be a masked entrainer effect.

Figure 3.21 shows one other example demonstrating that the entrainer effect on solubility depends on the solute involved since, together with the cosolvent, it determines the type and strength of interactions. The factor by which the solubility in CO_2 at 35 °C is increased by addition of 3.5 mol% methanol (expressed as the ratio between the solute mole fractions in CO_2/methanol and pure CO_2) varies widely between the four solutes in the following order: 2-aminobenzoic acid (ABA) > 2-naphthol (2-NOL) > acridine (ACR) > hexamethylbenzene (HMB). Because the polarizability of methanol is similar to that of CO_2, the effect of the cosolvent on the solubility of non-polar solutes such as HMB will be smaller than on polar solutes (see Fig. 3.21). Methanol, a hydrogen bond donor and acceptor, boosts the solubility of ABA to a greater extent than that of benzoic acid.

Masking of entrainer effect

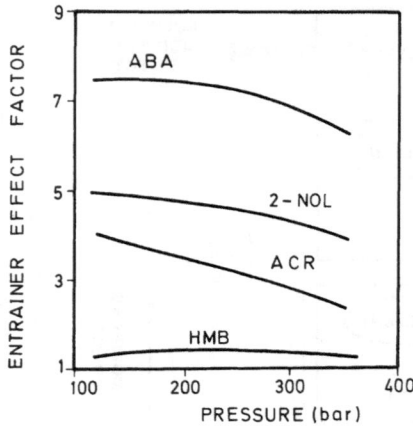

Fig. 3.21. Variation of the entrainer effect factor of 3.5 mol% methanol in SC CO_2 at 35 °C with pressure for various solutes. For details, see text. (Reproduced with permission of the American Chemical Society)

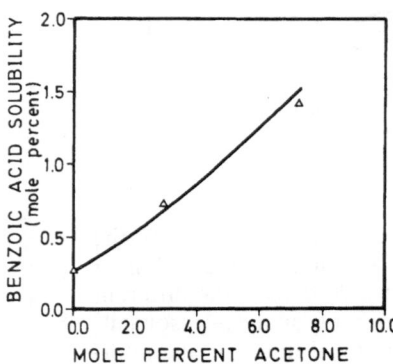

Fig. 3.22. Influence of the acetone content in SC ethane on the solubility of benzoic acid at 328 K and 182 bar. (Reproduced with permission of Elsevier Science)

This suggests that little or no zwitterion is formed, which is consistent with the extremely low dielectric constant of CO_2 (1.26 at 31 °C and 73 bar).

As can be seen in Fig. 3.22, which shows the variation of the solubility of benzoic acid in supercritical ethane containing different mole percentages of acetone at 328 K and 182 bar, the solubility is markedly dependent on the cosolvent concentration.

3.4.5.1 Clustering

Because of attractive intermolecular forces, a host of supercritical solvent molecules cluster around the solute molecules. The solvent "crowds" thus formed around solute molecules make up macroscopic clusters consisting **Macroscopic** of many coordination spheres. The size of such clusters can be estimated **clusters** from partial molar volume data.

The solvent clusters formed around the solute in mixed supercritical fluids are cosolvent-rich [21]. The local composition in supercritical CO_2 and cosolvent around an infinitely dilute solute can be estimated from solvatochromic measurements.

Table 3.10 shows the local compositions (as mole fractions) of cosolvent around the solute (Phenol Blue) in the first coordination sphere for several cosolvents at various concentrations and pressures. The liquid cosolvents included are quite different: acetone (non-protic, dipolar), methanol and ethanol (protic), and octane (non-polar, fairly polarizable). In all cases, the local cosolvent composition exceeds the bulk solution composition since the cosolvent interacts more strongly with the solute than does CO_2. The local composition shifts to the bulk value as the pressure is raised. The decrease in the local cosolvent concentration with increase in pressure is more marked for n-octane than for the polar cosolvents. This suggests that preferential solvation through hydrogen bonding is probably less

Table 3.10. Local cosolvent compositions around Phenol Blue (calculated from spectroscopic data obtained at 35 °C)

Overall conc. (%)	Pressure (bar)	Cosolvent local mole fraction (%)[a]			
		Acetone	Methanol	Ethanol	n-Octane
1.0	80	7.5	6.8	6.9	7.1
	85	4.2	3.7	3.8	3.2
	100	3.1	3.0	2.3	2.1
	200	2.5	2.6	1.8	1.5
	300	2.4	2.1	1.6	1.1
2.0	80	13.9	15.1	11.4	12.0
	85	8.2	9.6	7.0	7.4
	100	6.2	8.6	5.6	5.0
	200	5.0	7.3	4.3	3.0
	300	4.1	6.5	3.7	2.5
3.5	80	16.8	24.2	17.2	15.9
	85	11.1	18.2	12.9	11.7
	100	9.0	15.0	11.1	8.7
	200	7.1	12.0	8.3	4.6
	300	6.9	11.2	7.5	4.5
5.25	80	21.5	30.8	23.5	20.7
	85	16.1	24.8	18.8	16.4
	100	13.4	21.3	16.2	12.4
	200	10.1	16.8	12.4	7.3
	300	9.6	15.7	11.3	7.1

[a] In the first coordination sphere.

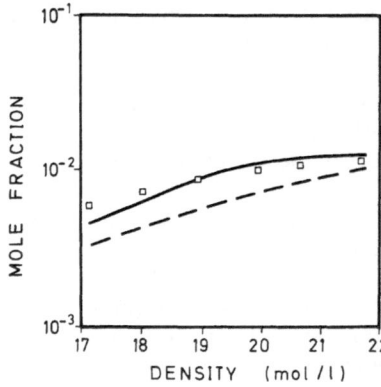

Fig. 3.23. Prediction of the solubility of acridine in SC CO_2 containing 3.5 mol% ethanol as a function of the SF density at 35 °C. (— AVDW-DDLC model; --- HSVDW model, □ experimental data). (Reproduced with permission of the American Institute of Chemical Engineers)

markedly dependent on compressibility than is that arising from non-polar forces.

Any theoretical model for prediction of solubility in modified supercritical fluids should therefore take the following into account: (a) that the solute interacts more strongly with the cosolvent than it does with CO_2; and (b) that its concentration in the solute local environment is increased as a result.

The Augmented van der Waals/density-dependent local composition (AVDW–DDLC) model [21] allows one to predict the effect of cosolvents on solubility with greater accuracy that other, earlier methods, because it takes account of both effects. Figure 3.23 shows experimental solubility values for acridine in CO_2 containing 3.5 mol% methanol at 35 °C and various densities. It also shows the theoretical solubility isotherms predicted by the HSVDW and AVDW–DDLC models, the latter of which is clearly closer to experimental facts.

3.4.5.2   Effect of the Cosolvent on Selectivity

The selectivity of a non-polar supercritical solvent towards polar solids relative to non-polar solids can be substantially increased (by one order of magnitude) by addition of small proportions of a cosolvent (e.g. methanol) without decreasing the dissolved solute concentration. Such a significant selectivity enhancement is feasible with solutes containing highly polar functional groups (e.g. hydroxyl, amino, carboxyl), which can take part in strong solute–cosolvent interactions (e.g. hydrogen bonding). The cosolvent can be used to increase selectively the solubility of solutes capable of interacting with it through dispersion, dipole–dipole and acid–base forces, with respect to other, non-interacting solutes.

Selectivity in SF applications is determined by the volatility of the solids involved and solute–cosolvent interactions in the supercritical phase. A non-polar supercritical solvent can be made highly selective towards similarly volatile compounds with different functional groups by addition of a small amount of an appropriate cosolvent. **Increasing selectivity**

The selectivity of the benzoic acid–hexamethylbenzene couple in SC CO_2 can be increased by using methanol as cosolvent. The two solutes have a similar vapour pressure and hence are similarly soluble. The solubility of benzoic acid in CO_2 is substantially increased on addition of a small proportion of methanol (Fig. 3.21). Thus, 3.5 mol% methanol raises the selectivity by a factor of 2.5. The yield, defined as the summation of individual solubilities, is also increased by 165% notwithstanding the substantially increased selectivity. On the other hand, selectivity decreases with increase in the yield through addition of a cosolvent, as is the case with the hexadecanol/octadecane/CO_2 system with acetone as cosolvent [33]. Hence the significance of choosing an appropriate solvent, one that exerts a greater entrainer effect on the species of interest than on any other – otherwise the selectivity towards such a species will be decreased in favour of the yield.

The entrainer effect of methanol is also responsible for a substantial increase in the selectivity of other systems such as 2-aminobenzoic acid/anthracene and 2-naphthol/anthracene. Thus, methanol has little effect on the solubility of anthracene since it is hardly more polarizable than CO_2. Molecular interactions in the form of hydrogen bonding between methanol and 2-aminobenzoic acid or between methanol and 2-naphthol are the key to the greatly enhanced selectivity of these systems. Selectivity is pressure-independent, whether a modifier is present or absent, over the range 120–350 bar, notwithstanding the fact that the HS-VDW model prediction that it should increase with increasing pressure in the presence of a modifier, as can be seen in Figs 3.24A and B, which show the experimental and HS-VDW-predicted variations with pressure of the ratio of mole fractions of the solutes (2-aminobenzoic acid/anthracene and 2-naphthol/anthracene, respectively) in the absence and presence of 3.5 mol% methanol. The presence of such a small proportion of the alcohol increases the selectivity by a factor of 2–10 (at a pressure of 120 bar) and the yield of the 2-aminobenzoic acid/anthracene system by 4.5.

3.4.6 Entrainer Effect of a Second Solute

The solubility of an analyte can be increased via the entrainer effect of a second, usually more readily soluble solute, with which it is coextracted. The solubility of the more soluble compound in the binary system is

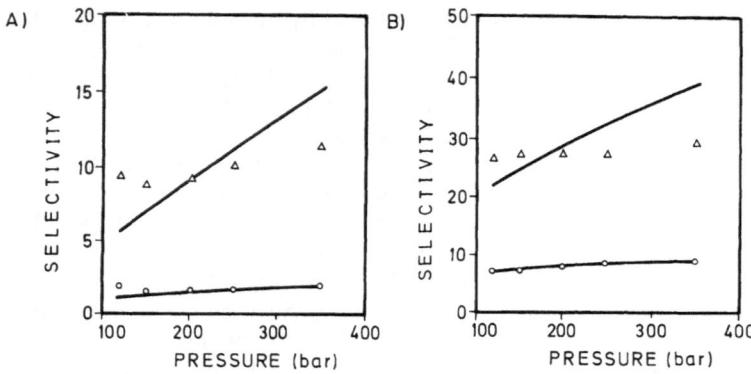

Fig. 3.24A,B. Influence of pressure on SFE selectivity in the presence (△) and absence (○) of a modifier. Straight lines correspond to theoretical predictions. **A** 2-Aminobenzoic/anthracene system. **B** 2-Naphthol/anthracene system. For details, see text. (Reproduced with permission of the American Chemical Society)

virtually the same, whereas that of the less soluble compound in the ternary system is increased as a result. Figures 3.25A and B show the solubility isotherms for phenanthrene and naphthalene at 308.2 K in a binary (one solid + one solvent) and a ternary system (two solids + one solvent), respectively. As can be seen, the solubility of naphthalene is scarcely altered by the presence of phenanthrene (less soluble), whereas that of phenanthrene is significantly higher in the presence of naphthalene, which occurs at a higher concentration [10].

Kurnik and Reid [10] measured the solubility of naphthalene(20%)–phenanthrene(75%), benzoic acid(280%)–naphthalene(107%) and phenanthrene(−10%)–2,3-dimethylnaphthalene(−10%) in SC CO_2 – the figures in brackets represent the maximum possible increase in the solubility of one component in the ternary system relative to its solubility in the binary system.

The solubility of a solute in a ternary system increases relative to that in the binary system proportionally to the solubility of the other solute in the supercritical fluid. For example, naphthalene is much more readily soluble in CO_2 than is phenanthrene. Therefore, naphthalene in a ternary system will increase the solubility of phenanthrene by 75% whereas its own solubility will be raised by only 20% as a result. Naphthalene and benzoic acid are both highly soluble in CO_2, so their solubility is increased by over 100% by virtue of their mutual presence.

Triglyceride mixtures   This trend has also been observed in triglyceride mixtures [15]; the solubility of trimyristin (MMM) and trilaurin (LLL) is scarcely affected by the presence of tripalmitin (PPP), whereas the solubility of this is increased

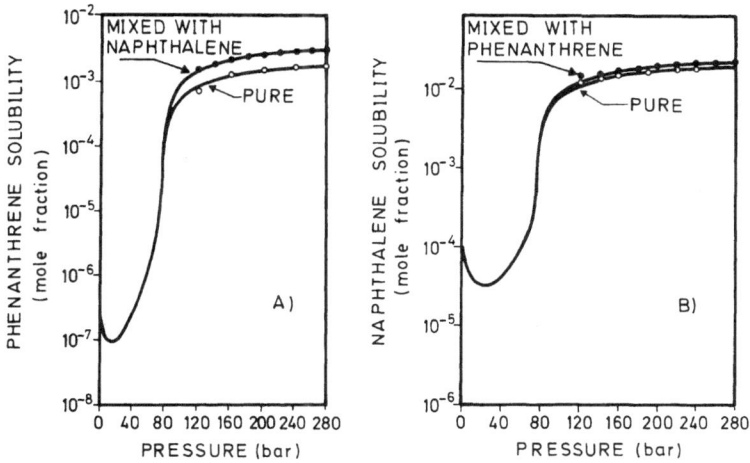

Fig. 3.25A,B. Variation of the SC CO_2 solubility of (**A**) phenanthrene and (**B**) naphthalene, both in pure and mixed form, with pressure at 308.2 K. (Reproduced with permission of Elsevier Science)

by the presence of both MMM and LLL, which is logical, taking into account their concentrations: that of PPP is lower than that of MMM and LLL by at least one order of magnitude. The increased solubility of PPP can be ascribed to the fact that the other, more soluble accompanying triglyceride (LLL or MMM), is present in a much higher concentration, so it exerts a favourable entrainer effect on the solubility of the other component (PPP).

When the interaction between a solute, s, and the new solute, s', is stronger than that between solute s and the solvent (e.g. because the new solute is more readily polarizable or forms hydrogen bonds), the solubility of solute s is increased by each molecule of s' that replaces a solvent molecule in the fluid phase.

3.4.7 Solubility near a UCEP

Normally, a solubility isotherm rises sharply near the critical pressure of a pure solvent, the rise being more gradual at high pressures. If the solubility of a solute continues to increase at high pressures, then one may consider the occurrence of a UCEP for the mixture under the prevailing conditions.

The solvent power of a fluid near a UCEP is appreciably higher than near the critical point of the pure fluid. The UCEP represents a critical

Solubility isotherm

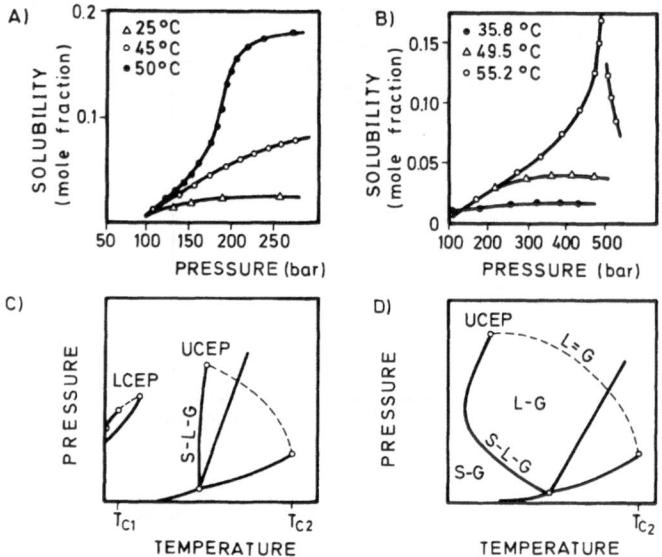

Fig. 3.26A–D. A,B Solubility isotherms and **C,D** phase diagrams for the naphthalene/ethylene **A,C** and biphenyl/CO_2 system **B,D**. For details, see text. (Reproduced with permission of the American Chemical Society)

mixing point in the presence of excess solid. The solubility of a solute in a supercritical fluids become increasingly sensitive to pressure and temperature changes as the pressure approaches that at the UCEP. The pressure required to work near a UCEP varies widely depending on the mixture components. Phase behaviour also varies widely at high pressures, even for very similar binary mixtures such as those of naphthalene–ethylene and biphenyl–CO_2. Both naphthalene and biphenyl are heavy, non-volatile solutes with a melting point well above room temperature (80.2 and 69.5 °C, respectively). On the other hand, ethylene and CO_2 are both gaseous under normal conditions and feature mild critical conditions (9 °C and 49.7 atm, and 31.05 °C and 72.8 atm, respectively). Figures 3.26A and B show the solubility isotherms for the naphthalene/ethylene and biphenyl/CO_2 systems, respectively, and Figs 3.26C and D show their respective $P-T$ projections. A comparison of the figures reveals a similarity between the isotherms of naphthalene at 50 °C and that of biphenyl at 55.2 °C. Both rise sharply with small pressure changes near 175 and 460 atm, respectively. On the other hand, the solubility increase for naphthalene from 45 to 50 °C in an isobar at around 200 atm is much greater than that observed from 25 to 45 °C. Similarly, the solubility of biphenyl at a constant pressure of around 450 atm increases much more markedly from 49.5 to 55.2 °C than between 35.8 and 49.5 °C. The sensitivity of the solubility of naphthalene and biphenyl to small pressure and/or temperature changes near 175 atm and 50 °C, and 460 atm and 55 °C, respectively, is a result of the nearness of the UCEP for the naphthalene/ethylene (52.1 °C, 174 atm) and biphenyl/CO_2 system (55.1 °C, 469.0 atm) [8]. The amount of solid dissolved in the supercritical fluid can be quite significant, with mole fractions above 0.15; 15 mol% naphthalene in ethylene or biphenyl in CO_2 corresponds to around 45% and 38%, respectively, by weight. However, at a temperature near that of the UCEP and pressures above this critical point, the solubility behaves rather differently for the two systems. Thus, at 55.2 °C and a pressure above 460 atm, the solubility of biphenyl in supercritical CO_2 decreases sharply with small pressure elevations; on the other hand, at 50 °C and a pressure above 174 atm, the solubility of naphthalene in SC ethylene increases with small pressure increases up to a limit. The disparate solubility of these two systems near the UCEP can be explained on the basis of the $P-T$ projection of the three-phase s-l-g curve. Unlike the naphthalene/ethylene system, the $P-T$ projection of the s-l-g curve for the biphenyl/CO_2 system goes by a temperature minimum. Consequently, the isotherm at 55.2 °C for the biphenyl/CO_2 system is representative of a liquid–gas(SF) equilibrium up to 469 atm and a solid–gas(SF) equilibrium above that pressure. On the other hand, the isotherm at 50 °C for the naphthalene/ethylene system is representative of a solid–gas(SF) equilibrium at any pressure and does not intercept the s-l-g curve of the system at any point.

3.5 Transport Phenomena

Transport phenomena most often determine the duration of a supercritical fluid extraction, whereas solute solubility determines its scope of application. Broadly speaking, the extraction of analytes from a given matrix involves three basic steps that are schematically illustrated in Fig. 3.27 and described below:

(1) The supercritical fluid is brought into contact with the surface and inside of the matrix, thereby starting the separation process.

(2) Retained analytes are removed by sweeping or displacement from the active sites of the matrix as a result of the higher affinity and/or concentration of the SF or modifier molecules. This is immediately followed by their dissolution (solvation) in the supercritical fluid.

(3) The analytes are transported from the inside of the matrix to its surface by essentially diffusional forces, and outside the matrix by primarily convective forces provided the extraction is performed in the **Diffusional forces convective forces**

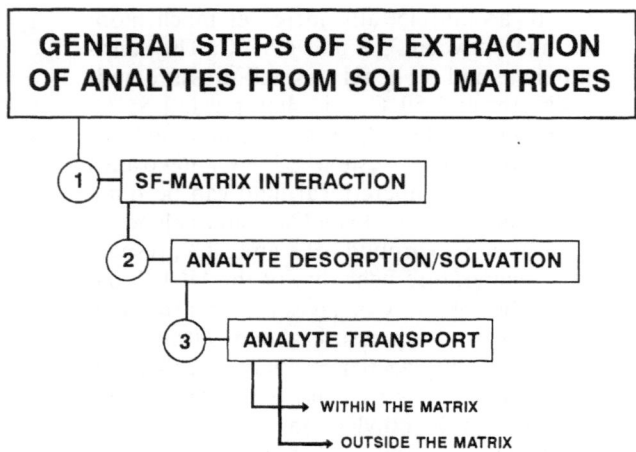

Fig. 3.27. Diagram showing the major steps involved in a supercritical fluid extraction

dynamic mode. In dealing with this transport, one should bear in mind the static fluid film formed around the matrix particles.

Most of the discussion of SFE has been focussed on solubility and pressure and temperature conditions in relation to solubility. However, solubility is rarely a limiting factor for extractions – particularly dynamic extractions. In many instances, the analyte is present in small amounts in the matrix and the concentration of the compound of interest in the supercritical fluid is below its solubility limit during the extraction. The fluid and conditions used are chosen so as to ensure that the analyte is soluble enough. The diffusion coefficient of an analyte in a supercritical fluid is higher than that in a liquid solvent and, obviously, a porous solid. In addition, the analyte is continuously exposed to a stream of pure solvent in dynamic extractions, so a potential saturation with the analyte should pose no hurdle to the extraction, however insoluble it may be. On the other hand, movement of the analyte to the bulk fluid is effected not only by diffusion, but also – and more significantly – by convection.

The rate of the overall extraction process is thus determined not only by the rate of the first or second step, but also by that of mass transfer from the matrix to the fluid.

One salient feature of SF extractions is that most of the compound is removed from the matrix in a fairly short time at the beginning of the process, after which, however, the extraction rate is considerably reduced; thus, extracting 99% of the solute takes 10 times as long as extracting 50%. This universal phenomenon may originate from different sources, namely: (a) differences in the extent of interaction between matrix and analyte arising from occurrence of the analyte in various forms or the presence of variously active sites in the matrix; (b) differences in the availability of analytes in the solid matrix arising from their location (internal or superficial) or distribution (even or uneven). These differences detract from the efficiency of industrial extractions and hinder quantification in analytical-scale extractions.

The situation varies depending on whether the desorption of adsorbed species or transport of the analyte from within the matrix particles to the outside is the rate-determining step. Below are discussed both possibilities.

3.5.1 Desorption of Adsorbed Species

The matrix surface and the analyte can interact weakly (through dispersive forces) or by chemisorption through electrons shared by the adsorbent and the analyte. Small geometry, basicity or acidity differences between molecules can result in substantial differences in the interaction of the matrix surface with various molecules. These phenomena operate differently for each analyte and sample owing to chemical differences between molecules, matrices and matrix–analyte interactions. The extraction conditions can be selected in such a way as to emphasize differences between interferences and analytes in a complex sample in order to achieve a selective separation.

Analytes partition between the adsorbent surface and the supercritical fluid where they are dissolved according to a partition or distribution coefficient, K.

A mass balance for the matrix–analyte system gives rise to

$$m\frac{d\theta}{dt} = -qc \tag{13}$$

where θ and c denote the concentration of the analyte species in the solid phase and fluid, respectively, m the sample weight and q the fluid flow-rate. At low concentrations, in the absence of intraparticular constraints, θ can be assimilated to the product of the distribution coefficient times the concentration (the distribution or partition coefficient, K, is the ratio of the analyte concentration in the solid to that in the solution, $K = \theta/c$). The solution to Eq. (13) yields the following expression for the relative concentration of adsorbed species:

$$\frac{\theta}{\theta_0} = \exp\left(-\frac{q}{mK}t\right) \tag{14}$$

where θ_0 and θ are the concentrations of the species in the initial solid ($t = 0$) and at time t, respectively.

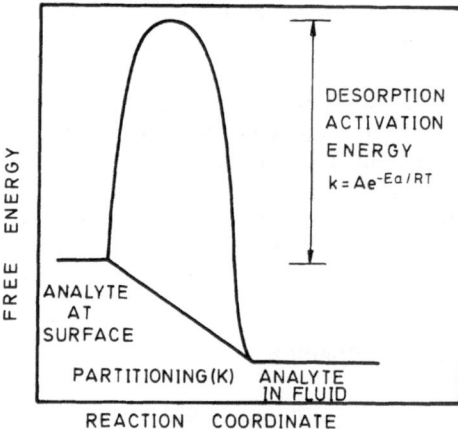

Fig. 3.28. Variation of the free energy during the desorption step in supercritical fluid extraction. (Reproduced with permission of the American Chemical Society)

However, immediate attainment of this equilibrium may be hindered by the kinetics of surface adsorption–desorption. Figure 3.28 shows the free energy for a general chemical desorption process. Desorption can be thermodynamically favourable ($\Delta G < 0$), but also kinetically unfavourable (i.e. pass by an energy maximum). The energy barrier involved, E_a, determines the kinetics of the process through the rate constant

$$k = A \exp\left(-\frac{E_a}{RT}\right) \tag{15}$$

where E_a is the activation energy, A the Arrhenius frequency or preexponential factor, R the gas constant and T the absolute temperature.

Energy barrier   Desorption from a solid surface, like any chemical reaction, is limited by an energy barrier. The energy barriers associated with weakly adsorbed species are fairly low, so they can readily be overcome thermally. In many cases, the analytes are chemically adsorbed on the surface matrix so they are desorbed via activated complexes. On the other hand, the high desorption activation energy of strongly adsorbed species must first be lowered by having the solvent molecules selectively interact with the matrix–analyte complex in order to cleave the link and release the analyte. Under these circumstances, the chemical and/or structural properties of the solvent are more significant than its solubility parameter, so the extraction can be substantially improved by altering the composition of the supercritical fluid, whether by using one of a different chemical nature (with pure fluids) or adding a small proportion of a modifier. For example, in the extraction of native polychlorinated dibenzo-*p*-dioxins (PCDDs) and dibenzofurans (PCDFs) from fly ash, CO_2 and N_2O – two structurally similar substances with a virtually identical solubility parameter – gives rise to rather different results. Thus, while CO_2 does not remove native PCDDs from their matrix, N_2O extracts them quantitatively [34].

3.5.2 Diffusion in the Solid

Molecular diffusion is the primary driving force of mass transfer within solid or small-pore size matrices. The rate of this process is determined by such factors as the coefficient of diffusion of the solute in the solid, the way the solute distributes in the solid (uniformly or non-uniformly between the surface and the inside), the solid porosity and its surface area to volume ratio (i.e. its geometrical shape and particle size). **Molecular diffusion**

The diffusion coefficient is inversely proportional to the square root of the solute molecular weight. Extraction can be favoured by the presence of a solvent capable of swelling the solid (e.g. a polymer) in order to facilitate diffusion of the analyte. Some porous samples accommodate organic colloids in their pores, so the analyte has to diffuse through the colloid as well. The presence of a modifier can be advantageous in this context as it is bound to alter the internal pore environment.

Below the two basic models used to explain dynamic extraction processes are discussed.

3.5.2.1 The Spherical Model

This models relies on the following assumptions:

a) The matrix particles are all spheres of the same size and the extractable material is evenly distributed within them at the beginning of extraction.
b) The rate at which the supercritical fluid is passed through the particles is always high enough to ensure that the concentration of extracted compound in the fluid is close to zero.
c) The compounds to be extracted move through the matrix in a diffusional-like fashion.

Even though real samples fail to meet the first assumption – whether because of their particle shape or size/concentration distribution – the model has been validated by the experimental results obtained for a wide variety of samples [35].

The mathematical solution to the diffusion equations for a compound present at an initially uniform concentration in a solid sphere of radius r immersed in a fluid where the compound is continuously maintained at a zero concentration is similar to that for heat conduction in a hot sphere immersed in a cold fluid [36,37], so it is called the *hot-ball model*. By expressing the equations of the previous model in terms of diffusion one obtains the following equation for the mass ratio, m, of extractable compound that remains inside the sphere after extraction time t, to the initial mass of extractable material, m_0:

$$\frac{m}{m_0} = \frac{6}{\pi^2} \sum_{n=1}^{\infty} \frac{1}{n^2} \exp\left(-\frac{n^2\pi^2Dt}{r^2}\right) \tag{16}$$

where D is the diffusion coefficient of the compound in the sphere. Equation (16) can be simplified by introducing a reduced time, t_r, which will be proportional to time ($t_r = \pi^2Dt/r^2$) for any system:

$$\frac{m}{m_0} = \frac{6}{\pi^2}\left[\exp(-t_r) + \frac{1}{4}\exp(-4t_r) + \frac{1}{9}\exp(-9t_r) + \cdots\right] \tag{17}$$

The solution to Eq. (17) is a summation of terms that decrease exponentially with increase in time. At long times, the last terms are negligible relative to the first. Therefore, a plot of ln (m/m_0) against time or a time-proportional quantity such as t_r will be linear at long enough times. Such is the case with the graph in Fig. 3.29A. The curve initially falls sharply and then more gradually until it becomes linear at $t_r \simeq 0.5$. The linear portion of the curve extrapolates to an intercept of about -0.5 at $t = 0$. The physical significance of the shape of the extraction curve lies in the fact that the process initially involves a massive concentration at the sphere surface, from which the solute rapidly diffuses to the fluid. As extraction develops, the surface concentration decreases, so a steep concentration

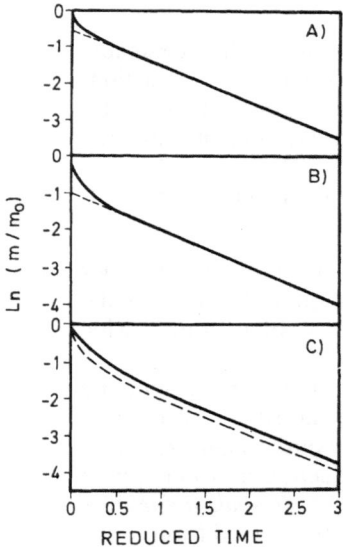

Fig. 3.29A-C. Prediction of the variation of the logarithm of the ratio of unextracted to initial mass as a function of reduced time. **A** Basic model. **B** Model including the effect of particle shape. **C** Model including the effect of solubility constraints (—) compared with the basic model (---)

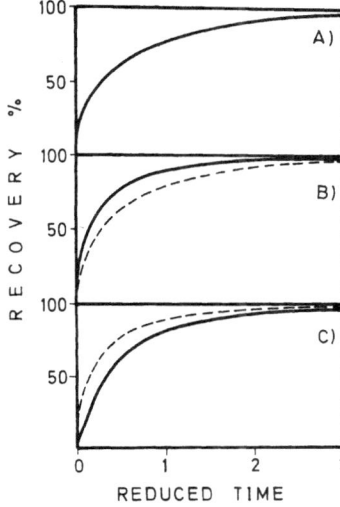

Fig. **3.30A–C.** Prediction of the variation of recovery as a function of reduced time. **A** Basic model. **B** Model including the effect of particle shape. **C** Model including the effect of solubility constraints (—) compared with the basic model (---)

gradient is established near the surface and diffusion, the rate of which is proportional to the gradient, is still quite fast. At long extraction times, the concentration profile across the sphere is much more gradual, so diffusion is much slower.

Even though the initial fall in the curve of Fig. 3.29A may seem of little significance relative to the overall graph, it represents the extraction of most of the compound. In fact, 63% of the solute is extracted during the time needed to reach $t_r = 0.5$. However, the time required to extract 99% of the material is such that $t_r = 4.1$. Figure 3.30 shows the whole extraction vs t_r plot predicted by this model, which reproduces the long interval of the extraction curve corresponding to very low extraction rates.

Soloving the expression defining the reduced time for t yields

$$t = \frac{r^2}{\pi^2 D} t_r. \tag{18}$$

t_r depends on a single variable: the percent recovery needed (based on the graph in Fig. 3.30, for example, $t_r = 4.1$ for 99% extraction). From Eq. (18) it follows that the time required for a given proportion of solute to be extracted is inversely proportional to the diffusion coefficient. Obviously, the easier diffusion within the solid (i.e. the higher D), the shorter will be the extraction process. Also, the smaller particle size is, the faster will be the process since t is directly proportional to the sphere radius squared. Thus, halving the sphere radius reduces the extraction time to one-fourth.

Extraction models based on non-spherical particle geometries provide curves similar to that in Fig. 3.29A. The initial curve fall is steeper for geometries with surface to volume ratios larger than that of a sphere. This effect is particularly outstanding in uneven, rough particles. One applicable curve is shown in Fig. 3.29B; the corresponding extraction vs t_r graph is depicted in Fig. 3.30B (the basic spherical model curve is also included as a dotted line for comparison).

The limiting effect on solubility is illustrated in Fig. 3.29C. The initial fall rate of the logarithmic curve is reduced, thereby delaying appearance of the linear portion and shifting the curve to the right relative to the prediction of the infinite dilution model, which is represented by a dotted line. Figure 3.30C shows the corresponding extraction vs t_r curves.

In real samples, analytes may not be uniformly distributed within the matrix, but at higher concentrations near the surface e.g. if they were newly adsorbed, cover the matrix or were displaced to the surface during grinding). The effect is similar to that of an uneven particle shape. In other cases, the analyte concentration near the surface has previously been reduced (e.g. by virtue of the leaching effect of rain water), so the curve has a gentler initial falling portion.

A comparison of the model predictions with experimental extraction data reveals an initial fall and a final linear portion in $\ln (m/m_0)$ vs time plots. The intercept of the straight portion is more negative than the theoretical value for a sphere (-0.5). The results can always be justified in terms of particle shape, solubility constraints and an uneven distribution of extractable components in the spheres.

Figure 3.31 shows the extraction of phenanthrene from a soil using SC CO_2 at 50 °C, a flow-rate of 0.45 ml/min and two different pressures: 400 and 180 atm. The experimental curves are identically shaped with those predicted by the theoretical model. The somewhat more negative experimental intercept may be the result of particle shape unevenness or a preferentially superficial distribution of the analyte.

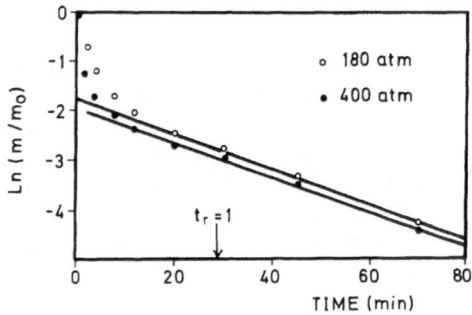

Fig. 3.31. Confirmation of the theoretical predictions of Fig. 3.29 for the SFE of phenanthrene from soils using SC CO_2 at 50 °C and two different pressures. For details, see text

The extraction of additives from polymers can be assumed to meet the model requisites quite strictly. Additives are uniformly distributed and the extraction process involves diffusion across the polymer. The sample particles, however, can have dissimilar shapes and sizes after they are ground to reduce their size so as to accelerate extraction. The slope of the linear portion obtained in the extraction of tinuvin-326 from a ground, sieved polypropylene sample with pure CO_2 are consistent with the model predictions. The slope provided by particles smaller than 0.6 mm in diameter is roughly five times greater than that yielded by larger particles (diameters between 0.6 and 1.2 mm); this corresponds to an effective particle diameter ratio of 2.3 between the two samples.

Polymer additive extraction

3.5.2.2 The Infinite Slab Model

The extraction model for thin film-shaped samples can also be derived from heat conduction equations by converting heat terms into diffusional terms [38]. In the extraction of a compound with a diffusion coefficient D which is uniformly distributed across an infinite film of thickness L, the ratio between the unextracted and initial mass at time t is given by

$$\frac{m}{m_0} = \frac{8}{\pi^2} \sum_{n=0}^{\infty} \frac{1}{(2n + 1)^2} \exp\left[-(2n + 1)^2 \pi^2 \frac{D}{L^2} t \right] \tag{19}$$

which, expressed in terms of the relative time, t_r, simplifies to

$$\frac{m}{m_0} = \frac{8}{\pi^2}\left[\exp(-t_r) + \frac{1}{9}\exp(-9t_r) + \frac{1}{25}\exp(-25t_r) + \cdots \right] \tag{20}$$

The $\ln(m/m_0)$ vs t plot predicted by this model is essentially identical with that of the spherical model, but the linear portion is reached more rapidly (Fig. 3.32A). The approximate equation for long times is

$$\ln\frac{m}{m_0} = -0.2100 - t_r \tag{21}$$

The theoretical extracted fraction vs time curve for an infinite slab is qualitatively similar to that of a sphere and also ends in a long tail (Fig. 3.32B). An overall 37% of solute is extracted during the initial interval defined by $t_r = 0.25$. However, 99% extraction is only achieved at $t_r = 4.4$, i.e. a 17-times longer time.

These theoretical models provide an extrapolation procedure for obtaining quantitative analytical information in a shorter time than that required for exhaustive extraction of the sample. For example, direct extraction of polymer pellets and use of an extrapolation method avoid the risk of additive losses in grinding.

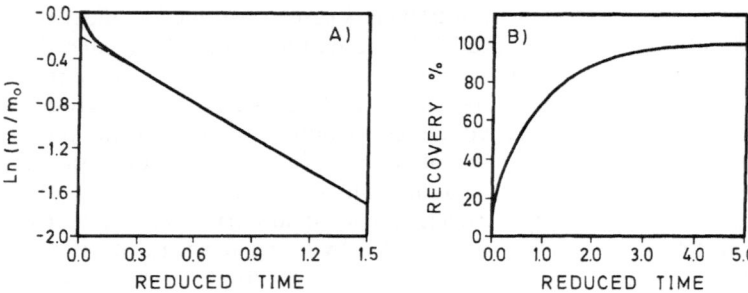

Fig. 3.32A,B. Theoretical performance of the infinite film slab in SFE. **A** Variation of the logarithm of the ratio of unextracted to initial mass and **B** recovery as a function of reduced time. (Reproduced with permission of the American Chemical Society)

The overall mass of compound in the sample, m_0, is given by

$$m_0 = m_1 + \frac{m_2^2}{m_2 - m_3} \tag{22}$$

where m_1 is the solute mass extracted over an interval at least as long as that encompassed by the initial non-linear portion of the solubility curve, and m_2 and m_3 are two masses extracted at two consecutive, identical subsequent times such that $t_3 - t_2 = t_2 - t_1$.

3.6 Factors Influencing Supercritical Leaching

Supercritical fluid extraction processes are affected by a host of experimental variables. In an SFE process, the analytes are (a) transferred from their host matrix to the supercritical fluid, (b) flushed from the extraction cell by the SF, and (c) collected for subsequent analysis. This section deals with variables which affect extraction proper (i.e. steps a and b), while those influencing collection and/or determination of the analytes are examined in Sect. 4.3.5.

Table 3.11 lists the principal variables influencing supercritical fluid extraction processes.

The large number and variety of factors on which SFE performance relies makes optimizing this type of process rather a difficult task. The **Simplex method** simplex method has traditionally been used to minimize the time required for development of extraction methods and increase the extraction efficiency [39]. Also, statistical procedures have been employed to identify the key parameters of extraction processes and minimize the number of experiments required for optimization [40,41].

Table 3.11. Experimental variables affecting SFE efficiency

Features of the fluid	**Dynamic factors**
– Nature of the fluid (polarity)	– Extraction time
– Pressure (density)	– Extracting fluid flow-rate
– Temperature	– Extraction cell (size, geometry, void
– Presence of a modifier	volume, stirring–US)
– Modifier concentration	
– Overall volume of extracting fluid	**Sample pretreatment**
	– Addition of liquids, solvents, derivatizing
Features of the solute	reagents, acids, etc.
– Type of analyte (volatility, polarity, MW)	– Addition of solids
– Concentration	
	Analyte collection mode
Features of the solid	(Chapter 4)
– Sample size	
– Particle size	
– Nature of the matrix (polarity and	
covalent binding to the analyte)	
– Presence of other extractable substances	
– Sample conditions (moisture, fat content,	
pH)	
– Encapsulation phenomena	

SFE variables can be altered for a twofold purpose, viz. improving the extraction efficiency and/or selectivity. This section discusses how such variables affect extraction efficiency, while Sect. 5.2.10 examines how differences in the extraction efficiency and/or collection of species can be exploited for boosting selectivity.

Variables can be adjusted in order to increase the extraction rate and/or the maximum amount of analyte that can be extracted. While increasing the extraction rate makes the process more expeditious, it is probably more interesting to improve the overall analyte recovery. In some cases, the analyte can only be recovered to a given extent however long the extraction is extended. The variables on which increased extracted levels rely are the most significant for optimization when quantitative extraction is sought.

Quantitative extraction

SFE variables can be optimized by determining the amount of analyte extracted over a preset interval at different values of the variable concerned or by running the whole cumulative extraction vs time curve for each variable value studied. The former choice is faster, but may lead to incorrect conclusions (Fig. 3.33). If measurements are made at a preset time t_1 shorter than that required for maximum efficiency (t_2 or t_{max}), the process involves comparing extraction rates, so no information on extraction efficiency can be obtained. Conversely, if measurements are made at $t > t_{max}$, one obtains information on the extraction efficiency but not on the extraction rate. Under the circumstances of Figs 3.33A and B, using t_1 to

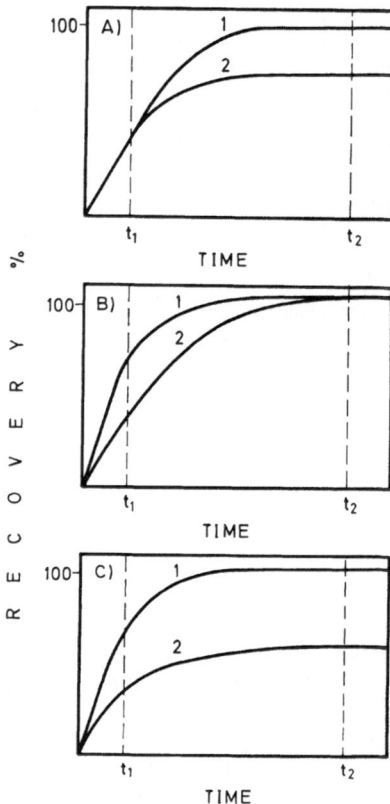

Fig. 3.33A–C. Optimization of SFE: study of the influence of a variable at two different values (1 and 2). **A–C** Graphs showing the kinetics of extraction in three common situations. Significance of using times t_1 and/or t_2 in the experiments

t_2 exclusively would lead to errors of opposite sign. On the other hand, under the conditions of Fig. 3.33C, any conclusions drawn from t_1 would be correct. Therefore, in studying the effect of a given variable, one should determine the recovery at both a relatively short time and a long enough time.

3.6.1 Properties of the Supercritical Fluid

The chief properties of supercritical fluids used as extractants (solvent power, penetrability, diffusivity, polarity, etc.) are discussed in Sect. 2.2.2. In practice, the performance of an SFE method relies on two essential properties of the supercritical fluid: its density, which is determined by its pressure and temperature, and its chemical nature (polarity).

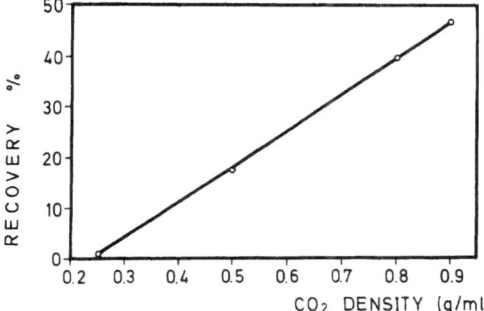

Fig. 3.34. Influence of SC CO_2 density on the recovery of pyrene added to silica after 10 min at a constant temperature of 40 °C. (Reproduced with permission of VCH)

The solvent power of an SF is a function of its density. No other separation technique allows the solvent power to be so readily altered by simply changing the physical conditions. The relationship between the solvent power and such fluid properties as pressure, temperature and density was described in Sect. 3.4. At a constant temperature, the density and hence the solvent power of an SF increases with increasing pressure. Optimization of an SFE procedure usually involves keeping the temperature constant and changing the pressure, though. **SFE optimization**

Figure 3.34 shows the influence of the CO_2 density on the extraction of pyrene added to silica on constancy of all variables except – obviously – the pressure [42]. As can be seen, the recovery is virtually linearly related to density throughout the range assayed.

Table 3.12 compares the results obtained in the extraction of analytes from a polluted soil at different pressures. As can be seen, full recovery of 4- and 5-ring PAHs required higher pressures than the rest [43].

As shown in Fig. 3.35, increased pressures improve the extraction efficiency of cholesterol added to glass wool. The minimum pressure required to achieve 98% recovery in 30 min is 17.7 MPa [44]. The efficiency with which diuron can be extracted from soils also increases with increasing pressure at the beginning of the process (Fig. 3.36) [45]. While the extraction rate falls as extraction develops, the overall amount extracted (average recovery = 70%) is statistically the same at three different pressures (110, 235 and 338 bar). At short extraction times, recovery can be boosted by using a higher pressure; at long extraction times, though, little is gained by increasing the pressure.

Occasionally, an increased pressure not only increases the solute solubility, but also facilitates its diffusion. Polymers adsorb large amounts of CO_2 under supercritical conditions (e.g. methyl polymethacrylate absorbs 25% CO_2). As a result, polymer pellets swell considerably, thereby favouring diffusion of the solute. Even though the diffusivity of dense gases is inversely proportional to pressure, CO_2 absorption increases with **Effect of pressure on diffusion**

Table 3.12. Effect of pressure on the extraction of PAHs from polluted soils[a]

Compound	Accepted concentration range (ppm)	Concentration (ppm)		
		250 atm	350 atm	400 atm
Naphthalene	24.2–40.6	23	23	25
Acenaphthylene	14.7–23.5	20	–	22
Acenaphthene	527–737	566	601	614
Fluorene	414–570	445	471	458
Phenanthrene	1270–1966	1682	1978	1911
Anthracene	373–471	357	439	400
Fluoranthene	1060–1500	1028	1459	1571
Pyrene	744–1322	703	1153	1269
Benzo[a]anthracene	214–290	74	235	284
Chrysene	271–323	74	251	314
Benzo[b,k]fluorantene	130–174	<1.0	107	155
Benzo[a]pyrene	80.1–114.3	<1.0	64	89

[a] Conditions:
SFE: 65 °C, 40 min, 0.9 ml/min.
GC: methyl 25 m × 0.2 mm i.d., from 60 °C (20 min) to 280 °C (30 min) at 7 °C/min;
Detection: MS.

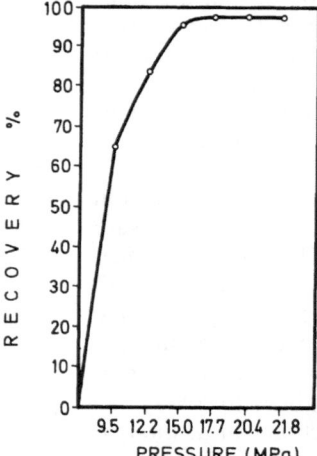

Fig. 3.35. Influence of pressure on the SFE efficiency of cholesterol added to glass wool using SC CO_2 at 45 °C for 30 min

it; therefore, the rate of solute diffusion through a polymer matrix can be raised by increasing the pressure. Figure 3.37 shows the variation of the proportion of additive extracted from poly(vinyl chloride) (PVC) as a function of pressure [46]. The curve rises sharply near the critical pressure and then levels off at around 40 MPa.

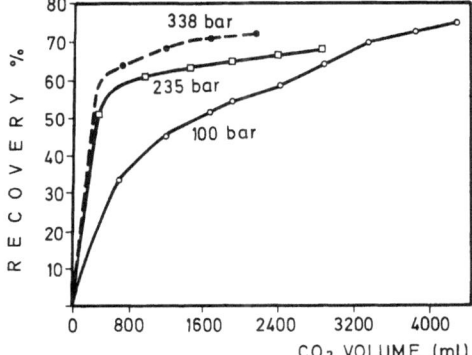

Fig. 3.36. Influence of pressure on the SFE of diuron from soils using 90:10 CO_2/acetonitrile at 100 °C and a flow-rate of 16.5 ml/min. (Reproduced with permission of Elsevier Science)

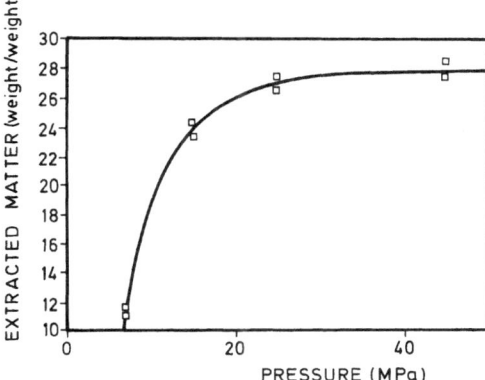

Fig. 3.37. Variation of the SFE efficiency of PVC additives with SC CO_2 as a function of pressure. (Reproduced with permission of the Royal Society of Chemistry)

Even though the analyte solubility in the supercritical fluid is a central factor of SFE, the supercritical fluid must also overcome analyte–matrix interactions occurring in some samples. Such interactions can readily be overcome by fluids of a higher polarity than CO_2. In such cases, diffusion in the solid matrix does not seem to be the extraction rate-determining step; otherwise there should be no change in the recovery rate on changing the nature of the supercritical fluid – unless the fluid induced a deep physical change in the matrix. The greater suitability of polar fluids is a result of their ability to interact with the matrix and efficiently compete with the analyte for its active sites. As can be seen in Fig. 3.38 for tetrachlorodibenzo-*p*-dioxins (TCDDs) and PAHs, SC N_2O enables more expeditious extraction of some samples than does SC CO_2. This is rather surprising taking into account the great similarity of the two fluids in terms of physical properties and solubility parameter; however, unlike CO_2, N_2O has a non-zero permanent dipole moment [47], hence the difference.

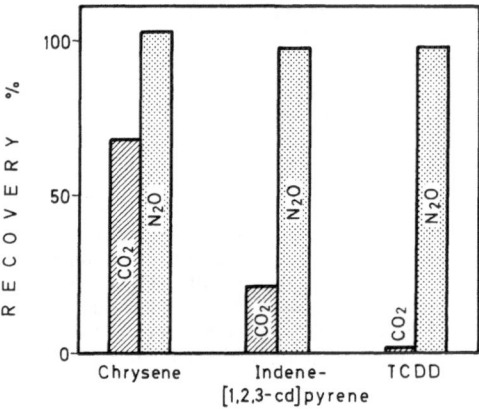

Fig. 3.38. Comparison of the extraction efficiency of various compounds with SC CO_2 and SC N_2O. For details, see text. (Reproduced with permission of the American Chemical Society)

Hildebrand solubility parameter   Choosing a supercritical fluid on the basis of its Hildebrand solubility parameter is of no use when different SFs are compared; it seems more appropriate in these cases to rely on the dipole moments of the fluids. Thus, based on its dipole moment, $CHClF_2$ should be more suitable than CO_2 and N_2O for extracting polar analytes – the opposite conclusion is drawn from a comparison of their solubility parameters, though. Table 3.13 compares the extraction of PCBs from a river sediment (SRM 1939) using various supercritical fluids [1]. Also, Fig. 3.39 compares the recovery of several PAHs achieved by using SC $CHClF_2$ and CO_2 with that obtained with dichloromethane (a liquid) and sonication [1]. As can be seen, SC $CHClF_2$ provides higher recoveries than CO_2.

Steroid extraction   Supercritical chlorodifluoromethane (Freon-22) has also been used for extraction and chromatography of polar compounds such as steroids [48]. This supercritical fluid provides significantly more efficient and much more expeditious extractions than does CO_2 in this respect. Figure 3.40 illustrates the extraction of oestrone added to glass wool using both fluids. Even though the extraction time with CO_2 was twice as long, the recoveries were much lower than those provided by Freon-22. In addition, raising the pressure to 18 MPa scarcely improved the recovery with CO_2, which peaked at only 16%. On the other hand, Freon-22 provided 100% recovery in only 15 min. Comparatively, exhaustive Soxhlet extraction with methanol took 7 h.

3.6.2 Properties of the Solid

Sample properties and extraction rate   Knowing how some sample properties (e.g. the matrix nature, porosity, moisture content, surface to volume ratio, size, etc.) affect the extraction rate is of paramount importance when the SFE technique is to be used for

Table 3.13. Comparison of Soxhlet extraction (NIST) and SFE by using $CHClF_2$, N_2O, CO_2 and methanol-modified CO_2 to recover PCBs from a river sediment (SRM 1939)

PCB	PCB concentration (µg/g), NIST	Recovery (%) ± sd			
		$CHClF_2$	N_2O	CO_2	$CO_2/MeOH$
2,4,4'	2.21 ± 0.10	63 ± 3	37 ± 2	36 ± 3	65 ± 6
2,2',5,5'	4.48 ± 0.06	83 ± 4	41 ± 2	38 ± 1	72 ± 5
2,2',3,5'	1.07 ± 0.12	108 ± 4	46 ± 1	44 ± 1	80 ± 4
2,3',4,4',5	0.51 ± 0.01	124 ± 2	63 ± 2	75 ± 10	97 ± 1
2,2',3,4,4',5'	0.57 ± 0.01	85 ± 0.5	61 ± 2	57 ± 3	96 ± 5
2,2',3,4',5,5',6	0.18 ± 0.01	91 ± 4	76 ± 7	89 ± 6	94 ± 7
2,2',3,3',4,4'	0.10 ± 0.01	88 ± 1	72 ± 7	65 ± 7	101 ± 7
2,2',3,4,4',5,5'	0.16 ± 0.01	104 ± 3	91 ± 7	90 ± 9	105 ± 9
2,2',3,3',4,4',5	0.11 ± 0.01	128 ± 7	71 ± 5	66 ± 7	99 ± 11
Average recovery (%)		97 ± 21	62 ± 18	62 ± 20	90 ± 14

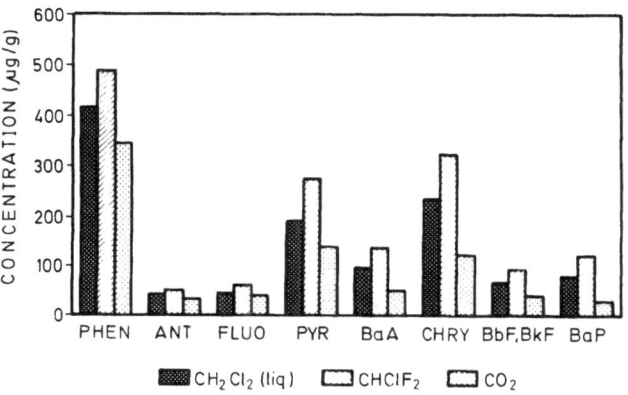

Fig. 3.39. Comparison of the extraction efficiency of various PAHs using a liquid solvent (CH_2Cl_2 plus sonication) and two different supercritical fluids (CO_2 and $CHClF_2$)

semi-quantitative screening purposes, where extraction is never complete. In order to be able to compare or correlate results, one must carefully control the above-mentioned sample variables, which is rather an arduous task. Under quantitative extraction conditions, controlling such variables is not so compelling since they only determine how drastic the experimental conditions need be (i.e. how long the extraction time or high the flow-rate must be, whether a modifier is to be used, etc.). It is therefore

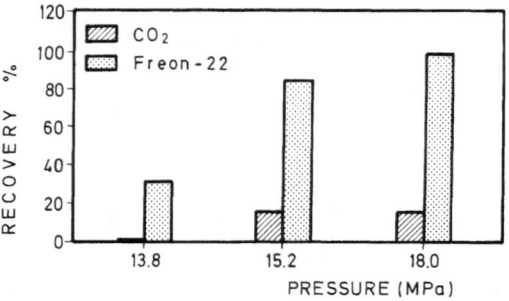

Fig. 3.40. Extraction with SC CO_2 for 30 min and SC freon for 15 min of oestrone added to glass wool at a constant temperature and three different pressures. (Reproduced with permission of Elsevier Science)

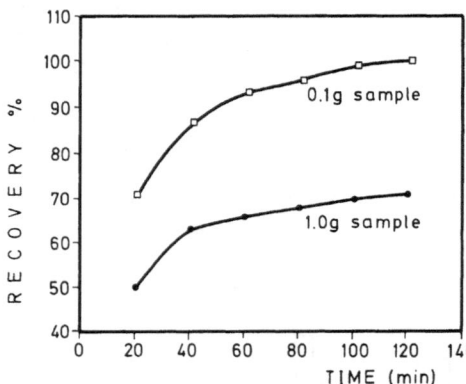

Fig. 3.41. Influence of the sample size of a river sediment and the extraction time on the efficiency of SFE of PCB-52 with SC CO_2. (Reproduced with permission of Springer International)

Sample pretreatment   customary to pretreat samples (e.g. by grinding, sieving, drying, lyophilization, mixing with a solid, adding a modifier, adjusting the pH) for this purpose.

Below is discussed the influence of the amount of sample, particle size and matrix composition on the extraction rate.

Increased amounts of sample call for proportionally increased supercritical fluid volumes. Thus, processing large amounts of sample requires using long extraction time or passing the fluid at a high flow-rate through the sample. Under equal conditions, recovery may decrease significantly with increasing amount of sample. Once the extraction conditions are **Large samples** optimized for large samples, the results are similar to those obtained for smaller samples. The sample volume and extraction time can have a great effect in those cases where the analyte concentration in the sample is quite high (at the parts per million level). Figure 3.41 shows the recovery of PCB-52 from a river sediment achieved with pure CO_2. As can be seen, recovery was quantitative after 50 min dynamic extraction of 100 mg of sample containing several PCBs at concentrations of a few parts per

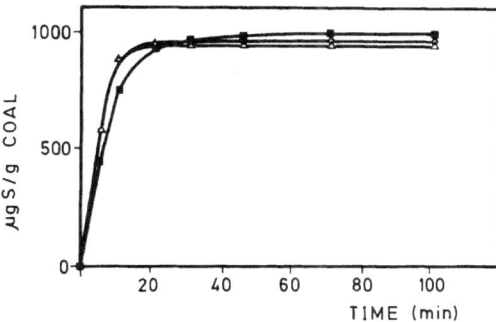

Fig. 3.42. Influence of the size of carbon particles (■ 60 mesh, 250 μm; ○ 100 mesh, 149 μm; △ 200 mesh, 74 μm) on the extraction of elemental sulphur with SC CO_2 containing 10% methanol

million. However, quantitative extraction from a 1 g sample could not be achieved even after 120 min. This may be the result of PCB saturation in the CO_2 [49].

As noted in Sect. 3.5.2, the extraction rate increases with decreasing particle size and increasing sample surface area to weight ratio. The effect of particle size is illustrated in Fig. 3.42, which shows the kinetics of extraction (with CO_2 containing 10% methanol) of elemental sulphur from a ground sample sieved through different mesh under an argon atmosphere [50]. Figure 5.12 shows the percent recovery of a volatile analyte separated by stepwise extraction from various polymer sheets of different thicknesses. Extraction was faster from the thinner sheets, which also had higher area to weight ratios [51]. Solid samples with a non-porous matrix can be lyophilized and ground in order to increase their surface area and facilitate extraction. Table 3.14 shows how grinding a frozen material can shorten the extraction time for polymers [46]. In addition, reducing particle size solves other problems derived from sample heterogeneity. **Increasing surface area of samples**

Recovery is markedly dependent on the nature of the matrix; as a result, the best supercritical fluid in terms of analyte solubility may not provide the highest possible recovery. Differences in the extraction efficiency between samples under identical extraction conditions arise from the different ability of the matrices to interact with the analytes. Such differences in the extraction of analytes also arise from the material on which the analytes are supported (Fig. 5.21) [52] and the addition method used [53]. The presence of functional groups at the matrix surface or its components and their ability to bind to the analytes, the organic matter and moisture content, and encapsulation phenomena are a few of the matrix-dependent factors that determine the ease with which analytes can be extracted.

The sample ability to covalently bind to the analytes is a function of its (bio)chemical activity. 2,4-Dichlorophenol can readily be removed from straw by SFE as the analyte can hardly bind to the matrix. Straw consists

Table 3.14. Extraction of whole and ground PVC pellets

Sample	Pressure (MPa)	Temperature (°C)	Time (min)	Total extracted material[a] (% w/w)
Ground with frozen material	35–45	45	60	20.7
Whole pellets	35–45	45	60	7.2
Whole pellets	45	95	230	21.7

[a] Total extractable additives in PVC: 28.5% (w/w) (manufacturer's formulation) and 30.0% (w/w) (liquid extraction overnight).

almost solely of inactive structural tissue (lignin, chiefly), so it only acts as a weakly adsorbent surface for analytes. This is not the case with matrices **Pesticide** treated in the field, where some of the pesticide bound to lignin in the **extraction** plant may remain in the matrix [54].

Hydrophobic matrices facilitate penetration of SC CO_2; on the other hand, too high a moisture content can make the analytes inaccessible to this supercritical fluid since it is water-immiscible. Some matrix components may partition undesirably, avoid retention in traps or hinder analysis of extracts.

Figure 3.43 shows the kinetics of extraction of pyrene and benzo[a]-pyrene from natural sludge, both as collected (45% moisture) and after air-drying (2% moisture), with CO_2 at 400 atm and 50 °C. The rate of extraction of low-molecular weight PAHs such as phenanthrene, fluoranthene and pyrene is significantly increased if the sample is previously dried – the recoveries shown in Fig. 3.43 are corrected for the drying weight loss. However, the rate of extraction of PAHs of higher molecular weight (e.g. benzo[a]pyrene) is hardly altered by drying [1].

High moisture Real samples with high moisture contents can plug restrictors as a result **content** of the water they contain freezing at restrictor tips. This can be overcome by raising the restrictor temperature at the expense of losses of the more volatile analytes.

Removing fat from the sample can hinder analyte collection and entail clean-up of the extract prior to analysis.

3.6.3 Properties of the Solute

The polarity, volatility and molecular weight of an analyte determine its solubility in supercritical fluids. Thus, the less volatile and more polar a given analyte is, the less readily soluble in pure CO_2 it will be. On the other hand, analytes of a high molecular weight containing polar groups call for high fluid densities and modified SC CO_2.

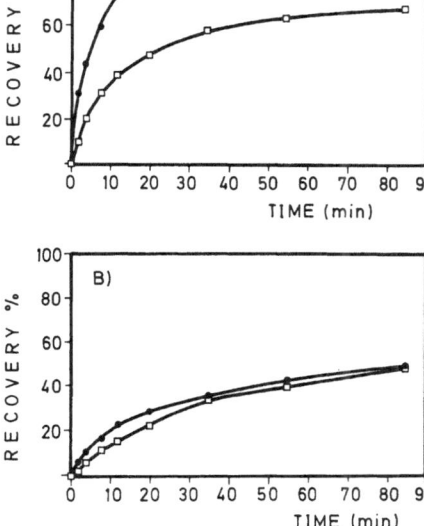

Fig. 3.43A,B. Influence of the moisture content (● 2%, □ 45%) in a sewage sludge on the SFE with SC CO_2 of (**A**) pyrene and (**B**) benzo(a)pyrene at 400 atm and 50 °C. (Reproduced with permission of the American Chemical Society)

The first step in an SFE process involves establishing the experimental conditions under which the analytes are soluble in the extraction medium. For this purpose, samples not containing the analytes are spiked with them. **Spiked samples** However, recoveries from samples where the analytes occur naturally are usually lower than those from spiked samples as a result of more significant matrix interactions. Thus, the average recovery of PCBs from sewage sludge is only 30% under conditions where the analytes are fully extracted from spiked samples [49]. Therefore, optimal conditions for quantitative extraction of analytes cannot be directly extrapolated to real samples, but must be readjusted in order to offset the adverse effect of solute–matrix interactions on recovery. Table 3.15 illustrates the influence of the form in which analytes are present in certified reference materials on their recovery [55]. While spiked samples allow quantitative recovery, those initially containing the analytes hardly afford recoveries above 30% under favourable conditions (a more dense SF in the presence of a modifier).

The highest possible solubility is achieved when the solubility parameter of the extracting SF is similar to that of the solute. The extraction conditions must be chosen in such a way that the solute is maximally soluble in the fluid whenever a large amount of solute (a major constituent) is to be extracted. However, extracting trace amounts of analytes is possible at

Table 3.15. Influence of the form in which analytes (PCBs) are present in solid (soil) samples (MCR) on recovery in the extraction with SC CO_2

Conditions				Average recovery (%)	
T (°C)	P (atm)	Density (g/ml)	Time (min)	Real material (BCR)	PCB-spiked material (NIST)
40	200	0.840	30	25	
40	150	0.781	30		99
60[a]	300[a]	0.830	30	30	

[a] 100 µl of methanol previously added to the sample.

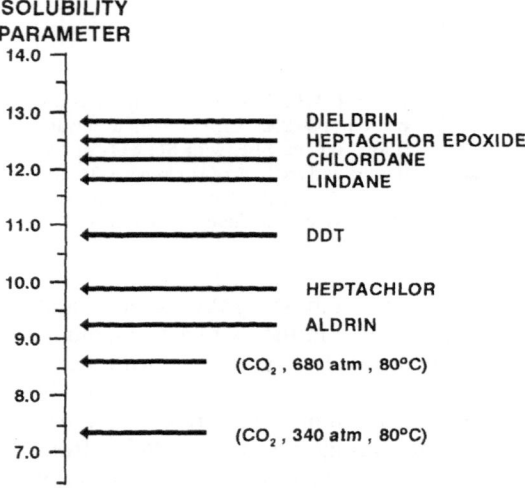

Fig. 3.44. Solubility parameters for various organochlorine pesticides and SC CO_2 under two different conditions. (Reproduced with permission of Preston Publications)

much lower pressures, i.e. by using supercritical fluids with lower solubility parameters than that of the analyte. Figure 3.44 shows the δ values for various organochlorine pesticides and supercritical CO_2 at two different pressures. As can be seen, the δ value of SC CO_2 is much lower than those of the pesticides, yet is high enough for trace amounts of these analytes to be dissolved [55].

Effect of analyte molecular weight   The efficiency of extraction of PAHs from some samples (sludge, sediments, soils, fly ash, urban air particulates) usually decreases with increase in the analyte molecular weight. Figure 3.45 shows the kinetics of extraction of various PAHs from silica using SC CO_2 [42]. The initial extraction

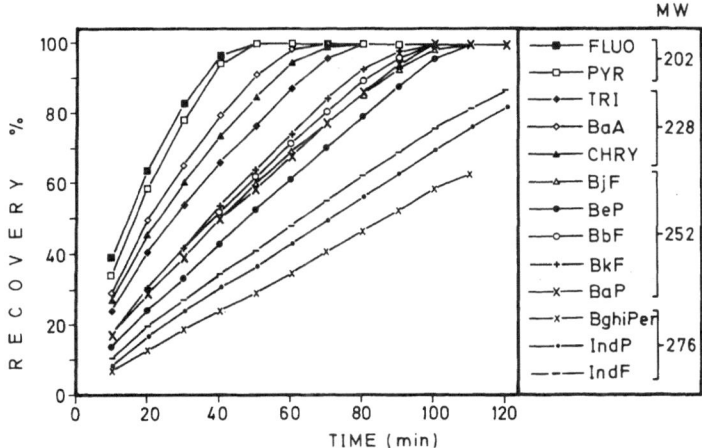

Fig. 3.45. Influence of the molecular weight of PAHs added to silica on their kinetics of extraction with SC CO_2 at 40°C and $\rho = 0.9$ g/ml. (Reproduced with permission of VCH)

rate is much higher for low-molecular weight PAHs than it is for their heavier congeners. This trend is consistent with the CO_2 solubility of PAHs, which decreases with increasing molecular weight of the solute. While the extraction efficiency achieved with $CHClF_2$ does not depend on the molecular weight of PAHs [1], those of PCBs with CO_2 and N_2O usually increase with increase in the MW of the solute [1] (Table 3.13).

3.6.4 Presence of a Modifier

The extraction efficiency of polar organic analytes, whether in polar or non-polar matrices, which cannot be quantitatively extracted with CO_2 can be increased by adding an organic modifier to the supercritical fluid in order to expand the range of extractable analytes using non-polar SFs such as CO_2. In fact, a small proportion of a polar cosolvent added to CO_2 **Polar** increases its polarity. Not surprisingly, cosolvents are used in 70–90% of **cosolvent** all SFE applications because they shorten the extraction time.

The mechanism by which a modifier exerts its effect depends on the type of matrix and form in which the analytes occur in it. Thus, a modifier can act in four basic ways (Fig. 3.46), namely:

a) By increasing the analyte solubility in the supercritical fluid through interaction with the solute in the fluid phase.
b) By facilitating desorption of the analyte from the matrix through interaction with the bound solute or the matrix active sites.

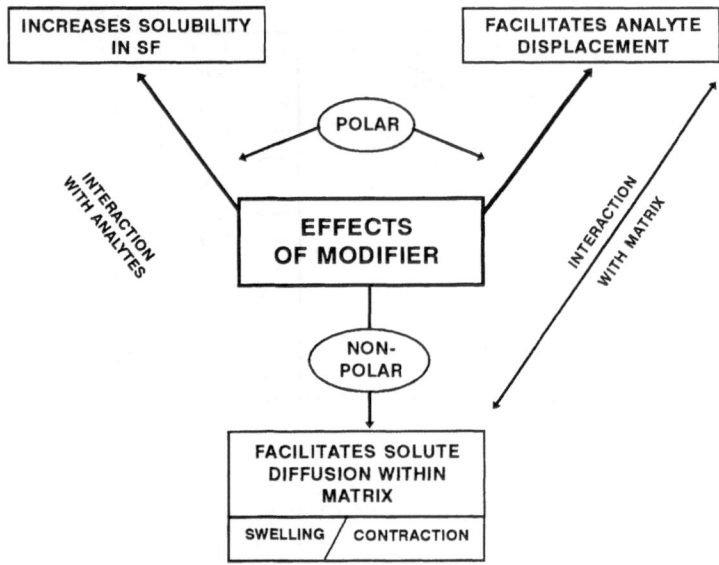

Fig. 3.46. Aspects of SFE under the influence of modifiers

c) By favouring diffusion of the solute within the matrix. The resulting recovery will be proportional to the extent of swelling of the material. Diffusion of the solute and penetration of the fluid are both favoured by swelling of the matrix under the influence of a modifier.

d) By hindering diffusion of the solute within the matrix through contraction, which results in diminished recovery. Such is the case of methanol in the extraction of monomers from Cellophane used as cigarette packet wrapping [56].

Mechanisms a and b require using a polar modifier, whereas c and d can result both from polar and non-polar modifiers.

The chief variables involved in the action of modifiers are their nature, concentration, way of addition, and imbibition and equilibration times.

Modifiers can be added to SFs in a static or dynamic fashion. In choosing either, one should consider such factors as the mechanism of action, available instrumentation and fluid costs.

Static addition   The simpler choice is static addition, whereby a given volume of modifier (between 20 and 200 µl, depending on sample size) is directly added to the sample held in the cell prior to extraction. The main disadvantage of this mode is that, as the supercritical fluid starts to circulate though the sample, the modifier is swept from the extraction cell, so the matrix is brought out of contact with the modified SF. The modifier concentration in the fluid decreases gradually during extraction, so the entrainer effect on analyte solubility also decreases with time.

Dynamic addition of the modifier is more effective than static addition since the modifier is continuously passed through the sample. The former can be carried out by using premixed fluid reservoirs or mixing the pure components in the propulsion system, whether by placing the liquid modifier in the cylinder of a syringe pump or by using two pumps (one for the fluid and the other for dispensing the required amount of modifier). Mixing in the extractor system affords greater flexibility in mixture preparation [i.e. the type and proportion of cosolvent(s)] and results in lower fluid costs than does using premixed fluid reservoirs, where the modifier may separate from the supercritical fluid at some time. **Dynamic addition**

When the modifier is added via the extraction cell it is important to bring the sample into contact with the modifier over a long enough interval (as long as the imbibition time) or carry out a previous static extraction in the presence of modifier in order to have it replace the analyte at the matrix active sites or swell the matrix.

Table 3.16 shows the organic solvents most frequently used as SC CO_2 modifiers. Methanol is the most common by far and is replaced with isopropyl alcohol, less aggressive, if the matrix is attacked by the first. Methylene chloride is specially suitable for extracting PAHs and PCBs from tissues and heavy hydrocarbons from soils. Acetone is commonly used to extract PCBs from soils and sediments. Toluene, both pure and mixed with methylene chloride, is usually employed to extract heavy hydrocarbons. These cosolvents are normally used in a 5–10% proportion with SC CO_2. Methanol at concentrations between 10 and 20% is frequently used for extracting pharmaceutical compounds. This alcohol is also very effective for expanding polymers in order to extract additives. Some hydrocarbons including n-hexane, n-heptane and n-octane have also been used as CO_2 modifiers for the isolation of fatty acids [57], as has ethyl acetate for isolating triglycerides [58]. **Commonly used modifiers**

The solubility of a solute in a liquid solvent is a measure, though not necessarily an accurate one, of the liquid efficiency as modifier in the solute extraction. For example, the solubility of diuron in methanol is not significantly different from its solubility in acetonitrile – the polarity and dielectric constant of which are somewhat higher than those of methanol –

Table 3.16. Most commonly used CO_2 modifiers

Methanol	Acetonitrile	Toluene
Propanol	Formic acid	Methylene chloride
Tetrahydrofuran	Acetone	n-Hexane

Table 3.17. Effect of the type of modifier on the extraction efficiency for sludge/fly ash using on-line coupled SFE/GC[a]

Compound	Recovery (%)						
	Benzene	Methanol	Propylene carbonate	Hexane	Formic acid	Acetonitrile	CO_2
Ethylbenzene	95	80	96	72	78	81	74
Cumene	96	79	96	70	76	79	72
2-Chloronaphthalene	92	82	93	66	70	83	66
1,2,4-Trimethylbenzene	96	84	96	70	78	85	71

[a] Conditions:
SFE: 225 mg of sludge/fly ash, 375 atm, 50 °C, 10 min (static), 7 min (dynamic), 50 µl of modifier.
GC: 30 × 0.25 mm i.d. DB-WAX from 30 °C (7 min) to 310 °C at 7 °C/min, FID.

yet the recoveries achieved by using acetonitrile as modifier are somewhat lower (69% vs 81%) [45]. Ideally, one should study the behaviour of each system towards different types of solvents including those which form hydrogen bonds (e.g. methanol), those which do not but are polar (e.g. dichloromethane), and non-polar solvents (e.g. *n*-hexane).

Table 3.17 compares the extraction efficiency for various analytes achieved by using different modifiers [59].

Figure 3.47 shows the beneficial effect of methanol used as modifier in the extraction of a variety of analytes that cannot be extracted quantitatively with SC CO_2. As can be seen, the presence of the alcohol resulted in significantly increased extraction efficiency in all cases, and even allowed the quantitative recovery of a linear alkylbenzenesulphonate (an ionic substance) [47].

Even though the modifier often increases the solubility of analytes, in most cases SFE cosolvents are intended to cleave bonds involving analytes that are chemically adsorbed onto their matrices. The extraction of caffeine from coffee is the most representative example of the use of modifiers in SFE. Caffeine in coffee is very readily soluble is supercritical CO_2 (around 1 mg/g at 200 atm and 40 °C). However, SC CO_2 does not allow efficient extraction of caffeine from coffee beans or tea leaves. By adding water to the matrix, however, the caffeine can be isolated almost quantitatively from it. The water functions to swell the matrix, which facilitates diffusion of the caffeine outside the vegetable tissue. Simultaneously, caffeine–tannin complexes within the vegetable cell structure are hydrolysed as a result [60].

Caffeine extraction

DDT extraction from soil

In the extraction of DDT from soil, 60–70% of the insecticide is removed by SC CO_2 within 10–20 min; however, extending the extraction time beyond that limit has no beneficial effect on recovery. In fact, part of

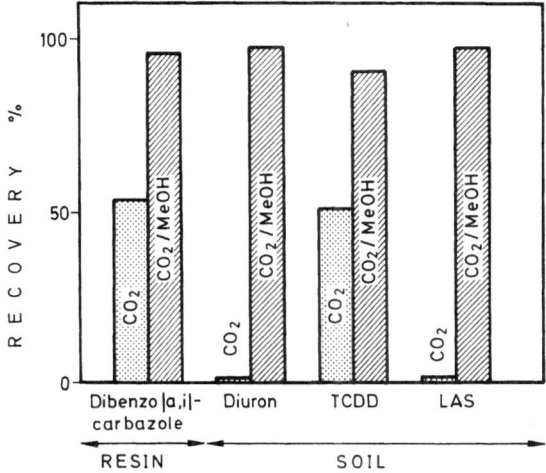

Fig. 3.47. Comparison of the efficiency achieved with SC CO_2 and methanol-modified CO_2 in the SF extraction of various compounds from a natural and a synthetic samples under the same operational conditions. (Reproduced with permission of the American Chemical Society)

the analyte is strongly bound to the soil, so it cannot be extracted into SC CO_2 however long the extraction is allowed to continue. A mixture of CO_2 and 5% toluene under the same extraction conditions does not improve matters much; however, CO_2 containing 5% methanol allows around 95% of the pesticide to be extracted within 5 min [61].

Acid pesticide extraction

Acid pesticides such as 2,3-D or dichlofop can only be recovered from glass wool to the extent of 30–70% using dry CO_2 at 20 MPa and 313 K. Water-saturated CO_2 under the same conditions raises the recovery to 70–100%. The extraction efficiency for other pesticides is independent of the water content of the supercritical fluid [62].

Fatty and resin acids extraction

Fatty and resin acids (FRAs), pollutants present in river sediments near paper pulp plants which are lethal to fish, have traditionally been extracted by the Soxhlet technique. While pure CO_2 does not allow these pollutants to be extracted, a mixture of this fluid with methanol or formic acid (or both) serves the purpose quite well. The effect of each modifier on the recovery of the principal FRAs is shown in Fig. 3.48. Recoveries are referred to the Soxhlet recovery, arbitrarily taken as 100%. While palustric acid degrades during extraction, its recovery is 2.5 times higher than that provided by Soxhlet extraction [63].

Figure 5.4B shows the influence of the modifier concentration on the extraction of diuron [45]. As can be seen, the extraction efficiency in-

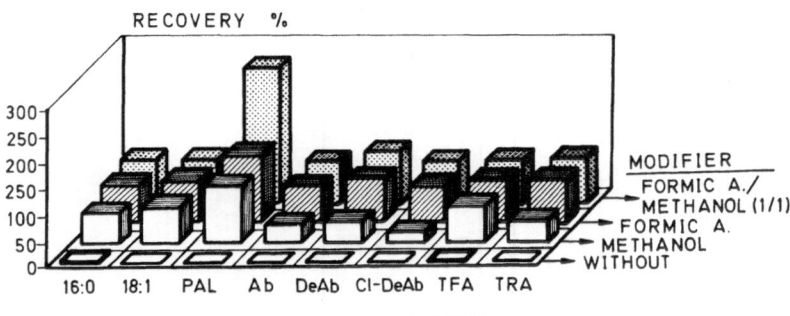

Fig. 3.48. Effect of various modifiers of SC CO_2 on the recovery of fatty (16:0, 18:1) and resin acids relative to their Soxhlet extraction (100%). Pal = palustric acid. Ab = abietic acid. DeAb = dehydroabietic acid. Cl-DeAb = chlorodehydroabietic acid. TFA = total fatty acids. TRA = total resin acids. (Reproduced with permission of Hewlett–Packard)

creases with increase in the modifier content in the supercritical fluid. Using the modifier and SF in premix reservoirs may give rise to irreproducible fluid delivery if the modifier concentration is too high.

Figure 3.49 shows the effect of a cosolvent in terms of type, amount and static equilibration time on the recovery of diuron and linuron from sassafras soil [64]. As can be seen in Fig. 3.49A, the recovery increased with increase in the amount of methanol added from 100 to 200 µl. However, increasing the modifier volume to 300 µl had almost no further effect. In addition, organic additives exhibit a different selectivity. Thus, ethanol is more effective than methanol for removing linuron from soils (Fig. 3.49B). Figure 3.49C shows the effect of including a static extraction interval immediately after methanol is added. Initial and overall recoveries improve as a result of this equilibration step.

Benzocaine extraction

Modifiers allow the working density and hence the working pressure used to be reduced. Thus, a volume of 150 µl of methanol in the extraction cell ensures quantitative extraction of benzocaine from pharmaceutical preparations at a CO_2 density of only 0.35 g/cm^3. In the absence of methanol, the density has to be raised to 0.65 g/cm^3 in order to obtain a comparable recovery [65]. Pure CO_2 at 400 atm does not allow dibenzo-p dioxins or polychlorinated dibenzofurans to be extracted from fly ash. While adding 10% methanol as modifier does not improve matters, 10% benzene is enough to ensure recoveries of around 95%, which are similar to those afforded by N_2O. The volume of benzene used can be reduced to about 0.5 ml if it is added directly to the extractor together with the sample prior to the extraction with pure CO_2, thereby avoiding the need to prepare a 10% supercritical mixture [34].

The recoveries obtained with methanol-modified CO_2 are comparable to those afforded by other, more polar pure fluids. As can be seen in

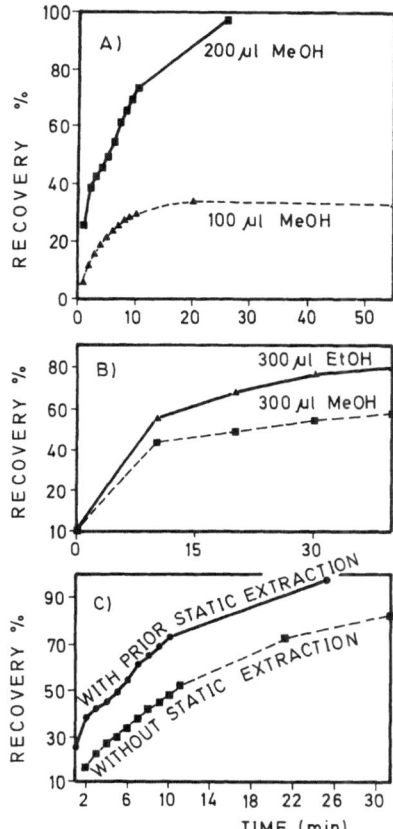

Fig. 3.49A–C. Study of the kinetics of extraction of (**A**,**C**) diuron and (**B**) linuron from soil in terms of (**A**) the amount of modifier added, (**B**) its nature and (**C**) the extraction mode used. For details, see text. (Reproduced with permission of Preston Publications)

Table 3.13, the average recovery of PCBs from a river sediment with CO_2/methanol was 90% vs 97, 62 and 62% achieved with pure $CHClF_2$, N_2O and CO_2, respectively [1].

The chief problems derived from the use of modifiers arise from their potential incompatibility with some detectors used in the subsequent determination of the analytes, as well as from impeded analyte collection through diminished retention at traps.

Modifier detector incompatibility

3.6.5 Additives

Solute–matrix interactions in real samples are of a great significance with a view to obtaining high recoveries, which often requires altering the initial matrix. Extraction recovery and selectivity can be improved by

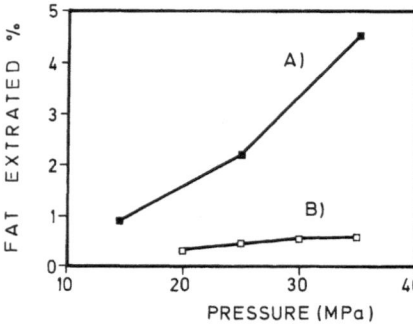

Fig. 3.50. Effect of an adsorbent mini-column packed with basic alumina on the variation of the amount of fat extracted from a biological material as a function of pressure. (A) Without column. (B) With column. (Reproduced with permission of Springer International)

adding solid or liquid substances to samples before the SFE proper is started. Some of the available treatments were mentioned in the previous section (modifiers) or are commented on in the next one (derivatizing reagents). This section is devoted to other types of substances added to samples in order to improve the extraction efficiency.

Some solids including desiccating salts, diatomaceus earth, sand and glass wool can be added to liquid or semi-solid samples with a high moisture content, whether to absorb the water or to disperse the sample in order to increase the surface exposed to the supercritical fluid [66,44]. In this way, the extraction efficiency is improved and potential problems arising from water circulating through the SFE system are avoided.

Selective adsorbents

Extraction of lipid material under the conditions required for extraction of analytes entails cleaning up extracts prior to analysis in order to avoid damaging the chromatographic system or the detector. Basic alumina is a selective adsorbent for removal of fat from lipid extracts. It increases the SFE selectivity towards Arochlor 1260 in its extraction from biological lipid matrices [67]. Samples are previously dried with anhydrous sodium sulphate and an amount of alumina as large as twice that of the sample is placed at the outlet of the extraction chamber or in a column after the extraction cell. The results thus achieved are shown in Fig. 3.50.

When the extraction pretreatment involves adding a small liquid volume to the extraction cell, this can readily be done with the aid of a micro-syringe. For this purpose, the syringe needle is inserted as deep as possible into the extraction cell, after which the solution is continuously delivered as the needle is slowly pulled out of the cell. In this way, the liquid is dispersed throughout the sample.

Water added to a sample can favour extraction of the analyte (e.g. caffeine from coffee beans, Fig. 3.51) or hinder it (e.g. benzocaine from tablets [65]). Moisture can favour sample aggregation. In addition, the presence of CO_2-immiscible liquids such as water in the sample may give rise to undesirable partitioning equilibria. If the analyte is an ionizable

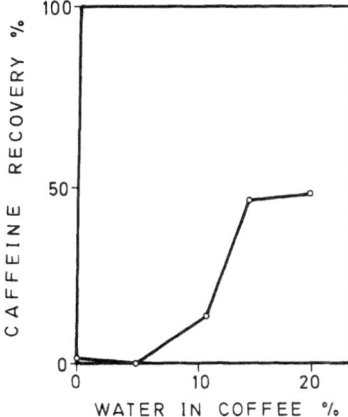

Fig. 3.51. Influence of the moisture content of coffee on the recovery of caffeine with SC CO_2. (Reproduced with permission of Elsevier Science)

species, its solubility in SC CO_2 will be dramatically reduced by the presence of water.

Moisture in a sample also plays a major role in the extraction of resin and fatty acids in sediments with modified CO_2 [63]. Recoveries are maximal for samples with a 5–10% moisture content. Lyophilized sediments provide analyte recoveries that are 25–40% lower than those afforded by the as-collected samples. However, adding 5% water to a dry sediment enables quantitative the extraction at a subsequent stage.

If a significant fraction of the analyte to be extracted is bound to the matrix, sample hydrolysis may be required. This is often the case with pesticides in plant materials. Non-extractable bound residues are formed in plants by synthetic processes involving binding of the pesticide and/or its metabolites to carbohydrates (as a β-glucoside) and aminoacids. Release of 2,4-dichlorophenol from vegetable tissue during SFE was assayed by using several sample pretreatment procedures including direct injection of cosolvents such as ethanol and isooctane or acid and basic aqueous solutions in the cell, and heating the sample at different temperatures prior to extraction. The most substantially enhanced extraction efficiency with no detriment to selectivity was achieved by using 100 µl of 17% phosphoric acid at 100 °C for 4 h.

A strong acid treatment may partly destroy the sample matrix, thereby releasing the analytes. An acid pretreatment of samples to remove dibenzo-*p*-dioxins and polychlorinated dibenzofurans from fly ash enables quantitative recovery with SC CO_2, otherwise only feasible with more polar fluids such as N_2O or a modifier. In fact, the extraction efficiency soars from 9% to 100% as a result of the acid treatment [34].

Acid pretreatment

3.6.6 Derivatization

Supercritical fluid extraction provides an expeditious means for the quantitative extraction of a host of non-polar organics (e.g. those suitable for gas chromatography) from a variety of matrices. However, quantitative extraction of polar or anionic analytes calls for addition of organic modifiers to CO_2. In situ chemical derivatization under supercritical conditions is an alternative to increasing the extraction efficiency for polar and ionic species. In addition, it conditions analytes for their subsequent chromatographic determination (Fig. 3.52). The functional group that is subjected to derivatization can belong to the analyte or the sample matrix, as shown in Fig. 3.53.

As a result of derivatization, one can obtain less polar substances than the analytes which therefore lend themselves more readily to extraction. On derivatization, the analyte polar groups (hydroxyl, carboxyl) are converted to other, less polar functions (ether, ester, silyl) which make the derivative more readily soluble in supercritical CO_2.

The increased extraction efficiency of reactive compounds can be the result of dissociative or associative derivatization mechanisms. In a dissociative mechanism, the adsorbed analyte is first desorbed from the active sites it occupies in the matrix and then dissolved in the SF for conversion into a less polar, more readily solvatable derivative. On the other hand, in an associative mechanism, derivatization is performed with the analyte adsorbed at the matrix active sites. The non-polar derivative thus obtained is subsequently desorbed from the matrix into the SF.

In addition to increasing the SF solubility of polar compounds, derivatization reagents may compete with analytes and displace them from the active sites of the matrix. Such compounds act both as polar modifiers and

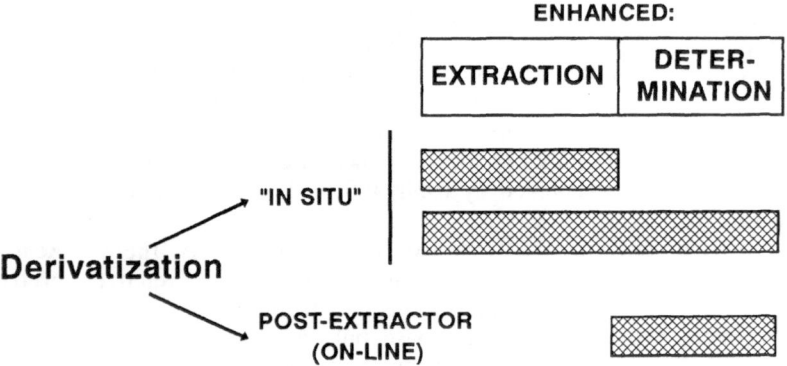

Fig. 3.52. Beneficial effects of the two derivatization modes on SFE

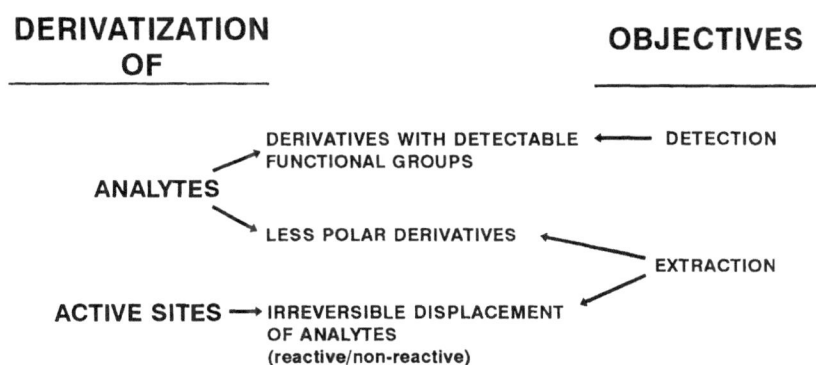

Fig. 3.53. Aspects of SFE influenced by derivatization and objectives of its use

derivatizing reagents. The extraction of non-reactive substances can thus also be improved via a displacement mechanism. Active sites are the target of the derivatizing reagent, which removes the intact (unreacted) analyte molecules from them. Derivatization at the matrix active sites has the advantage over polar modifiers of being irreversible (i.e. the analyte cannot be readsorbed after it is displaced from the matrix).

In situ derivatization/SFE is implemented in two steps. First, the analytes are derivatized under static conditions and then the transformed analytes are recovered by dynamic SFE. The derivatizing reagent is added to the extraction cell holding the sample. Over a short interval called the derivatization time, the analytes undergo simultaneous derivatization and extraction under static conditions, after which they are subjected to dynamic SFE.

The derivatizing reagents most commonly employed in SFE are those typically used in GC derivatization. One of the most widely used derivatization reactions is that of silanization, whereby an active hydrogen from a functional group is replaced with a silyl group (a trisubstituted silicon atom). Substituting an active hydrogen by a silyl group has major effects on the solute, namely: it decreases its polarity and ability to form intermolecular hydrogen bonds, and increases its volatility and thermal stability. Trimethylchlorosilane (TMCS) and hexamethyldisilazane (HMDS) are the two most frequently used silanizing reagents. In alkylation reactions, an active hydrogen is replaced with an alkyl group. The reagents most commonly used for this purpose are low-molecular weight hydrocarbon halides (bromides and iodides). Esterification reactions are also frequently used to derivatize acid (carboxyl) groups by means of alcohols. Thses reactions are rather slow, so they must be accelerated with acid catalysts such as boron

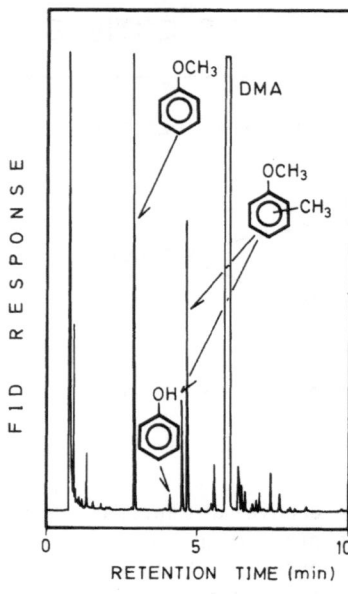

Fig. 3.54. In situ derivatization of phenol in waste water by using TMPA prior to SFE and GC/FID detection. DMA = N,N-dimethylaniline, a product of TMPA decay. For details, see text. (Reproduced with permission of the American Chemical Society)

trichloride. If the alcohol contains halogen atoms, the resulting derivative is suitable for sensing with an electron capture detector.

A mixture of HMDS and TMCS but no solvent directly added to the sample allows simultaneous silanization and extraction. The extract contains the trimethylsilyl esters and ethers of acids and alcohols, respectively, in addition to other, underivatized substances, extraction of which is also improved through displacement from the matrix active sites [68].

The in situ extraction and derivatization of resin and fatty acids (RFAs) as pentafluorobenzyl esters using a solution of pentafluorobenzyl bromide (PFBBr) in acetone and triethylamine rather than the usual modifier improves their GC/ECD determination. The recoveries thus achieved are roughly three times lower than those obtained by off-line coupled SFE/ derivatization. The recovery is increased (to 60%) by increasing the extraction time from 10 to 60 min. Even if in situ extraction–derivatization provides no quantitative results, it can still be used for rapid semi-quantitative screening at a high sampling rate [63].

Trimethylphenylammonium hydroxide (TMPA) in methanol has been used under SC conditions for the derivatization and quantitative extraction of polar organic compounds from a variety of real samples such as soils, sediments, waste water, adsorbent discs and micro-organisms [69]. As can be seen from Fig. 3.54, phenol can be efficiently derivatized (>98%) to anisole. The extract also contains cresols as methyl ethers. The reagent,

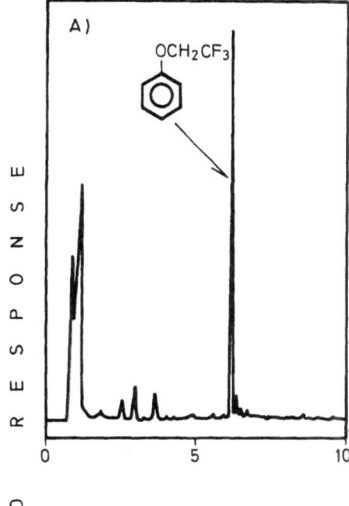

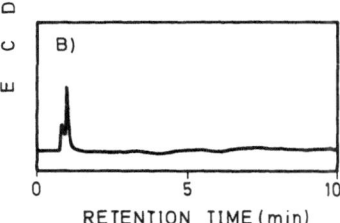

RETENTION TIME (min)

Fig. 3.55A,B. In situ derivatization of phenol in waste water by using (**A**) a 1:4 TMPA/2,2,2-trifluoroethanol mixture and (**B**) pure TMPA prior to SFE and GC/ECD detection. (Reproduced with permission of the American Chemical Society)

1 ml of 20% TMPA in methanol, is added to the sample, 1 ml of waste water, which is previously mixed with 1 g of sand to increase its surface area. The derivatization–extraction is carried out in two steps (static and dynamic), each lasting 15 min.

The selectivity and sensitivity of derivatization/SFE can be improved by introducing detector-specific functional groups (e.g. by formation of halides for GC analysis with an electron capture detector). Figure 3.55 shows the chromatograms obtained by derivatization/extraction of phenol from waste water using either a 1:4 mixture of 20% TMPA and 2,2,2-trifluoroethanol or TMPA alone [69].

The derivatization efficiency depends on the matrix components involved. In choosing the reagent concentration and derivatization time for in situ derivatization–extraction one must take into account the presence of potentially reactive components in the matrix. As can be seen in Table 3.18, 2,4-dichlorophenoxyacetic acid (2,4-D) can be efficiently recovered as its methyl ester from sand on derivatization and dynamic extraction

Selectivity and sensitivity optimization

Table 3.18. Recoveries obtained by derivatization/SFE of 2,4-dichlorophenoxyacetic acid (2,4-D) added to two matrices

Matrix	Reagent	Derivatization time (min)	Recovery (%) ± sd[a]
Sand	1.5% TMPA	5	98 ± 6
River sediment	1.5% TMPA	5	23 ± 2
River sediment	20% TMPA	5	63 ± 15
River sediment	20% TMPA	15	92 ± 4

[a] Each value represents a triplicate extraction.

for 5 and 15 min, respectively. The recovery obtained for 2,4-D spiked sediment samples under the same conditions is much lower. Analyses of the extracts, which contained 4% organic carbon, showed the presence of methyl esters of carboxylic acids of probable biological origin. Recoveries can be further improved by increasing the reagent concentration and derivatization time. Occasionally, stepwise derivatization is required [69].

Selectivity can be accomplished by careful selection of the derivatizing reagent. The derivatization–extraction of 2,4-D from agricultural soils in the presence of Dicambra is selective if a BF_3/methanol mixture is used as derivatizing reagent instead of TMPA in methanol (Fig. 3.56). The subsequent derivatization of the BF_3/methanol extract with TMPA provides a derivatized Dicambra peak, which indicates that underivatized Dicambra is extracted into CO_2/methanol [69].

When derivatization follows extraction, the aim is usually to obtain derivatives that are easier to determine (Fig. 3.52). On-line derivatization coupled to SFE under supercritical conditions has been used to determine **Fatty acids in oilseeds** the fatty acid composition of oilseeds [70]. Triglycerides are transesterified to methyl esters over a solid catalyst following extraction into supercritical CO_2. The catalyst consists of methanol-presoaked alumina and is placed after the sample in the extraction cell (Fig. 3.57).

3.6.7 Temperature

The temperature is an experimental variable of special interest to SFE as it affects every extraction step: desorption, diffusion and dissolution. The relative significance of each step depends on the particular matrix and analyte. Supercritical fluid extraction must be done at a temperature at least 10 °C higher than the critical temperature of the fluid. The maximum usable working temperature may be limited by thermolabile components of the matrix.

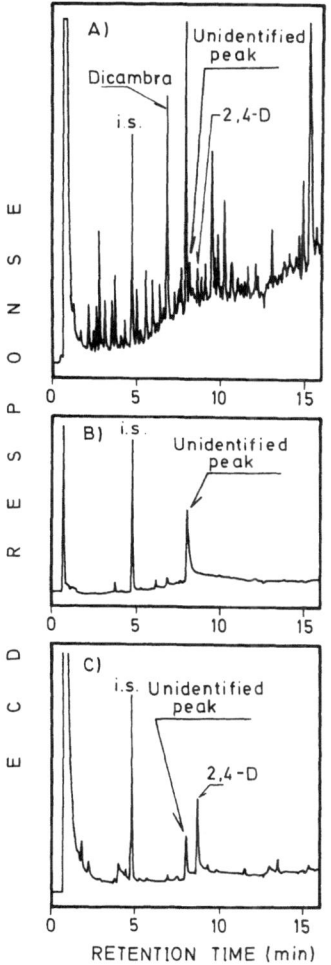

Fig. 3.56A–C. Influence of the type of reagent used on the selectivity of derivatization prior to SFE. Determination 2,4-D and Dicambra in agricultural soils. Derivatizing reagent: (**A**) TMPA/methanol, (**B**) derivatization blank, (**C**) BF₃/methanol. i.s. = internal standard. (Reproduced with permission of the American Chemical Society)

The solute volatility and diffusivity increase with increasing temperature; however, the density of a supercritical fluid at a given pressure (and hence its solvent power) decreases with increasing temperature (Fig. 3.58). In addition, the temperature plays a major role in desorption kinetics, which may determine the extraction rate of adsorbed analytes.

The solubility parameter for CO_2 and solutes decreases as the temperature is raised. At high pressures, a decrease in the solubility parameter for the solute can be more significant than one in that of the solvent.

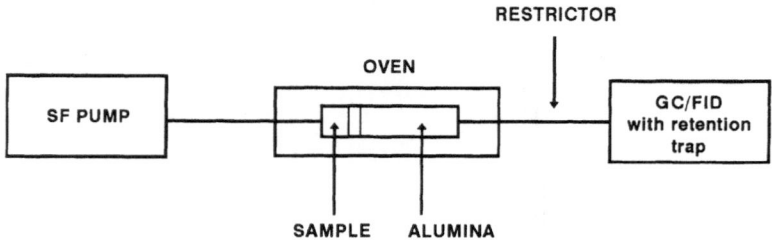

Fig. 3.57. Supercritical fluid extractor including a column packed with alumina in methanol for on-line derivatization of extracts. (Reproduced with permission of Springer International)

EFFECT OF TEMPERATURE IN SF EXTRACTION

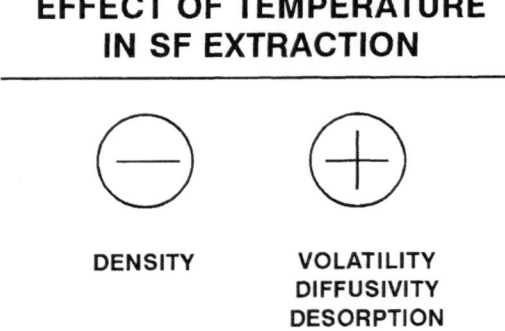

Fig. 3.58. Beneficial and adverse effects of temperature on various aspects of SFE

Solubility–pressure isotherms intercept at around 30–40 MPa, so solubility increases with temperature at higher pressures. The increased solubility achieved in extractions performed at high pressures may be partly responsible for the increase in the extraction efficiency with increasing temperature.

High-temperature extractions also at a high SF density occasionally require too high pressures for the extraction system to withstand.

The rate of diffusion of the solute through the sample matrix and within CO_2 can also be increased by raising the temperature. In addition to the increased solute diffusivity, an increase in the CO_2 diffusivity can facilitate absorption of the fluid in a polymer and hence diffusion of solutes across it. In extracting additives and side products from polymers, a temperature elevation (under high enough solubility conditions for all surface molecules to be dissolved) can be used to boost diffusion to the surface, which is the rate-determining step of the process.

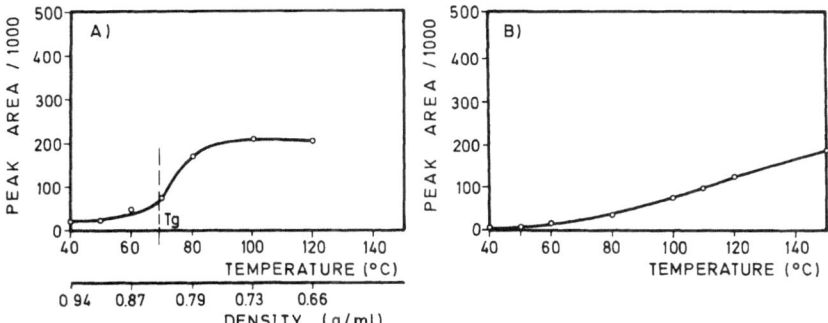

Fig. 3.59A,B. Influence of temperature on the structure of some samples subjected to SFE. Extraction of a cyclic trimer of PET with SC CO_2 at a constant pressure of 360 bar (**A**) and a constant density $\rho = 0.5 \, g/ml$ (**B**). T_g = polymer glass transition temperature. For details, see text. (Reproduced with permission of Pergamon)

Increased temperatures can induce structural changes in some matrices **Structural** (e.g. synthetic polymers) that may affect diffusion. Polymeric matrices are **changes** quite interesting because their physical properties, oligomer and additive diffusion included, can change as the polymer goes through different transition temperatures (e.g. in passing from a glass state to an elastic state to a molten state). Figure 3.59A shows the amount of cyclic trimer, the chief low-molecular weight component of poly(ethyl terephthalate), extracted at 360 bar and several temperatures [71]. The S-shaped curve obtained can be ascribed to changes undergone by the polymer. At low temperatures, increasing temperatures result in increased diffusion of oligomers within the matrix and the high solvent power of the extracting fluid allows quantitative dissolution of the oligomers as they reach the surface. At the glass transition temperature, T_g, diffusion increases abruptly, so the amount of oligomer extracted also increases sharply above T_g. At $T > T_g$, where oligomer diffusion in the matrix is quite high, the extracted material calls for a high solvent power and flow-rate for effective, expeditious removal. The slight decrease in the amount extracted at 120 °C is probably the result of a decrease in the solvent power. In this temperature range, the solvent power becomes the major limiting factor for the amount of oligomer extracted. A plot of amount extracted against temperature at a constant density does not show the jump at T_g typically observed in the previous curves (Fig. 3.59B). The glass transition temperature, the location of which is markedly dependent on pressure and absorbed SF molecules, changes in constant-density extractions. The extraction tem- **Adverse effect** perature must be lower than the melting point, otherwise melting will **of melting** reduce the sample surface and hence the extraction rate.

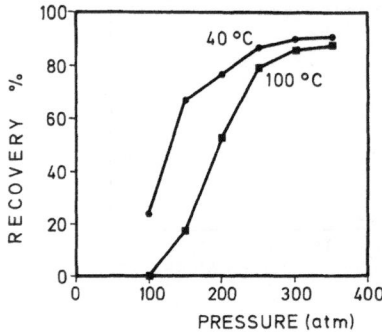

Fig. 3.60. Special case where increasing temperatures detract from SFE efficiency: variation of the efficiency of extraction with 95:5 SC CO_2/methanol of mebeverine alcohol (MEBOH) from ODS cartridges as a function of pressure at two different temperatures. (Reproduced with permission of the American Chemical Society)

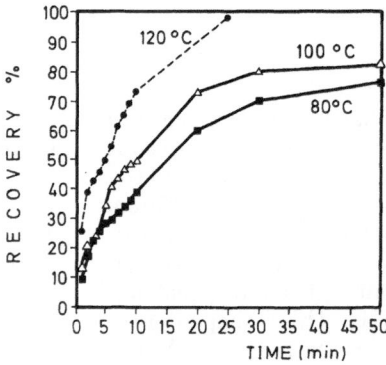

Fig. 3.61. Influence of temperature on the kinetics and efficiency of extraction of diuron from soils (1 g of sample) with SC CO_2 at a density $\rho = 0.6$ g/ml. Static pre-extraction time = 10 min. (Reproduced with permission of Preston)

Figure 3.60 shows the effect of pressure and temperature on the recovery of mebeverine alcohol (MEBOH) from solid-phase ODS extraction cartridges by SFE for 10 min. As can be seen, MEBOH recovery increased with pressure up to a constant value of around 87% near 300 atm. Increasing the extraction temperature from 40 to 100 °C decreased the recoveries obtained over the same pressure range. This diminished recovery was the likely result of the decrease in the supercritical fluid density at high temperatures [72].

As can be seen in Fig. 3.61, diuron recovery from sassafras soil with modified CO_2 increases with increasing temperature at a constant density, which reflects a potential increase in the solute volatility [64]. Figure 5.8A shows the kinetics of extraction of diuron from soil with CO_2 containing 10% methanol at a constant pressure and two different temperatures (75 and 100 °C). The higher temperature provides the higher recovery (81% vs 48%) [45].

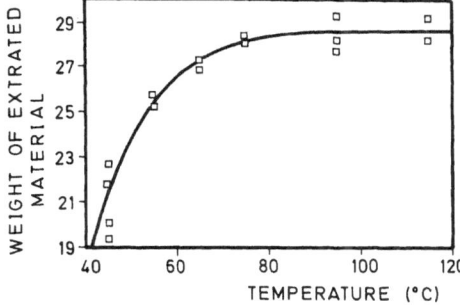

Fig. 3.62. Influence of temperature on the SFE of additives from PVC at a constant pressure of 45 MPa and a flow-rate of 5 ml/min for 30 min. (Reproduced with permission of the Royal Society of Chemistry)

Figure 3.62 shows the influence of temperature on the extraction of PVC additives [46]. As can be seen increased temperatures have a favourable effect on the extraction efficiency, as usual in SFE.

3.6.8 Dynamic Factors

In addition to extracting the analytes, supercritical fluids serve one other major purpose: transferring them from the extractor to the collection vessel or chromatographic system. In order to ensure efficient transport of **Transportation** the analytes, the volume of the extraction cell, the extraction time and the **of analyte** flow-rate of the supercritical fluid must be carefully combined in order that the flow-cell may be vented as many times as required. The CO_2 volume passed through the cell in one extraction should be at least ten times higher than that of the cell. This is usually accomplished by using an increased flow-rate and a miniature extraction cell that can be frequently vented without lengthening the extraction time unduly.

When the analytes are present at a much lower concentration than their solubility in the SF, dynamic extraction seemingly offers little advantage over static SFE. However, if the extraction rate is limited by the analyte solubility in the SC, dynamic extraction (and high flow-rates) will obviously **Flow-rate** excel over static SFE.

With some samples, the flow-rate is the most influential factor on recovery [49]. The higher the flow-rate used is, the more analyte is extracted over a given interval. However, as can be seen in Fig. 3.63, the rate of extraction of fluoranthene from 2.5 g of soil sample at flow rates between 0.15 and 1.2 ml/min scarcely changes above 0.3 ml/min [73]. The optimal extraction conditions cannot be established on the basis of the number of times the extraction cell is vented because the extraction of most analytes is kinetically limited. Increasing the flow-rate is occasionally impractical as it may decrease the analyte collection efficiency or, in

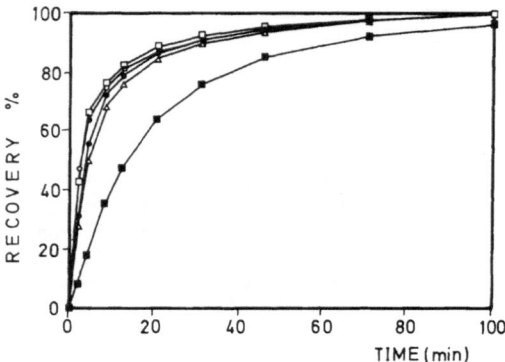

Fig. 3.63. Influence of the SC CO_2 flow-rate on the SFE of fluoranthene from a soil sample (2.5 g) under constant remaining operating conditions. (■ 0.15, ● 0.3, △ 0.5, □ 0.7 and ○ 0.9 ml/min). (Reproduced with permission of Chapman & Hall)

coupled SFE/GC, cause the analyte to disperse through deposition on a longer column stretch. In addition, very high flow-rates result in greater pressure drops across the extraction cell, which decrease the extraction efficiency [45]. Such drops depend on how tightly the sample is packed in the extraction cell.

If the inner volume of the extraction cell is reduced (e.g. by introducing a piece of inert material into it), the extractant that passes through it will be renewed more often. The effect of the cell void volume (the difference between the cell volume and the sample volume) varies dramatically depending on whether the extraction is dynamic or static. In static extractions, a high void volume favours extraction as it allows more fluid to be held in the cell and analytes to be more extensively solvated as a result –

Unlimited diffusion unlimited diffusion is assumed. This may be the root of an unexpected phenomenon observed in changing the amount of sample while keeping the cell size constant [74]. The variation of the amount of detergent fragrances (highly CO_2-soluble) that is extracted as a function of sample size (Fig. 4.7) goes by a maximum rather than continuously rising in response to the greater amounts of extractable compound logically present in larger samples. This extraction involves a static extraction period comparable in length to that of dynamic extraction (5.0 vs 6.1 min); also, the cell volume (1.5 ml) does not change and, even though the amount of extractable compound increases with sample size, the void volume also decreases as a result. Conversely, if the cell is too large for the sample volume used in dynamic extraction, incomplete extraction may result by effect of the flow velocity. Flow across the extraction cell is laminar, so there is no normal component to the general flow. In addition, the flow

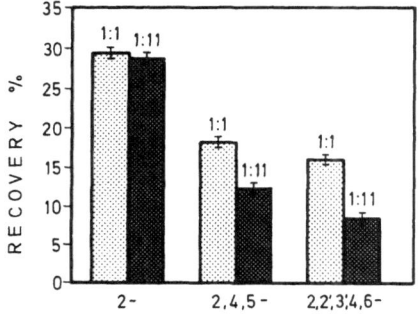

Fig. 3.64. Influence of the cell geometry (two different inner diameter to length ratios, 1:1 and 1:11) on the SFE efficiency for various PCBs retained in ODS sorbents. (Reproduced with permission of Pergamon)

velocity near the cell walls is close to zero even if the linear velocity is not negligible. Therefore, if the sample is deposited on the cell walls, the carrier flow velocity may be inadequate to sweep the analytes. This may be why small amounts of sample in large cells provide poorer results than in smaller cells.

The linear velocity is probably the most influential factor related to cell size. The higher the velocity is, the faster should be the extraction. In fact, small-scale extractions proceed at an adequate rate with velocities of around 1 cm/min.

The cell and flow orientations have minimal effects on the extraction rate except at very low flow-rates (e.g. 0.15 millilitres of liquid CO_2 per minute in the pump head), where the system void volume hinders extraction.

On a constant cell inner volume, the effect of the cell geometry on the SF extraction rate for some real samples (e.g. highly homogeneous, sieved soil samples and heterogeneous lemon peel samples) is negligible provided elongated cells of small internal diameter or wide, short cells are used [75]. Otherwise, recovery decreases with decreasing inner diameter to length ratio [76,77]. When analytes are to be extracted from solid adsorbents, the dimensions of the extraction cell used can have a marked effect – occasionally comparable to that of the SF density – on recovery [77], depending on the analyte and, especially, the solvent used. The recovery of PCBs from octadecylsilane (C_{18}) adsorbents increases significantly with increasing cell width (i.e. with increase in the inner diameter to length ratio); however, such an effect of the cell geometry is not observed with Florisil adsorbents [76]. Figure 3.64 compares the recoveries of various PCBs from C_{18} using two extraction cells of different geometry: a short, wide cell (9.9 mm × 9.9 mm i.d., 1:1 i.d./L ratio) and a longer, thinner cell (50.0 mm × 4.4 mm i.d., 1:11 i.d./L ratio). The differential effect of C_{18} has two possible explanations. Thus, the analytes may be reretained by C_{18} after extraction, as in supercritical fluid chromatography, thereby slowing down the extraction and decreasing the recovery. Also,

Geometry of extraction cell

the higher linear velocity (up to five times) of SC CO_2 in the 1:11 cell may hinder transfer of the analyte from the adsorbent to the supercritical fluid stream.

The recovery of PAHs from C_{18} can be increased by a factor greater than two on increasing the inner diameter to length ratio from 1:20 to 1:1, depending on the particular analyte. The recovery increment obtained by changing the cell dimensions increases roughly linearly with number of fused rings in PAHs [77].

As noted earlier, the variation of recovery with the extraction time conforms to a curve such as those in Fig. 3.33. The extraction rate decreases with time, as reflected in the slopes of the curves, which tend to zero with time. The amounts extracted per unit time decrease as extraction develops up to a time after which recovery remains virtually unchanged at 100% or less.

The time required for quantitative extraction tends to increase from spiked samples to real samples. For example, complete extraction of cholesterol from egg yolk takes 60 min compared to only 30 min for quantitative extraction of this substance added to glass wool [44].

References

1. Hawthorne SB, Langenfeld JJ, Miller DJ, Burford MD (1992) Anal. Chem. 64:1614
2. Nielen MWF, Sanderson JT, Frei RW, Brinkman UATh (1989) J. Chromatogr. 474:388
3. Houben RJ, Janssen HGM, Leclercq PA, Rijks JA, Cramers CA (1990) J. High Resol. Chromatogr. 13:669
4. Wallace JC, Krieger MS, Hites RA (1992) Anal. Chem. 64:2655
5. Johnston KP, Eckert CA (1981) AIChE J. 27:773
6. Chrastil J (1982) J. Phys. Chem. 86:3016
7. Sadek PC, Carr PW, Doherty RM, Kamlet MJ, Taft RW, Abraham MH (1985) Anal. Chem. 57:2971
8. McHugh MA, Seckner AJ, Yogan TJ (1984) Ind. Eng. Chem. Fundam. 23:493
9. Kurnik RT, Holla SJ, Reid RC (1981) J. Chem. Eng. Data 26:47
10. Kurnik RT, Reid RC (1982) Fluid Phase Equilib. 8:93
11. Schafer K, Baumann W (1988) Fresenius Z. Anal. Chem. 332:122
12. Maxwell RJ, Hampson JW, Cygnarowicz-Provost M (1991) LC–GC 9:789
13. Larson KA, King ML (1986) Biotechnol. Prog. 2:73
14. Mitra S, Chen JW, Viswanath DS (1988) J. Chem. Eng. Data 33:35
15. Banberger T, Erickson JC, Cooney CL, Kumar SK (1988) J. Chem. Eng. Data 33:327
16. van Leer RA, Paulaitis ME (1980) J. Chem. Eng. Data 25:257
17. McHugh M, Paulaitis ME (1980) J. Chem. Eng. Data 25:326
18. Johnston KP, Ziger DH, Eckert CA (1982) Ind. Eng. Chem. Fundam. 21:191
19. Dobbs JM, Wong JM, Lahiere RJ, Johnston KP (1987) Ind. Eng. Chem. Res. 26:56
20. Dobbs JM, Johnston KP (1987) Ind. Eng. Chem. Res. 26:1476
21. Kim S, Johnston KP (1987) AIChE J. 33:1603
22. Dandge DK, Heller JP, Wilson KV (1985) Ind. Eng. Chem. Prod. Res. Dev. 24:162

23. Francis AW (1954) J. Phys. Chem. 58:1099
24. Hildebrandt JH, Scott RL (1950) The solubility of non-electrolytes, 3rd edn. Reinhold, New York
25. Giddings JC, Meyers MN, McLaren L, Keller RA (1968) Science 162:67
26. Barton AFM (1983) CRC Handbook of solubility parameters and other cohesional parameters. CRC, Boca Raton, Florida
27. King JW, Friedrich JP (1990) J. Chromatogr. 517:449
28. Ikushima Y, Goto T, Arai M (1987) Bull. Chem. Soc. Jpn. 60:4145
29. Peng DY, Robinson DB (1976) Ind. Eng. Chem. Fundam. 15:59
30. Walsh JM, Ikonomou GD, Donohue MD (1987) Fluid Phase Equilib. 33:295
31. van Alsten JG, Hansen PC, Eckert CA (1984) Supercritical enhancement factors for nonpolar and polar systems. AIChE Annual Meeting, San Francisco
32. Schmitt WJ, Reid RC (1986) Fluid Phase Equilib. 32:77
33. Brunner G (1983) Fluid Phase Equilib. 10:289
34. Alexandrou N, Pawliszyn J (1989) Anal. Chem. 61:2770
35. Bartle KD, Clifford AA, Hawthorne SB, Langenfeld JJ, Miller DJ, Robinson R (1990) J. Supercrit. Fluids 3:143
36. Carslaw HS, Jaeger JC (1959) Conduction of heat in solids. Charendon, Oxford, 233
37. Crank J (1975) The mathematics of diffusion. Clarendon, Oxford, 89
38. Bartle KD, Boddington T, Clifford AA, Cotton NJ, Dowle CJ (1991) Anal. Chem. 63:2371
39. Campbell E (1992) Optimization of operating parameters used in supercritical fluid extraction. Pittcon'92, New Orleans
40. Otero Keil Z (1992) Optimization of SFE by statistical methods. Pittcon'92, New Orleans
41. Lopez Avila V, Dodhiwala NS, Beckert WF (1990) J. Chromatogr. Sci. 28:468
42. Tena MT, Luque de Castro MD, Valcarcel M (1993) Lab. Rob. Autom. 5:255
43. Dolata LA, Levy JM, Rosselli AC, Ravey RM (1992) Making SFE Work for environmental applications. Pittcon'92, New Orleans
44. Ong CP, Ong HM, Li SFY, Lee HK (1990) J. Microcol. Sep. 2:69
45. McNally ME, Wheeler JR (1988) J. Chromatogr. 447:53
46. Hunt TP, Dowle CJ, Greenway G (1991) Analyst 116:1299
47. Hawthorne SB (1990) Anal. Chem. 62:633A
48. Li SFY, Ong CP, Lee ML, Lee HK (1990) J. Chromatogr. 515:515
49. David F, Verschuere M, Sandra P (1992) Fresenius J. Anal. Chem. 344:479
50. Louie PKK, Timpe RC, Hawthorne SB, Miller DJ (1992) Determination of elemental and organic sulfur in coal using supercritical fluid extraction (SFE) and pyrolysis/SFE. Pittcon'92, New Orleans
51. Schmidt S, Blomberg L, Wannman T (1989) Chromatographia 28:400
52. Yang J, Colvin BA, Bonanno AS, Griffiths PR (1992) SFE/SFC/FT-IR Identification of compounds in and on polymeric matrices. Pittcon'92, New Orleans
53. Wuchner K, Ghijsen RT, Brinkman UATh, Grob R, Mathieu J (1993) Analyst 118:11
54. Thomson CA, Chesney DJ (1992) Anal. Chem. 64:848
55. King JW (1989) J. Chromatogr. Sci. 27:355
56. Sandra P, private communication
57. Ikushima Y, Saito N, Goto T (1989) Ind. Eng. Chem. Res. 28:1364
58. Ikushima Y, Hatakeda K, Ito S, Saito N, Asano T, Goto T (1988) Ind. Eng. Chem. Res. 27:818
59. Levy JM, Rosselli AC, Khorassani MA, Dolata LA, Storozynsky E, Boyer DS, Ravey RM (1992) Modifiers: the ultimate answer in SFE. Pittcon'92, New Orleans
60. Andersen MR, Swanson JT, Porter NL, Richter BE (1989) J. Chromatogr. Sci. 27:371
61. Dooley KM, Kao CP, Gambrell RP, Knopf FC (1987) Ind. Eng. Chem. Res. 26:2058

62. Schafer K, Baumann W (1989) Fresenius Z. Anal. Chem. 332:884
63. Hewlett Packard Application no. 228–154
64. Wheeler JR, McNally ME (1989) J. Chromatogr. Sci. 27:534
65. Hewlett Packard Application no. 228–117
66. Damian J, Myer L, Liescheski P, Tehrani J (1992) Supercritical fluid extraction of organic analytes from aqueous media and wet matrices. Pittcon'92, New Orleans
67. Johansen HR, Becher G, Greibrokk T (1992) Fresenius J. Anal. Chem. 344:486
68. Hills JW, Hill HH Jr, Maeda T (1991) Anal. Chem. 63:2152
69. Hawthorne SB, Miller DJ, Nivens DE, White DC (1992) Anal. Chem. 64:405
70. King JW, France JE, Snyder JM (1992) Fresenius J. Anal. Chem. 344:474
71. Kuppers S (1992) Chromatographia 33:434
72. Liu H, Cooper LM, Raynie DE, Pinkston JD, Wehmeyer KR (1992) Anal. Chem. 64:802
73. Westwood SA (ed.) (1993) Supercritical fluid extraction and its use in chromatographic sample preparation. Blackie Academic & Professional, London
74. Hewlett Packard, Application no. 228–145
75. Burford MD, Langenfeld JJ, Hawthorne SB, Miller DJ (1992) Effect of extraction cell shape, flow rate and solvent collection parameters on supercritical fluid extraction (SFE) efficiencies. Pittcon'92, New Orleans
76. Furton KG, Lin Q (1992) Chromatographia 34:185
77. Furton KG, Rein J (1991) Chromatographia 31:297

4 The Analytical-Scale Supercritical Fluid Extractor

4.1 Introduction

Chapter 1 is concerned with supercritical fluid extraction (SFE) in terms of the analytical process, with emphasis on the significance of its preliminary operations. Chapter 2 describes the physico–chemical properties of supercritical fluids and reviews their applications. Finally, Chapter 3 deals in systematic fashion with solid–supercritical fluid extraction (leaching) and the most influential factors on its efficiency, throughput and selectivity.

This Chap. 4 is concerned with practical aspects of SFE such as the basic elements and functioning of a supercritical fluid extractor. It also deals with the different possible operational modes and connections (on-line and off-line) between the extractor – which is not an instrument, but an apparatus – and the sensing system used. The wide use of gas (GC), supercritical fluid (SFC) and liquid (HPLC) chromatographs in this context is a result of the need to discriminate between the analytes in the extract (the selectivity of SFE generally allows "families" of analytes rather than individual components to be isolated). The advantages and disadvantages of the various available alternatives are compared on the basis of a number of examples. Finally, the features of commercially available extractors are described and compared in order to provide the reader with useful guidelines for choosing the most suitable extractor in each case.

4.2 The Supercritical Fluid Extractor: A Broad View

Cost of apparatus

In contrast with the widespread belief among the uninitiated, an analytical-scale supercritical fluid extractor can readily be assembled in the laboratory in a fairly short time and for less than $100 000. The error most frequently made by inexperienced researchers in this field is probably using too many switching valves, transfer lines and other types of mechanism. Broadly speaking, the less sophisticated set-ups provide the better results, are the less polluting and give rise to smaller losses. However, the complexity of some samples and/or specific analytical requirements may occasionally call for a highly sophisticated system performing like a commercially available extractor.

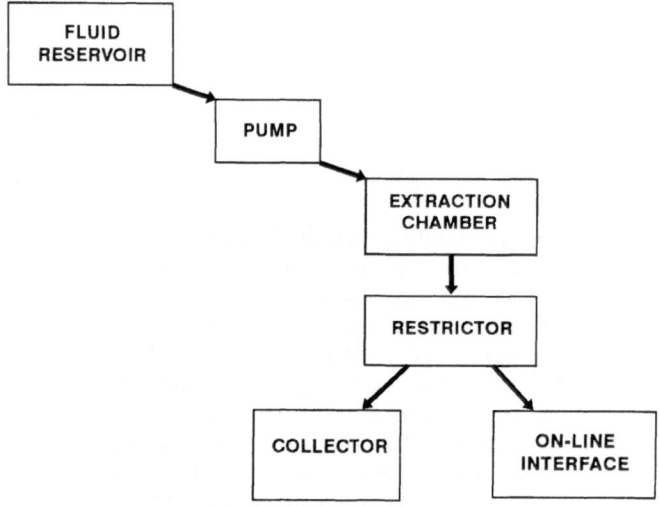

Fig. 4.1. Basic scheme of a supercritical fluid extractor

Figure 4.1 depicts the basic elements of an analytical-scale extractor, namely:

a) A source of high-purity supercritical fluid.
b) A high-pressure pump delivering the fluid at constant, readily controllable pressure and flow-rate. Even though the pumps typically used in supercritical fluid chromatography can also be used in SFE, there are some inexpensive alternatives (because extractions are usually performed at a constant pressure, none of the sophisticated pressure–density ramp controllers used in the chromatographic technique are required here).
c) A cell or chamber holding the sample, thermostated with the aid of an appropriate device. The extraction chamber can be purchased from manufacturers or built from tight-fitting tubes (e.g. empty chromatographic columns or pre-columns). The cell temperature can be controlled by placing it in a chromatographic oven or a straightforward heating tube.
d) A restrictor intended to maintain the required pressure in the extraction chamber.
e) A system for collection of extracted components after the SF is depressurized, connected with the extraction chamber via the restrictor. Extracts are usually collected by bubbling the depressurized fluid through a container holding a few millilitres of an appropriate solvent, a cryogenic trap or a cartridge packed with adsorbent material.

While supercritical fluid extraction is a preparative technique and requires no coupling to others, some special samples and/or analytes make it necessary or desirable to use it as a preliminary step to one or more separation technique(s) (usually gas, liquid or supercritical fluid chromatography [1,2]).

4.3 Basic Elements of a Supercritical Fluid Extractor

Figure 4.1 shows the essential elements of a supercritical fluid extractor, namely a fluid reservoir, a propulsion system, an extraction chamber, a depressurization system and a collection system, which are described in detail below.

4.3.1 The Fluid Reservoir

4.3.1.1 Description

Substances that can be used as supercritical fluids (SFs) are usually commercially available in steel, aluminium and stainless steel cylinders whose capacity and content weight vary over the ranges shown in Table 4.1. The pressure within one such cylinder is high enough for its contents to be liquid whichever state they may take under normal conditions. The purity of SFs must meet the specific requirements of each applications, as noted in Sect. 3.3. Cylinders are supplied as standalone elements or parts of combined units, the number and nature of which depend on the complexity of the extraction system to be fitted to the cylinder. Sealings for connections between the cylinder and the SF delivery port must be of Teflon or Kel-F since other materials swell, deteriorate or are partly dissolved on contact with most SFs.

Metal cylinders

Connections seals

Table 4.1. Size and weight of CO_2 tanks typically used in SFE and SFC

Approximate dimensions (diameter × height, cm)	Approximate weight of contents (kg)
23 × 130	25
23 × 65	12
18 × 85	11
16 × 49	4
10 × 44	2

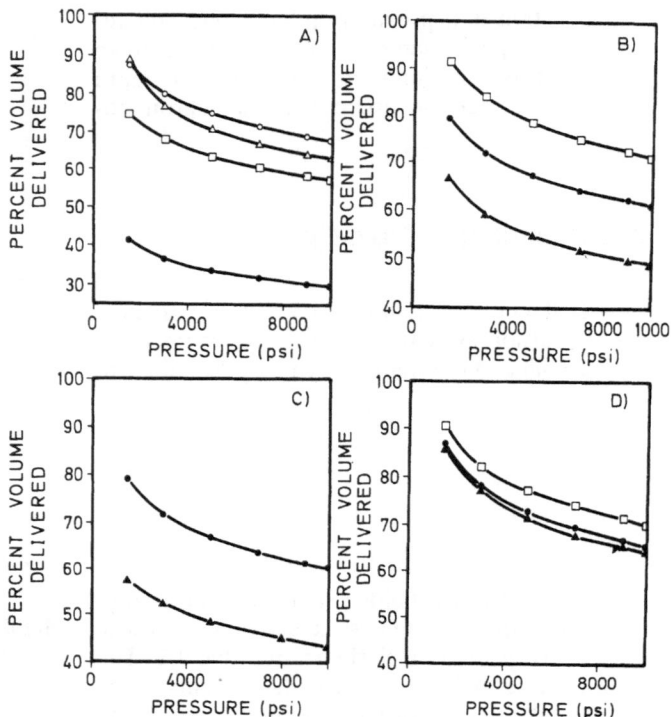

Fig. 4.2A–D. Variation of the CO_2 delivery/pressure ratio as a function of the housing cylinder. **A** Cylinder with the dip tube at the bottom and the pump on the workbench (○ without helium; pump cooled to 18 °C with water, 2.2 min warm-up time after loading; ▲ cylinder with about 1500 psi helium pressurization; □ pump cooled to 18 °C with water, (2 min warm-up time after loading; ● without helium, no pump cooling, 2 min warm-up time after loading). **B** Cylinder with the dip tube and pump at the bottom (□ pump cooled at 18°C with water, 2 min warm-up time; ● pump cooled to 18°C with water, 2 min warm-up time; ▲ without cooling, 2 min warm-up time). **C** Standard cylinder (no dip tube), standing, with pump on the bench; pump cooled to 18 °C with water. (● 22 min after loading; ▲ 2 min after loading). **D** Inverted, off-ground cylinder delivering CO_2 by syphoning (□ pump cooled at 18 °C, 22 min after loading; ● no cooling, 22 min after loading; ▲ no cooling, 2 min after loading). (Reproduced with permission of ISCO)

The features and usage of an SF cylinder affect the efficiency of the high-pressure system used to propel it along the extractor and hence extraction proper. Irrespective of their constituent material, cylinders can be supplied with or without a dip tube, and with or without a helium head space. Helium is scarcely soluble in most SFs and their mixtures, so it minimizes the amount of liquid or gas that is wasted as a result of line cooling or pressure drops along lengthy transfer lines. Finally, cooling of the pump head, the time during which cooling is applied and the cylinder position (standing or inverted) in cylinders without a dip tube may have dramatic effects on the supply of SF to the extraction system.

Cylinders furnished with a dip tube, a pressurized head space and cooling for long enough to ensure equilibration provide the most regular SF delivery. Figure 4.2A shows the variation of the proportion of super-critical CO_2 delivered by a syringe pump at different pressures from a cylinder furnished with an extraction tube, a pressurized (around 1500 psi) or non-pressurized head space, and cooling for different times or no cooling, with the pump placed above the cylinder delivery port. Figure 4.2B shows a similar graph for cylinders with a non-pressurized head space and the pump located about 1 m below the cylinder output. As can clearly be seen, the cylinder with a dip tube, a pressurized head space, equilibration at an appropriate temperature and the propulsion system lying below the cylinder output provides the best efficiency. However, because the first two requisites are not met by some commercially available cylinders, the quality of SF delivery can be improved in two ways de-pending on the required pressure. Thus, with the cylinder upside down (i.e. the fluid emerging from the bottom), the gas formed in the cylinder free zone remains away from the outlet, from which it can be dispensed in a liquid state until the cylinder contents are exhausted. This is only the case with SFs that occur as gases at ambient temperature (e.g. CO_2). If the output of the inverted cylinder lies above the propulsion system, the resulting syphoning facilitates delivery of the SF, as shown in Figs 4.2C and D.

Regular delivery

Cylinder upside down

Syphoning

4.3.1.2 Connection to the Extractor

The interface between the cylinder and the extractor consists of several units, some of which are essential and others only occasionally required or desirable. Figure 4.3 depicts such units, viz. the connecting line, filters and check, safety, bleed and switching valves for a system using one or two supercritical fluids.

The *connecting line*, which joins the cylinder and extractor, consists of a stainless steel tube of 1.5 mm inner diameter. The line should be as short as possible in order to minimize conversion of the fluid into a gas before the pump is reached. If the cylinder cannot be placed near the extractor, it should be one with a pressurized head space.

Connecting line

Nevertheless, a *bleed valve* is the best choice when the connecting line cannot be made short enough. Bleed valves are usually placed immediately before the extractor, so that the fluid reaching them can be wasted until some liquid emerges from it, thereby avoiding the undesirable presence of gas in the pump.

Bleed valve

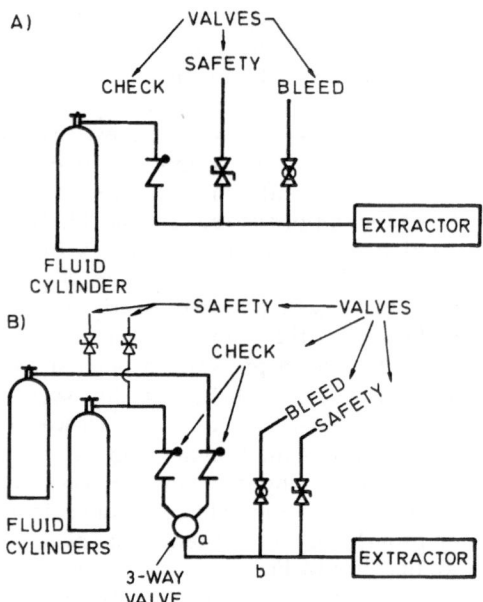

Fig. 4.3A,B. Ways of connecting the fluid cylinder(s) and the extractor

Safety valve   A *safety valve* or a similar device should normally be used in order to avoid pressures above those the system and material used can reasonably withstand. The valve is actuated when a preset pressure is exceeded in order to release fluid and hence any back-pressure arising from malfunctioning of some measuring device.

Check valve   A *check valve* attached to the line minimizes contamination of the fluid stored in the cylinder.

If the extractor is connected to several cylinders, then a three- or four-**Switching** way *switching valve* is required to supply the fluid required at each stage. **valve** In this situation, the *ab* segment in Fig. 4.3 should lend itself readily to venting and purging with fresh fluid. In order to avoid problems arising from the reactivity of some fluids such as NH_3, the interface should include a nitrogen purging line to minimize reaction in the transfer tube.

Filters   *Filters* of small pore size (5–10 µm) are occasionally employed prior to or after the high-pressure pump in order to remove any contaminating particles from the fluid. The presence of such particles, though, can be minimized or avoided by using cylinders of appropriate materials. **Cylinder inner** Aluminium cylinders provide better inner surfaces than do steel cylinders **surfaces** in this respect, especially if the supercritical fluid they contain includes a modifier. In fact, modified SFs should never be stored in steel cylinders because many modifiers react with steel to form organometallic complexes.

On the other hand, aluminium and stainless steel are not affected by most modifiers.

4.3.1.3   Functioning, Cautions

Maintaining the purity of a fluid to be used for SFE entails exercising some cautions in replacing cylinders. This should be done with as much care as one usually prepares samples. Specifically:

a) The product quality (SF grade) should be checked on the cylinder label.
b) The cylinder should be allowed to stand at the future working temperature for at least 24 h.
c) The connecting material required should be spotlessly clean and made ready on an also clean surface.
d) The operator should wear nylon gloves.
e) Before the cylinder is fitted, the fluid on-off valve should be rapidly turned three times until some liquid emerges. This can safely be done by switching the valve on and off rapidly, but gently. This flushes any deposited material in the valve, after which it must be wiped with a dry cloth before the cylinder is fitted to the system.
f) If a new tube is to be used as connecting line, then it should be of stainless steel and washed with an acid, rinsed with water and dried in an oven prior to fitting.
g) If an on-line filter is to be employed, it should be replaced with each new cylinder of fluid.
h) The extractor should be purged no fewer than six times in order to remove retained impurities by using a pressure of 350–400 atm and a temperature at least 20 °C higher than the working temperature.
i) In replacing cylinders, all open and shut connections should be wiped with a clean cloth. No solvents or lubricants that may penetrate into the system should be used for this purpose.

4.3.1.4   Measurement of the Cylinder Contents

The contents of CO_2, N_2O or any other compressed gas left in a cylinder can be determined by measuring the cylinder internal pressure, to which its contents are proportional. On the other hand, the amount of liquefied gas left in the cylinder cannot be measured through the pressure as the latter does not change proportionally to oscillations in the amount of gas inside the cylinder. **Internal pressure**

The simplest and most frequently used way of measuring the cylinder contents is by weighing. For this purpose, the cylinder is placed on a **Weighing** balance that continuously records weight variations and hence the amount of fluid consumed and that remaining in the cylinder.

Some manufacturers such as BB Nerike have developed several devices for direct, visual measuring of liquefied gas that provide an accurate indication of the cylinder contents irrespective of its internal pressure and temperature.

4.3.2   The Propulsion System

High pressure pumps   Propulsion units typically employed in SFE consist of a high-pressure pump similar to those used in HPLC or supercritical fluid chromatography, even though none of the pressure–density ramp devices used in SFC are usually needed here. In any case, the pumps must meet some requisites for optimal performance of the overall set-up, namely:

a) They should be functional over a wide pressure range (ideally, between 1000 and 10 000 psi). In addition, the working pressure should be constant and readily measurable.
b) They should provide reproducible, non-pulsating streams at flow-rates between 1.0 µl/min and 90 ml/min.
c) They should be chemically inert, particularly in those zones that are to come into contact with the supercritical fluid.

Other, desirable features of the propulsion system include rapid recovery (in a matter of a few seconds) on cell pressurization, expeditious loading (with syringe pumps) and compression to the working pressure, and the ability to continuously monitor the fluid flow-rate via a computer screen.

4.3.2.1   Single Systems

Commercially available pumps, be they stand-alone units or parts of extraction systems, are of either of two types: syringe or dual piston pumps.

Syringe pumps consist of a barrel that its loaded with fluid which is propelled by a piston at a preset rate with the aid of a gearing system and a step motor.

Dual piston pumps have two chambers with two pistons, an eccentric and a motor arranged in such a way that the piston positions are subject to a lag of 180°: while one chamber is pumping, the other is being loaded. Both chambers are provided with a highly viscous fluid that compresses and moves a steel diaphragm which isolates it from the fluid passage zone.

Based on the specifications of commercially available pumps, SFE operational conditions can be varied between the following ranges:

- Flow-rate: $1.0 \mu l/min$ to $90 ml/min$.
- Pump capacity: unlimited in piston pumps; up to $270 ml$ in syringe pumps.
- Precision: $\pm 1.0\%$.
- Operational modes: programmed pressure, density and/or flow-rate.

Dual piston pumps have the advantages over syringe pumps that they deliver a continuous supply of fluid (limited by the barrel size in syringe pumps) and allow easy fluid changeovers. Unlike piston pumps, however, syringe pumps deliver a non-pulsating flow.

If the fluid to be used as extractant is liquid at ambient temperature, then setting the working pressure required should pose no problem. However, if the fluid is a gas under normal conditions, it may call for a compressor in large-scale work (samples of over $50 g$), particularly when syringe pumps are used.

Part of the technical research into SFE is being aimed at improving propulsion systems. Figure 4.4A shows a computer-controlled pneumatic amplifier pump [3] that provides a nominal pressure amplification factor of $122:1$ and features an SF delivery volume of $10 ml$. The fluid to be pumped is loaded into the pump via two filters (one of silica gel and the other of Ambersorb XE-340 resin) and a stainless steel cooling reservoir. The output pressure is controlled by a transducer mounted on the high-pressure side of the pump. The signal obtained can be improved by using an optional analogue-to-digital converter. A pressure gauge allows a minimal pressure to be maintained on the low-pressure side. The control software, written in BASIC, is rapid enough to manage the small changes required to keep the fluid at a given pressure or create a gentle pressure ramp.

Figure 4.4B shows a system based on a simpler operation principle [4]. It consists of a 200-ml stainless steel container furnished with a safety valve accommodating a disc that breaks at $700 atm$ (Fig. 4.4B.1). The system works as follows: the container (1) is cooled below the critical pressure of the fluid by means of a cooling source (9) and is loaded with liquid from the high-pressure cylinder. In the next step, the container is heated, so the pressure inside it is raised and measured and controlled electronically with the aid of the transducer (3) coupled to the thermostat (10). The supercritical fluid is dispensed to the extraction chamber at an accurately controlled pressure. The cylinder can typically supply $50 l$ of CO_2 at ambient pressure, which is enough for $5 h$ worth of continuous extraction through a 20-μm restrictor with an output flow-rate of $150 ml/min$. If a continuous supply is required, the extraction system can be fitted with two containers in parallel: while the first container is being used, the other can be made ready for use and vice versa (Fig. 4.4B.2). This system can readily be miniaturized by replacing the high-pressure containers with

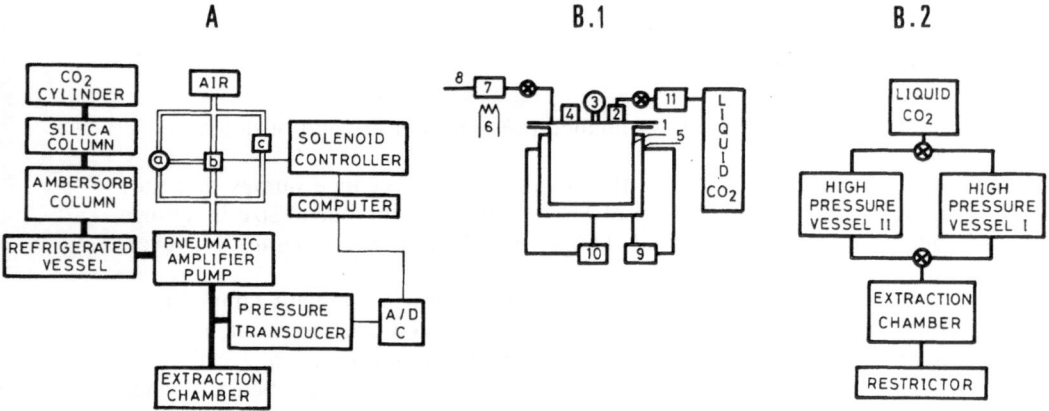

Fig. 4.4A,B. Non-commercially available propulsion systems. **A** Pneumatic amplifier pump (*a*, pressure regulator; *b* and *c*, large and small solenoid valve, respectively). **B.1** High-pressure collection system (1, high-pressure vessel; 2, safety valve; 3, electric pressure transducer; 4, electric temperature transducer; 5, heating/cooling jacket; 6, electric heater; 7, extraction chamber; 8, restrictor; 9, cold fluid source; 10, hot fluid source; 11, active carbon trap). **B.2** Continuous-supply fluid source based on two high-pressure vessels. For details, see text. (Reproduced with permission of Hucthig)

stainless steel tubing. In this case, the amount of SF each piece of tubing can supply is limited, but can be sufficient for the extraction of small amounts of sample, followed by direct deposition in a capillary column. Two tubes can be arranged in parallel for this purpose if needed.

4.3.2.2 The Need for a Dual Propulsion System: Use of Modifiers

As discussed at length in Sect. 3.4.5, cosolvents alter the polarity of supercritical fluids, thereby extending their typical application to non-polar analytes to typically polar compounds. Modifiers are thus commonplace in an SF extraction system. However, including a modifier in an SF set-up is not always easy. As can be seen in Fig. 4.5, there are several ways of introducing a modifier into an extractor, namely:

Introduction of modifier into extractor

a) By purchasing the supercritical fluid with the modifier in it, which detracts from the intrinsic flexibility of cosolvents (keeping a modest stock of SF cylinders containing different modifiers at various concentrations is fairly expensive).

b) By adding the modifier in the pump module during loading, which entails cleaning the module for each new modifier and concentration.

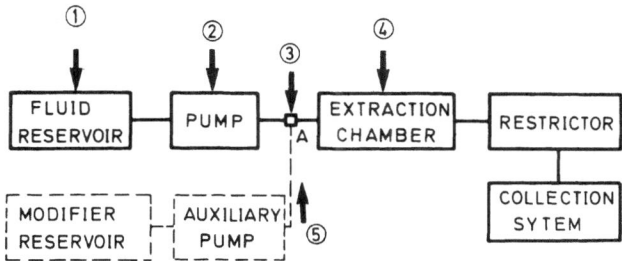

Fig. 4.5. Zones for introduction of a modifier into an SF extractor. A can be modifier or mixed fluid chamber. For details, see text

c) By using a modifier chamber between the pump and extraction chamber [5,6]. The chamber consists of a stainless steel cell holding the modifier that is placed within a chromatographic oven and through which the SF is pumped. This choice has the disadvantage that the modifier only acts during the first few moments in dynamic extractions.

d) By soaking the sample in the modifier, which can be rapidly flushed out of the extraction chamber by the SF, thereby loosing its effect under a dynamic regime.

e) By programmed addition to the SF with the aid of an auxiliary pump for delivery and a chamber between the pumps and the sample cell ensuring thorough mixing of the two phases prior to reaching the sample. Programmed addition of the modifier offers a number of advantages over the previous alternatives including the following: greater cleanliness, easier handling and higher reproducibility; easier addition and more thorough mixing than that achieved by adding a given modifier volume to the pump module during the loading operation; readier modifier changeovers; and a very simple pressure programme to synchronize the two pumps (SF and modifier).

Programmed modifier addition

Some manufacturers including Isco, Suprex, Computer Chemical Systems and Dionex market extractor ancillary modules for addition of modifiers which not only offer the above-mentioned advantages but allow mixing at fairly low pressures (around 2000 psi), thus ensuring thorough mixing, programming the addition in terms of mole percentages rather than volume percentages; switching between two types of modifiers via a valve without the need to replace cylinders; and simultaneous delivery to several cells in extractors performing parallel extractions.

4.3.3 The Extraction Chamber

4.3.3.1 General Features

The extraction chamber or cell is the compartment where the unknown sample is placed for subjection to the action of the supercritical fluid. This function can be served not only by commercially available units, but also by customized cells for specific purposes or even empty chromatographic columns or pre-columns. In any case, for one of these units to be efficiently used in an SFE process, it must meet several requisites, namely:

a) It should allow ready, convenient introduction of the sample.
b) It should be easy to shut without the need for any tools.
c) It should lend itself to sealing via software.
d) It should be chemically inert – SFE chambers are usually of porous stainless steel.
e) It should lend itself to ready, rapid mounting and disassembly from the system.
f) It should encompass a broad range of capacities (from 100 µl to 50 ml).
g) It should withstand high pressures (at least 700 atm).

Figure 4.6 shows a typical extraction cell consisting of a body, which determines its useful volume or capacity and is furnished with frit lids preventing sweeping of the unknown solid by the supercritical fluid, in addition to threaded sealing systems for tight fitting.

Even though cells can usually withstand higher pressures than the extraction system (up to 20 000 psi), they are generally fitted with a safety valve including a disc that bursts at a much lower pressure than that withstood by the cell (around 7000 psi), which protects the operator in the event of malfunctioning of the electronic or manual temperature controls.

4.3.3.2 Cell Size

The capacity of an extraction chamber loaded with an SF is dictated by the features of the sample to be extracted, so it can vary over a wide range (between less than 1 mg to a few hundred grams). Ideally, the sample size

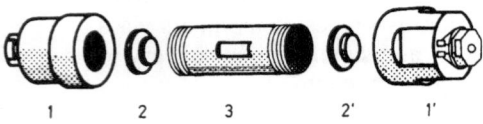

1   2   3   2'   1'

Fig. 4.6. Commercially available extraction chamber. (1,1'-manually fitted threaded cap; 2,2'-frit and seal unit; 3-cell body)

should be small enough to allow using as little SF as possible and achieving quantitative trapping of the analyte after extraction. However, this is not always possible, whether because the analyte of interest is present at a low concentration in the sample or because this is rather heterogeneous. As a result, manufacturers offer cells holding between 0.1 and 50 ml.

The smaller cells are preferentially used in on-line applications; because collection is quantitative, a small amount of analyte is sufficient for analysis, which is specially advantageous in dealing with expensive or scarcely accessible samples.

4.3.3.3 Sample Size to Cell Volume Ratio

The optimal sample size to cell volume ratio depends on whether extraction is performed statically or dynamically.

The pioneers of SFE concluded that expeditiousness was inversely proportional to the cell void volume on the assumption that the super-critical fluid not only dissolved the analytes but also swept them to the cell; therefore, the higher the cell void volume were, the larger number of cell volumes of SF would be required to separate/sweep the analytes quantitatively. This assertion has been corrected in view of the fact that, if the cell volume is inadequate, the SF carves channels in the sample matrix rather than permeates it (dynamic mode) or saturates it with analyte (static mode). This argument has been checked experimentally in the static mode by conducting identical extractions of 0.25, 0.5 and 0.6 g of a **Static mode** detergent in a 1.5-ml cell. The variation of the peak areas obtained by gas chromatographic separation followed by flame ionization detection for one of the extracted analytes (a fragance) with the amount of detergent are shown in Fig. 4.7. As can be seen, extraction was maximal for a sample

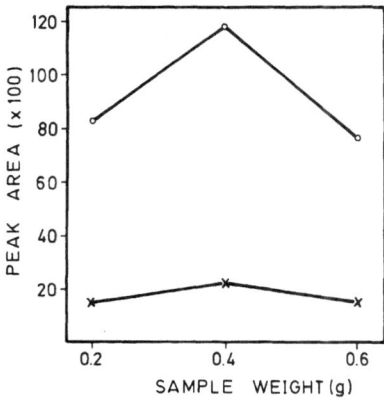

Fig. 4.7. Influence of the cell volume to sample size ratio on the extraction efficiency of two fragrance components at two different retention times: (○) 8.9 min and (+) 5.2 min. Cell volume = 1.5 ml

size of 0.5 g. Even though there is no irrefutable explanation for this phenomenon, the ratio of cell size to sample volume must logically influence the efficiency of SF extractions.

Dynamic mode   On the other hand, working in the dynamic mode requires considering additional aspects, as pointed out by Schmidt et al. [7], who assayed identical samples placed in cells of 0.1 and 0.3 ml capacity. The extraction yield of the larger cell was 12% lower than that of the smaller one; in addition, the former resulted in poorer precision than the latter ($\pm 6.0\%$ vs $\pm 1\%$ as rsd). This is partly due to the variation of the flow-rate with the cell size: 0.11 mm/s in the smaller cell and 0.22 mm/s in the larger one. The cell volume can be reduced to the desired value by packing it with an appropriate amount of inert material. Several manufacturers including Hewlett–Packard market cells furnished with additional packing material.

4.3.3.4   Cell Geometry

Extraction efficiency   The geometry of the extraction chamber is markedly influential on the extraction efficiency, as shown in experiments conducted with PAHs of variable molecular weights supported on C_{18} and two cells of the same capacity (about 0.8 ml) but different dimensions (1.0 cm $\times$ 1.0 cm i.d. vs 7.3 cm $\times$ 0.37 cm i.d., i.e. inner diameter to length ratios of 1:1 and 1:20, respectively). The recoveries afforded by the 1:1 cell exceeded those provided by the 1:20 cell by a factor dependent on the number of PAH rings of the congener in question, viz. 13, 33, 109, 137 and 206% for methoxychlor, pyrene, perylene, benzoperylene and coronene, respectively. As a rule, the factor increased with decreasing analyte extractability and was found to be a linear function of the number of fused rings of the PAHs studied. Figure 4.8A shows the recoveries achieved for two such PAHs.

The cell geometry also affects one of the most frequently controlled variables of the SFE process: density. In fact, the logarithm of the percent recovery increases linearly with the SF density. A regression analysis of data obtained by the least-squares method provided the following equations:

$$\log R_4 = 3.281d - 0.362 \quad (r = 0.999)$$

$$\log R_5 = 3.055d - 0.644 \quad (r = 0.999)$$

$$\log R_6 = 2.849d - 0.909 \quad (r = 0.994)$$

$$\log R_7 = 2.737d - 1.347 \quad (r = 0.999)$$

where Rs denote the percent recovery of pyrene (R_4), perylene (R_5), benzo(*ghi*)perylene (R_6) and coronene (R_7), and d the density of supercritical CO_2 at 100 °C. In contrast to the above-described effect of the

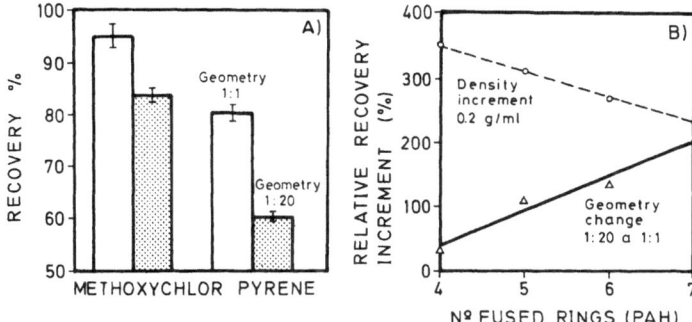

Fig. 4.8A,B. **A** Influence of cell geometry on the extraction efficiency of PAHs. **B** Influence of cell geometry and density on the relative recovery increment as a function of the number of fused rings of PAHs. (Reproduced with permission of Pergamon)

cell geometry, the recovery increment decreases with increasing analyte extractability. The ratio of the recovery increment to the density increment decreases roughly linearly with the number of rings of the PAH. Figure 4.8B reflects the opposing effects of the two variables.

4.3.3.5 Special Cells

Several authors have designed customized cells for special samples including liquid cells, microcells and gas sampling units.

While the main field of application of supercritical fluid extraction is that of solid samples, there are some applications to liquid – usually aqueous – matrices, which require altering a conventional extraction chamber in order to achieve a longer sample–extractant contact time. As shown in Fig. 4.9, the modification [9–11] usually involves inserting a capillary at each end of the cell in such a way that the SF enters it from the top, reaches the bottom, bubbles through the sample and leaves it through the lower capillary, at the end of which the SF, containing the extracted analytes, emerges. In this way, analytes can be extracted from liquid samples while the matrix is held in the chamber. However, the potential solubility of the liquid matrix in the SF must be taken into account since, for example, water is slightly (0.3%) soluble in SC CO_2, so it may freeze at the restrictor outlet.

Microcells (cells with a volume of a few microlitres) can be custom built **Microcells** from standard cells by removing or replacing the body with an altered bolt acting as sample chamber (Fig. 4.10A) or by bringing the male thread of

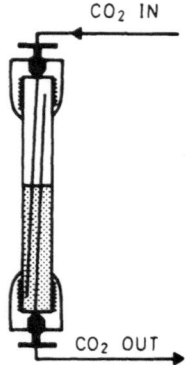

Fig. 4.9. Liquid extraction cell. (Reproduced with permission of the American Chemical Society)

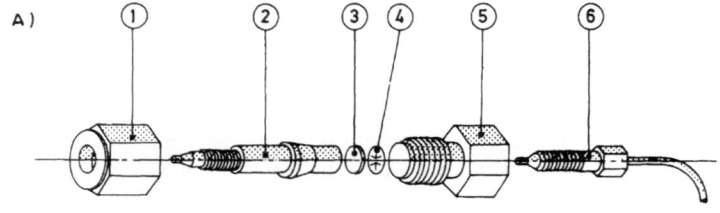

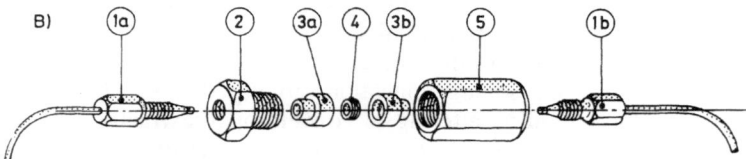

Fig. 4.10. A Mini-extraction cell ($V = 85\,\mu l$) (1, nut; 2, extraction chamber male bolt; 3, frit; 4, stream distributor; 5, bolt) **B** Micro-extraction cell ($V = 3-4\,\mu l$) (1a,b, male nuts with ferules; 2, female casings; 3a,b, female connection and frit holder; 4, frit). (Reproduced with permission of Pergamon)

the bolt into contact with the top of the frit disc; in this way, the space between the filter and the connector acts as the chamber [15] (Fig. 4.10B).

Lohleit et al. [16] developed a small cell for analysis of aerosols also functioning as a sampling unit. Figure 4.11 depicts the cell in its sampling (A) and extraction position (B). Air is passed through the system at a flow-rate of $1\,m^3/h$, particulates being removed by means of quartz fibre

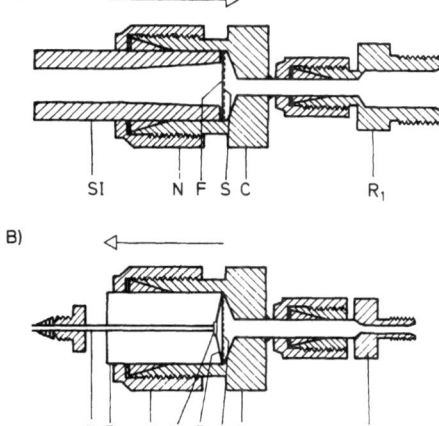

Fig. 4.11A,B. Mixed unit for sampling of organic particulates in air and extraction. **A** sampling position; **B** extraction position; SI, conical collection tube; N, nut; C, cap; S, sieve; F, filter; R-1 and R-2, joints; M, microfrit; B, GC connecting T-block. (Reproduced with permission of Elsevier Science)

filters. Potential contamination of the unit is avoided by preheating the whole system to 500°C. This ensures negligible blank extraction values.

4.3.3.6 Multi-Extraction Systems

The larger the sample size to be subjected to SFE is, the greater obviously is the amount of fluid to be passed through the cell if the extraction time is to be kept unchanged. However, increasing the flow-rate through the cell for this purpose is not always practical as it may reduce the collection efficiency. In this event, and also obviously when a fairly high throughput is required for various samples, it is customary to use several cells working simultaneously. Supercritical fluid extractor manufacturers include multi-cell options in their catalogues; cells can be used in numbers from 2 to 8 for parallel or sequential work, depending on the pump capacity and whether connection to the measuring instrument is made on-line or off-line.

Figure 4.12 shows three different ways of assembling a multi-extraction system with the aid of one or two switching valves. The system in Fig. 4.12A includes a single valve that switches to the appropriate SF from the pump and delivers it to the cells. After extraction (dynamic or static), which can be suited to each individual cell, the fluid is depressurized and collected in vials. Using two valves (Fig. 4.12B) allows the SF to be dispensed via one of them and the appropriate vial to be selected through the other – this latter, a 12-way valve, allows two fractions per sample to be collected. In the position shown in Fig. 4.12B, extracts are collected in

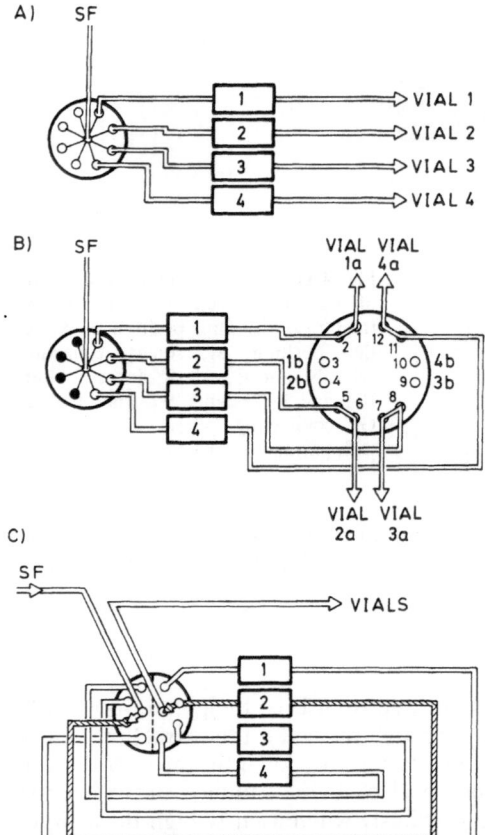

Fig.   4.12A–C. Multi-extraction systems (only four sample positions are shown, 1–4). **A** System with a switching/dispensing valve. **B** With two valves for sequential collection of two extracts from each sample. **C** With a single 10-way, 4-position valve for extraction by sequential recirculation of each sample

vials *a*; actuating the valve to the other position sends the extracts to vials *b*. This configuration can be used both statically and dynamically. Dispensing the SF and selection of the collection vial can both be entrusted to a single valve (Fig. 4.12C). This, in addition to implementation of the static and dynamic mode, allows recirculation of the fluid by connecting each cell outlet to its inlet. These systems are highly reproducible from cell to cell, as shown in Table 4.2 for various analytes sequentially extracted by using a 4-cell multi-extraction system coupled on-line to a gas chromatograph.

Table 4.2. Repeatability achieved by sequential usage of different cells in multi-extraction systems

Extracted analyte	Sample cell	Area (On-line SFE–GC system)	sd	rsd (%)
Naphthalene	1	975999		
	2	1095281	50543	4.92
	3	1007597		
	4	1032128		
Phenanthrene	1	1171944		
	2	1278659	50011	4.12
	3	1226372		
	4	1176554		
Pyrene	1	1764219		
	2	1866492	81292	4.35
	3	1226372		

4.3.3.7 Performance of Extraction Chambers

The wide gamut of commercially available extractors and individual units for supercritical fluid extraction offers a broad range of operating conditions and specifications which are summed up below.

a) Chamber working conditions:
   – Temperature: 35–200 °C.
   – Longest programmable extraction time: 6000 min.
   – Longest equilibration time: 60 min (or selectable).
b) Number of extraction chambers: 1–8.
c) Features of the chamber:
   – Material: stainless steel or PEEK/stainless steel.
   – Capacity: 0.15–50 ml.
   – Maximum recommended pressure: 700 atm.
   – Sealing: PEEK.
   – Filter pore size: 0.5–2.0 µm.
   – Fitting: with or without tools.
   – Leak sensor: Absent or incorporated into a computer-controlled feed-back system.

4.3.4 The Depressurization System

The pressure change from the supercritical conditions in the extraction cell to the prevailing atmospheric conditions is effected via an interface known

Restrictors as a *restrictor*, which also controls the flow-rate of the SF that is circulated through the cell. Restrictors are usually capillary tubes of fused silica, stainless steel or another metal that are 10–50 cm long and 10–40 µm in inner diameter. The restrictor is one of the key elements of the extractor as it is largely responsible for the success or failure of the extraction process.

4.3.4.1   Types of Restrictors

The type of depressurization system (restrictor) used is dictated by the working conditions involved in the extraction process, which in turn are determined by the nature of the sample and analytes.

Commercially available restrictors are of either of two types, viz. (a) fixed restrictors, which are manufactured in various designs (Fig. 4.13A), and (b) variable restrictors (Fig. 4.13B).

Fixed restrictors *Fixed restrictors* consist of narrow-bore, hollow or frit-packed tubes by which the pressure cannot be regulated, so density or pressure changes in the supercritical fluid can only be effected by altering the flow-rate of the propulsion system. Inasmuch as the flow-rate and pressure vary reciprocally in this type of restrictor, changing the former variable while keeping the latter constant entails replacing the restrictor with one of appropriate length and inner diameter. Figure 4.13A shows a profile view of the five most common types of fixed restrictors.

Linear or Fjelsted restrictor a) The linear or Fjelsted restrictor [17] is the simplest of all in terms of design and was the earliest applied to capillary SF chromatography. It consists of a short capillary tube (10–25 cm) of uncoated silica whose

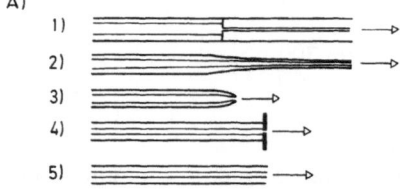

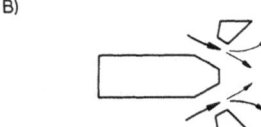

Fig. 4.13A,B. Types of restrictors. **A** Fixed (1 linear; 2 tapered; 3 integral; 4 pinhole; 5 frit). **B** Variable

inner diameter should be considerably smaller than that of the column. This type of restrictor is quite inexpensive, easy to replace and scarcely prone to plugging. On the other hand, because depressurization in SFC takes place throughout the tube, the solvent power of the mobile phase decreases along the restrictor, thereby hindering transfer of high-molecular weight compounds to the detector. In addition, cooling associated with depressurization may cause the solvent and some sample to condense on the walls, and the detector response to be distorted as a result.

b) The tapered or Chester restrictor [18] reduces the risk of solvent deposition as a result of one of its ends being tapered to an inner diameter of only 1–3 µm. This type of restrictor can be made manually with the aid of a flame, but is poorly reproducible. Still, depressurization takes place along a few centimetres of the tapered section, which, in addition, is extremely fragile. **Tapered or Chester restrictor**

c) The integral or Guthrie restrictor [19] can also be made in the laboratory by carefully heating a fused silica tube until it plugs – this requires some practice and perseverance. The sealed end is then ground to open a small hole whose diameter is adjusted to the desired value. With this type of design, pressure drops occur over a shorter stretch than in the previous two types of restrictor, so low-volatility compounds remain in the supercritical solutions until they are rapidly depressurized and dispersed in the gas phase.

d) The pinhole or Smith restrictor [20] also allows virtually instantaneous depressurization. It includes a very small orifice that is laser-drilled in a piece of metal foil. This type of restrictor was formerly used as an interface between SF chromatographs and vacuum systems, and is difficult to construct and fit.

e) The frit or Richter restrictor [21] is a compromise between those of Smith and Guthrie or those of Fjeldsted and Chester. It uses porous frit as packing for a fused silica capillary. The frit pores provide a myriad of "corridors" and reduce the risk of plugging as a result of the relatively short depressurization stretch involved.

Variable restrictors (Fig. 4.13B) allow the extracting fluid output to be regulated at will, so (a) the restrictor itself controls the pressure system and hence allows both the density and pressure of the fluid to be regulated, and (b) the pressure drop is instantaneous, which reduces the risk of plugging. In any case, if some plugging occurs, the time wasted as a result is minimal since the flow-rate can be rapidly raised to flush the restrictor and returned to its appropriate value for development of the extraction process.

4.3.4.2 Problems Arising from Depressurization

Experiments aimed at optimizing depressurization systems have shown that high-molecular weight substances can be more readily separated from the SF than smaller molecules, particularly if the restrictor is not properly heated. Mechanical restrictors frequently result in significant carry-over between samples and hence in poor reproducibility. The pressure regulator valves used for greater control of the extraction flow-rate are also a source

Carry-over   of contamination and carry-over between samples.

On the other hand, the solubility of a solute in a supercritical fluid is a function of pressure; therefore, a pressure drop can result in the less

Preventing   volatile analytes condensing on the restrictor walls or clustering and
restrictor   eventually plugging of the restrictor. There are several ways of avoiding or
plugging   eliminating restrictor plugging, some of which are described below.

The restrictor dimensions (length and inner diameter) determine the SF flow-rate and hence the extraction rate. If restriction is very strong (i.e. if the restrictor is very long and thin), the resulting recoveries are very low and the risk of plugging high. As a rule, large diameters result in greater efficiency as they allow more SF to pass through the cell [22]; however, too large a restrictor diameter can give rise to analyte losses through saturation of the collection system. For example, quantitative recovery of PAHs takes 3 times longer with a restrictor of $15\,\mu$m i.d. than it does with one of $25\,\mu$m i.d. [23]. An i.d. of $30\,\mu$m results in a slightly altered shape of the peak provided by an on-line coupled chromatograph, whereas an i.d. of $40\,\mu$m gives rise to markedly distorted peaks. By way of example, the flow-rates provided by restrictors of 20, 25 and $30\,\mu$m i.d. at a pressure of 300 atm are 150, 240 and 300 ml/min, respectively.

The restrictor can also be plugged intermittently or permanently by undissolved sample components swept by the SF. This can be avoided or corrected by sonicating the restrictor or placing a filter at the extraction chamber outlet [24]. Some authors prefer flushing the detector briefly by flooding it after each extraction. The washing liquid (solvent) to be used for this purpose obviously depends on the type of sample concerned (e.g. 2,2,4-trimethylpentane for vegetable samples [25]).

The restrictor dimensions not only determine the risk of plugging, but also have a decisive influence of other aspects of the extraction process. High flow-rates are usually avoided in SFE since fairly large samples can be extracted in relatively short times; however, this may cause the collection system to saturate. For example, using CO_2 at 400 atm and restrictors of 15, 30 and $40\,\mu$m i.d. typically provides SF flow-rates (measured as flow-rates of liquid CO_2 in the pump) of around 0.6, 1.0 and 1.6 ml/min, respectively, which are equivalent to 270, 450 and 750 millilitres of depressurized gas per minute. While the chromatograms obtained under the same chromatographic and extraction conditions with restrictors

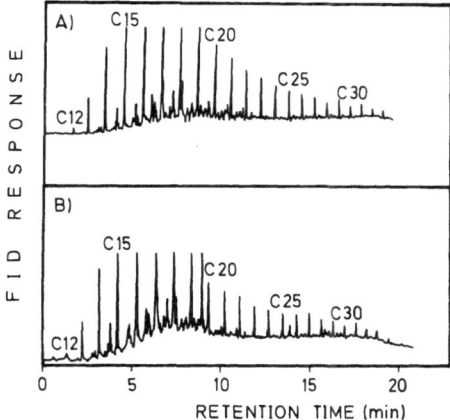

Fig. 4.14A,B. Influence of the restrictor diameter (**A** 25 μm; **B** 40 μm) on the shape of the chromatographic peaks provided by an on-lined coupled SFE/GC system in the analysis of an air-dried sample of petroleum-contaminated sediment. (Reproduced with permission of Preston)

of 25 and 30 μm i.d. are scarcely different, those provided by a restrictor of 40 μm have significantly broader peaks (Fig. 4.14).

Heating the restrictor is an efficient way of preventing plugging as it transfers kinetic energy to the molecules in contact with its walls and inhibits clustering. Depending on the amount and molecular weight of the extracted analytes, the heating temperature required may be as high as 250 °C, which may pose problems arising from decomposition or degradation of thermolabile analytes. This can be avoided or minimized by decreasing the restrictor inner diameter in order to ensure that the SF travels through it at a high speed, thereby reducing the residence time and the risk of degradation of its contents. Experiments on a sample containing such a thermolabile compound as dicumyl peroxide revealed that heating the sample at 135 °C in a closed vial for 30 min caused the analyte to degrade partly (Fig. 4.15A). If the same sample was extracted with an SF at 100 °C for 30 min using a restrictor heated at 250 °C, the analyte loss was very small or nil relative to the previous experiment (Fig. 4.15B); however, the loss was quite substantial if extraction at 150 °C was prolonged for 90 min with the restrictor at 250 °C.

Heating of restrictor

The risk of extracted analyte losses on depressurization depends on the restrictor suitability. Figure 4.16 shows the identical recoveries of oils from fatty appetizers obtained under the same extraction conditions (CO_2 at 80 °C and 600 atm, 600 ml/min gas flow-rate at the restrictor outlet) by halting the extraction at 15-min intervals for weighing or allowing it to proceed uninterruptedly for 60 min, with a single weighing at the end.

A laboratory-modified commercially available gas chromatograph equipped with a flame ionization detector effectively avoids plugging of the restrictor [26]. For this purpose, the detector is not used as such, but

Flame ionization detector

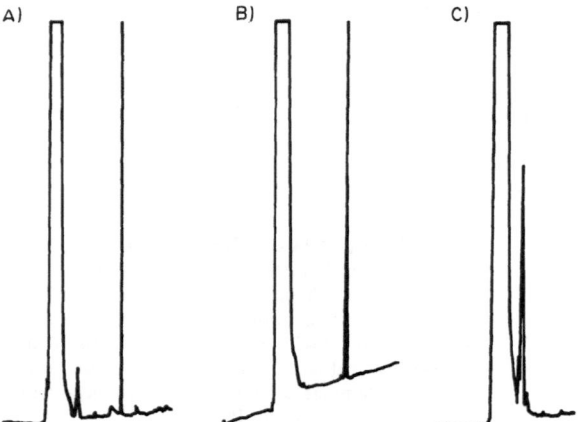

Fig. 4.15A–C. Comparative study of the degradation of a thermolabile compound (dicumyl peroxide) under different conditions. **A** Closed-circuit, no SF. **B** SFE system at 180°C for 30 min, with the restrictor heated at 250°C. **C** As B, but with heating at 150°C for 90 min

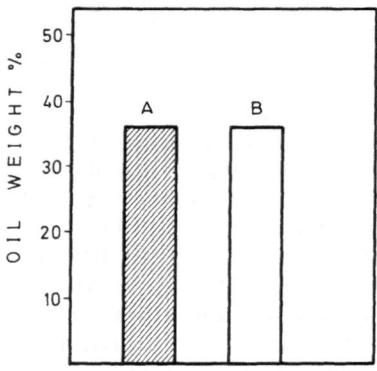

Fig. 4.16A,B. Percent weight of oil extracted from fatty appetizer samples in a four-step process with intervening depressurization **A** or in a single pass **B**. Total extraction time in both cases, 1 h

as a thermostated block onto which the extraction chamber is mounted, as shown in Fig. 4.17A. A plastic adapter not shown in the figure is placed above the detector flame orifice and the extraction chamber is positioned onto it. A nitrogen stream is introduced normal to the chamber outlet through the H_2 line in order to obtain a flow of hot gas through the restrictor tip, thereby avoiding deposition.

Restrictor material   The material of which the restrictor is made may also make it more or less prone to plugging. In this respect, metal restrictors are obviously better heat conductors than are fused silica restrictors. In addition, liquid

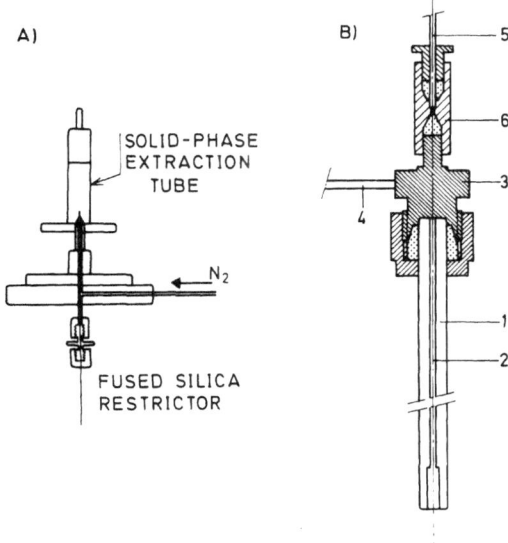

A)

SOLID-PHASE
EXTRACTION
TUBE

N₂

FUSED SILICA
RESTRICTOR

B)

Fig. 4.17A,B. Special restrictor designs used to avoid plugging. **A** Modified FID used as thermal conditioning system. **B** Restrictor with thermal protection and conditioning (1, glass tube; 2, fused silica capillary; 3, T-piece). (Reproduced with permission of Elsevier Science)

samples may activate fused silica under typical extraction conditions [27]. The fragility of silica restrictors can be overcome by wrapping them in special coverings that extend their service life to a few weeks. One such covering is a glass tube that fits tightly to a cryogenic/thermal desorption trap injector (Fig. 4.17B) [26]; the desorption oven also serves as an oven for the restrictor, thereby preventing condensation or precipitation of extracted components during depressurization. Inserting the carrier (helium gas) coaxially with the restrictor effluent enables working at a controlled pressure and prevents the SF from invading the helium line.

4.3.5 Collection Systems

4.3.5.1 Types

Separation of extracted species from the extracting SF can be achieved by (a) lowering the pressure in order to decrease the solubility, and (b) changing the temperature and hence the solubility of the dissolved analytes. Figure 4.18 shows the basic extractor arrangements used depending on whether collection is based on altering either variable. The most frequently employed analytical-scale technique for isolation of

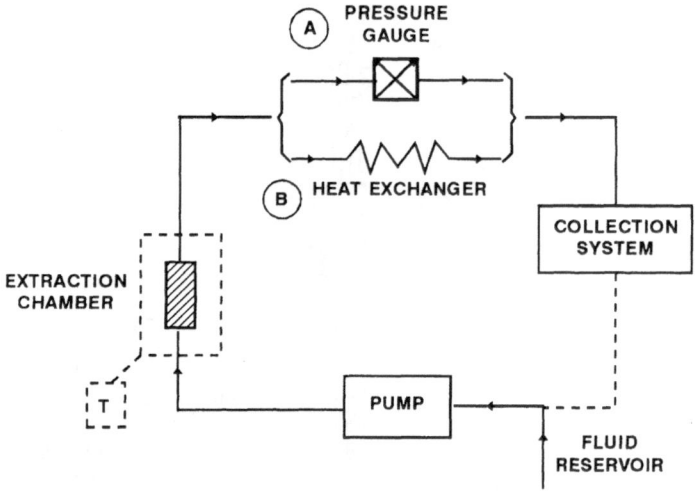

Fig. 4.18. Ways of separating the extract from the supercritical fluid. *A* By lowering the pressure. *B* By heat exchange. In industrial systems, the gas from the collection system is recirculated to reduce consumption

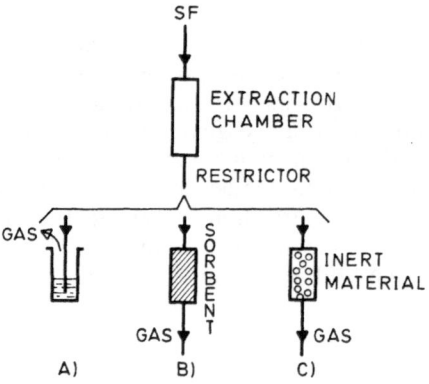

Fig. 4.19. Conventional collection systems. *A* Bubbling through a liquid. *B* Adsorption in an active material. *C* Cryogenic trapping in an inert material

analytes from an SF is depressurization, which can be accomplished in various ways (e.g. by bubbling through an appropriate solvent, adsorption onto a solid material and cryogenic trapping, Fig. 4.19.

Solvent bubbling
Solvent bubbling is possibly the simplest way of collection as it only requires the restrictor outlet to be inserted through the septum of a collection vial containing a few millilitres of solvent. The efficiency of this collection technique is affected by such factors as (a) the analyte solubility

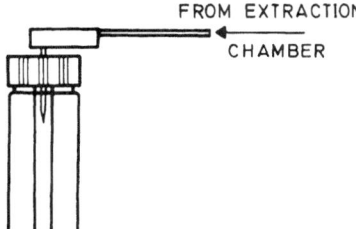

FROM EXTRACTION
CHAMBER

Fig. 4.20. Bubbling collection system patented by Dionex. (Reproduced with permission of Dionex–Lee Scientific Division

in the collecting solvent, which is a function of the solvent polarity; (b) the solvent volume used, which may nor not be close to its saturation value; (c) the size of the bubbles formed, which determine the exchange interface value; (d) the time of contact with the collector, along which mass transfer from the gas to the liquid phase takes place, which depends on the SF output flow-rate and the distance travelled by the bubbles; and (e) the solvent temperature, which has a decisive effect on solubility.

The temperature of the collector solvent is one of the most influential variables in this context as it determines to what extent it can evaporate or be swept by evolved gas bubbles. The temperature of the bubbling system is altered by the effect of the restrictor end (which is usually kept at a high temperature in order to avoid plugging through saturation) being immersed in the solvent, as well as cooling resulting from expansion of the SF.

The collection efficiency can be improved by altering several experimental variables (e.g. by using a small vial i.d. in order to increase the distance travelled by the bubbles in a given collector volume, or by placing a filter at the restrictor outlet so as to reduce bubble size). Decreasing the inner diameter of a restrictor is inadvisable as it increases the extraction time.

A patented system from Dionex (Fig. 4.20) solves many of the problems posed by this collection mode and ensures a high efficiency. On the one hand, the restrictor penetrates in the collector by only 0.5 cm in order to avoid heating the solvent; however, bubbling takes place in the bottom of the vial as a result of the presence of an internal cylinder, which maximizes the time of contact between bubbles and the collection solvent. On the other, the solvent is cooled under controlled conditions in order to **Cooling** increase its viscosity, which results in smaller bubbles and hence in an increased surface area to volume ratio and in more efficient transfer of analytes to the collecting solvent. Two additional advantages of cooling the collection vial and solvent are that volatility is thereby reduced and aerosols can be more effectively retained [24].

Collection on a sorbent material entails using an appropriate solid in the line or at the restrictor outlet. The material used for this purpose can be in

Table 4.3. Comparison of the efficiency of two collection systems

SF	TNG		DNG	
	AI	C_{18}	AI	C_{18}
Pure CO_2	71.1	89.6	2.2	2.9
98% CO_2 + 2% methanol	88.2	97.3	6.1	9.9

For details, see text.

a solid form, as a bed or as a column packing. In any case, this collection mode involves an additional step: desorption of the analytes from the adsorbent by elution with a small volume of solvent for subsequent determination or, alternatively, thermal desorption and sweeping by the eluent if an on-line coupled extraction/chromatographic system is used. Therefore, both steps (retention on the adsorbent material and removal from it) should be quantitative. If the separated analytes are of a different polarity, a mixture of adsorbents ensures adequate collection for all. Additional heating of the restrictor is usually required to avoid plugging through freezing.

Supercritical fluid modifiers affect adsorbent materials by changing or saturating their active sites and hence altering their efficiency. A clear example of the influence of a modifier on the collection efficiency depending on the particular adsorbent material used is shown in Table 4.3 for the extraction of dinitroglycerin (DNG) and trinitroglycerin (TNG) retained in granulated active coal with pure CO_2 and CO_2 containing 2% methanol, followed by collection in stainless steel (SS) or C_{18}. While deposition on stainless steel takes place by the sole effect of cooling and pointing the restrictor outlet at the metal balls, deposition on C_{18} is also aided by sorption of the solutes, so both analytes are more efficiently collected by this latter substrate.

Cryogenic trapping   Mulcabey et al. [25] carried out a more comprehensive study of the efficiency of various materials in the normal and reversed phase mode for off-line collection of hydrocarbons, phenols and their mixtures. For this purpose, they employed a special system equipped with a flame ionization detector (FID), of which only the temperature block (working T range = 50–100 °C) was used in order to control the collector temperature, which was measured by placing a thermocouple in the space bound by the restrictor and the collection solid phase. The set-up used to investigate the efficiency of the collection system is depicted in Fig. 4.21. A known amount of sample was introduced into the extraction system via the 50-µl loop of an injection valve. The system was calibrated by filling the loop with the sample of interest, unloading it in a 2-ml autosampler vial, adding

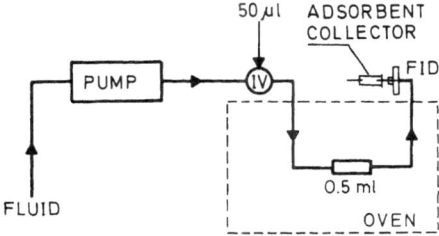

Fig. 4.21. Special system used to study the efficiency of various adsorbent materials. For details, see text. (Reproduced with permission of the American Chemical Society)

Table 4.4.A. Efficiency of trapping of a test mixture by various polar adsorbent phases using CH_2Cl_2 as conditioner and eluent

Compound	Diol trap	Silica trap	Amino trap
Acetophenone	71.1 ± 7.0	80.2 ± 4.4	69.3 ± 4.4
N,N-Dimethylaniline	72.1 ± 7.6	83.2 ± 4.6	73.9 ± 4.0
n-Decanoic acid	27.6 ± 6.1	0 ± 0	0 ± 0
2-Naphthol	95.6 ± 7.0	84.8 ± 6.6	28.7 ± 6.5
n-Tetracosane	93.4 ± 5.4	83.4 ± 4.7	79.2 ± 5.5

Table 4.4.B. Efficiency of trapping of a test mixture by various polar adsorbent phases using CH_2Cl_2 (A) and CCl_4 (B) as conditioner and eluent

Compound	Octyl trap		Phenyl trap	
	A	B	A	B
Acetophenone	70.0 ± 10.1	81.9 ± 8.5	74.6 ± 7.1	88.1 ± 3.7
N,N-Dimethylaniline	68.3 ± 7.2	81.5 ± 5.2	73.2 ± 7.7	87.2 ± 5.5
n-Decanoic acid	43.2 ± 18.8	1.9 ± 3.2	55.2 ± 16.7	17.4 ± 2.5
2-Naphthol	86.1 ± 12.5	31.9 ± 24.7	90.9 ± 7.0	77.2 ± 5.6
n-Tetracosane	80.7 ± 6.7	87.5 ± 3.2	91.5 ± 5.5	84.7 ± 4.7

an internal standard and determining the analytes by gas chromatography. The values thus obtained were taken as 100% for each component. A void volume of 0.5 ml in the system ensured efficient mixing of the analytes with the SC CO_2 rather than having them swept to the trapping zone. C_{18}, C_8 and phenyl reversed phases provided hydrocarbon collection efficiencies in the range 94–100%. The collector was kept at − 20 °C. Retention took place by cryogenic trapping and adsorption. The less volatile hydrocarbons (C_{18}–C_{32}) emerging from the restrictor were retained on the walls of solid-phase extraction (SPE) tubes or the surface of the packing material. The

Table 4.5. Comparison of the frit trapping efficiency achieved with various conditioning and eluting solvents

	A	B	C	D
Conditioning	3 ml of CH$_2$Cl$_2$	3 ml of CCl$_4$	3 ml of CH$_2$Cl$_2$	3 ml of CCl$_4$
	3 ml of CH$_3$OH	3 ml of CH$_3$OH	3 ml of CH$_3$OH	3 ml of CH$_3$OH
Elution	2 ml of CH$_2$Cl$_2$	2 ml of CH$_2$Cl$_2$	2 ml of CCl$_4$	2 ml of CCl$_4$
Analyte:				
Acetophenone	82.3 ± 4.0	88.0 ± 5.6	89.7 ± 6.4	93.3 ± 3.8
N,N-Dimethylaniline	79.1 ± 5.9	89.5 ± 6.5	89.0 ± 6.3	92.8 ± 3.8
n-Decanoic acid	51.5 ± 17.2	74.1 ± 13.4	71.6 ± 13.0	97.5 ± 10.7
2-Naphthol	81.9 ± 7.0	95.1 ± 2.7	81.6 ± 8.7	82.1 ± 3.1
n-Tetracosane	84.5 ± 7.8	86.0 ± 2.8	92.3 ± 10.9	91.5 ± 4.8

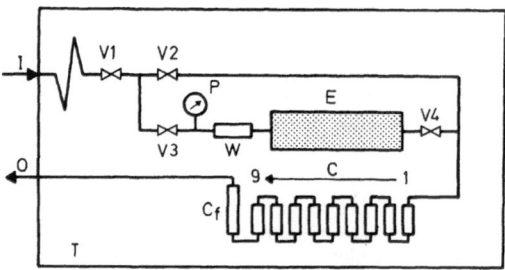

Fig. 4.22. Multi-collection system consisting of nine serially arranged columns. I, SF inlet; V$_1$–V$_4$ valves; P, pressure gauge; W, water saturation column; E, extraction chamber; C$_1$–C$_9$ trapping columns; C$_f$ final column; O, fluid outlet; T, thermostated bath. (Reproduced with permission of Springer)

more volatile compounds (C$_{10}$–C$_{16}$) were trapped by adsorption. The phenyl trap proved to be less efficient than the alkyl traps owing to weaker interactions. Thus, the C$_{10}$ hydrocarbon was trapped by 80% on the phenyl support and by 100% on the C$_{18}$ support. The sorbents used to isolate the analytes from a test mixture also proved to be variably efficient (Table 4.4A). The normal phases assayed in the study were diol, silica and amino. The reversed-phase recoveries obtained as a function of the eluent used are given in Table 4.4B. As can be seen, the recovery was affected by the type of adsorbent and eluent used. The polyethylene frits used to preserve packing of the adsorbent material were checked for their contribution to the overall trapping process. Table 4.5 shows the results obtained. The properties of the solvents used for conditioning of the adsorbent and elution, and the order in which they were employed, proved to have a more marked effect on each particular analyte than did the features of the analyte itself [28,29].

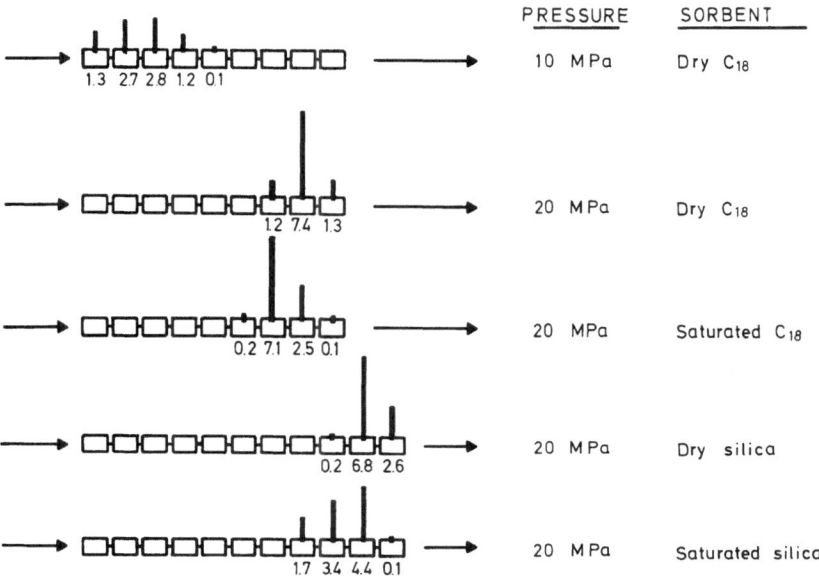

PRESSURE   SORBENT

10 MPa   Dry C$_{18}$

20 MPa   Dry C$_{18}$

20 MPa   Saturated C$_{18}$

20 MPa   Dry silica

20 MPa   Saturated silica

Fig. 4.23. Influence of pressure, type of adsorbent and presence/absence of moisture on the partitioning of 10 µg of linuron added to glass wool in the serially arranged collection columns of Fig. 4.22. (Reproduced with permission of Springer)

One way of ensuring 100% collection of extracted analytes is using several serially arranged columns (Fig. 4.22) packed with the same or a different material. The experiments performed by Schäfer and Baumann using between 9 and 11 RP18 or SI60 columns (particle size, 3 µm) revealed a decisive influence of pressure on the collection efficiency: the sequential number of the column where the greatest amount of product was collected was found to depend on the pressure in the extraction system (Fig. 4.23). The adsorbent material and the presence or absence of moisture proved to be less markedly influential [30].

Serially arranged columns

Cryogenic trapping involves cooling down the extraction mixture until the SF expands and the analytes deposit. The trapping temperature depends on whether the analytes are to be isolated from the fluid or this is to be liquefied and the collection vessel sealed in order to avoid losses of the analytes that crystallize partially or form aerosols during cooling. When the temperature of the cryogenic trap is very low (close to 0 °C), the restrictor must be heated in order to avoid the formation of two phases. As in collection on solid adsorbents, an extra preparative step is also required where some analyte may be lost. In liquefying cryogenic traps, the liquefied fluid must be evaporated and the analytes collected in an

Cryogenic trapping

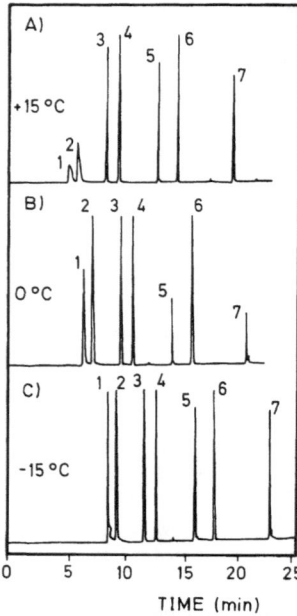

Fig. 4.24A–C. Influence of the cryogenic trap temperature on collection of amines extracted with N_2O from a Celite matrix and introduced on-line into a gas chromatograph equipped with a flame ionization detector. 1, heptylamine; 2, N-methylheptylamine; 3, 2-ethylaniline; 4, tributylamine; 5, dodecylamine; 6, diphenylamine; 7, octadecylamine. (Reproduced with permission of the American Chemical Society)

appropriate solvent. Sampling in the liquid phase is hindered by low temperatures resulting in nucleation and crystallization of the less volatile analytes; on the other hand, too high a temperature may result in losses of the more volatile compounds. Figure 4.24 shows the efficiency of collection of seven amines in a cryogenic trap at different temperatures prior to their GC determination [31].

A comparison between the two types of cryogenic collection systems (open and sealed) reveals that the choice is crucial, not only with highly volatile analytes. The analytes shown in Table 4.6 were collected in an open system by means of a long-necked flask cooled to 0 °C and a sealed system with the flask cooled by using liquid nitrogen. Since the analytes were scarcely volatile at 0 °C, the most likely origin of the low recoveries obtained was the formation of aerosols. Since high temperatures (e.g. 150 °C) result in the occurrence of two CO_2 phases at the working pressure (410 bar), formation of small solute particles ($<0.2\,\mu m$) must prevail at the higher temperatures and give rise to low efficiencies. The higher recoveries provided by methanol-modified CO_2 can be ascribed to the formation of larger liquid methanol and hence more readily deposited drops during expansion. As a rule, the sealed system was more efficient than its closed counterpart [32].

Table 4.6. Recovery of six analytes from XAD-2 resin by using various supercritical collection and extraction methods

Analyte	% Recovery				
	Collection system[a]		Extraction conditions[b]		
	Open	Sealed	CO_2 (150 °C)	CO_2 (50 °C)	CO_2/CH_3OH (80:20 mol/mol) (150 °C)
Chrysene	2.2	75	2.2	25	660
Benzanthrene	2.5	79	2.5	28	62
1-Nitropyrene	1.8	83	1.8	24	29
Dibenzo[a,i]carbazole	6.3	95	6.3	10	65
Coronene	8.2	81	8.2	6.5	62
Rubrene	0	25			

[a] Extraction conditions: CO_2 at 150 °C and 400 bar, around 250 ml of liquid.
[b] In the open operational mode. Extraction at 410 bar with 250 ml of liquid.

4.3.5.2 Special Collection Systems

Pariente et al. [33] developed an unusual collection procedure for use prior to on-line IR spectroscopic analysis (Fig. 4.25A). The restrictor output lies above a KBr window onto which the extract is deposited. The expansion of supercritical CO_2 on leaving the restrictor has a cooling effect that is strong enough for atmospheric steam to be condensed and drip on the KBr window. This is avoided by immersing the collection zone in dry air-purged container.

Aqueous samples may pose additional problems arising from the increased moisture content of the SF stream emerging from the extraction chamber, which may partly or fully plug the restrictor. This problem, faced in collecting the n-hexane extract in the extraction of cholesterol from egg yolk and blood serum [11], can be overcome by placing a thermostated (100 °C) stainless steel cell (15 cm × 4 mm i.d.) between the extraction chamber and the restrictor (Fig. 4.25B).

4.3.6 Thermostating

Supercritical fluid extraction calls for strictly controlled, accurately reproducible temperatures in the different units in order to ensure that the process takes place under optimal, fully reproducible conditions. For this reason, commercially available extractors are typically equipped with an oven similar to those used by gas chromatographs which allows the

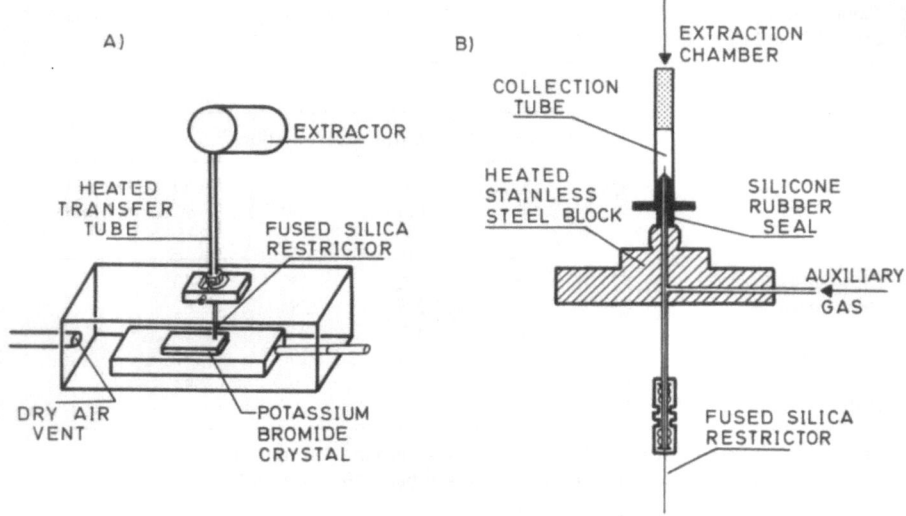

Fig. 4.25A,B. Special collection systems. **A** Direct deposition on a KBr pellet for IR determination. **B** System positioned between the extraction chamber and the restrictor to avoid plugging of the latter. (Reproduced with permission of the American Chemical Society)

temperature of the sample chamber, restrictor, restrictor outlet and collection system to be programmed, and the propulsion unit to be cooled if needed. Proper functioning of the overall system requires the temperature to be controlled to within ±0.01 °C. Customized extractors usually include a chromatographic oven where the sample cell is placed, all other units being subject to no temperature control or furnished with simpler heating devices.

The working temperature of the extraction cell usually ranges between 50 and 100 °C (see Tables 5.15–5.19), even though commercially available apparatuses offer working ranges encompassing temperatures between at least 35 and 150 °C.

As noted earlier, the restrictor temperature depends on the SF used, the presence or absence of modifiers, the features of the analyte and even those of the sample matrix – particularly if the sample is liquid.

The temperature of the collection unit is central when volatile analytes are involved. If it is too low, the extract can be trapped as a result. Also, water from wet solid samples or liquid samples may build up and form crystals when a collection procedure other than bubbling through a solvent is used.

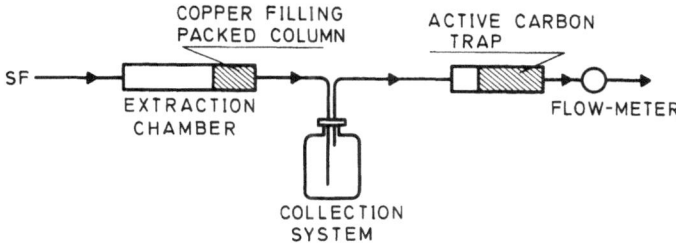

Fig. 4.26. Systems for removal of sulphur (copper filing packed column) and collection of volatile compounds (active carbon trap). (Adapted from [33] with permission of Pergamon Press)

As a rule, it is desirable to use as low as possible a temperature in the collection unit and keep the restrictor at a high enough temperature to avoid precipitation within it as a result of solubility limitations of the SF.

Some special thermostating systems designed for specific applications were commented on above in dealing with each extractor component.

4.3.7 Ancillary Components

Many of the ancillary devices added to the basic scheme in Fig. 4.1 were discussed in describing the different essential units of the SF extractor, namely:

a) Safety, check and switching valves, and filters for SF delivery units.
b) Compressors, modifier or mixing chambers fitted to propulsion systems (Fig. 4.5).
c) Switching valves fitted to the inlet and/or outlet of an extraction cell in multiextraction or recirculation systems.
d) Special components required for on-line coupling of an instrument, the features of which depend on the instrument concerned and are discussed in Sect. 4.6.

By way of example, consider a special component that can be required for dealing with complex samples (e.g. soils and sediments, which usually **Complex** contain high proportions of sulphur). Such samples entail using a trap to **samples** retain the undesirable substance(s) between the extraction chamber and the restrictor. Figure 4.26 shows the special system used by Pyle and Setty [34] to remove sulphur coextracted with the analytes of interest from soil samples. It consists of a typical 3 cm × 4.6 mm i.d. HPLC pre-column packed with 2 g of copper filings.

Samples containing volatile components that can be extracted into the SF and subsequently released with it to the laboratory atmosphere can be processed by passing the SF through an active carbon trap after expansion of the fluid and separation of the analytes (Fig. 4.26).

4.4 Extraction Modes

The contact between the supercritical fluid and sample from which extraction takes place can be established in a static or dynamic fashion. In the *static mode*, where the sample is soaked in SF, the fluid flow is halted over **Static mode** a given interval after which it is propelled to the collection vessel, where **contact** analytes are concentrated (Fig. 4.27A). This penetration period is specially useful when the analytes cannot be expeditiously removed from the matrix. As a rule, slow extractions and/or dense matrices call for the static mode, which is typical of off-line systems.

Dynamic mode In the *dynamic mode*, the SF is pumped through the sample into the **contact** collection vessel or interface and onto the measuring instrument (Fig. 4.27B). The extracting fluid is passed once through the sample and driven to the collection unit in the off-line mode. Coupling the extractor on-line to a chromatograph or detector requires using an interface suited to the

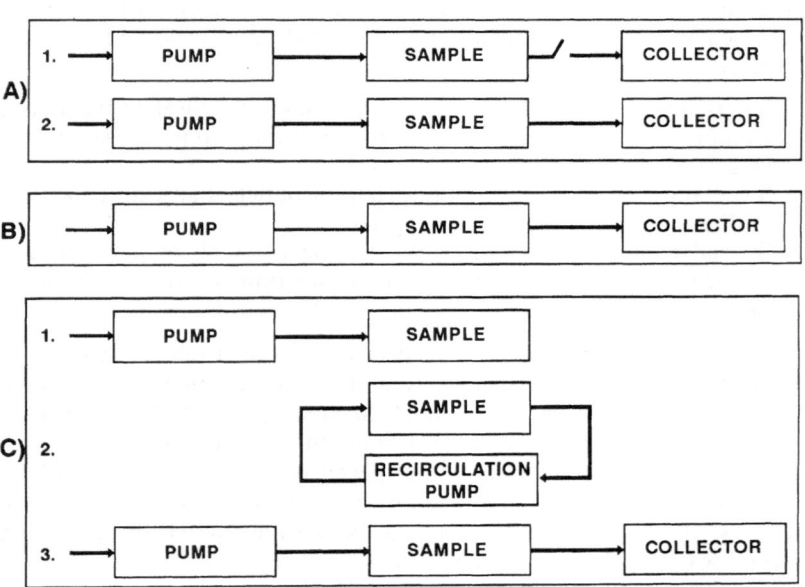

· **Fig. 4.27A–C.** SFE modes. **A** Static. **B** Dynamic. **C** Recirculation

instrument in question. There are a host of possible variants in this respect since the SF containing the extracted analytes can even be directly fed to a detector (UV, flame, ionization) and subsequently wasted. At preset intervals, the SF containing the extract can be driven to other devices by means of a switching valve (dynamic sampling). Selective (dynamic) fractionation can be accomplished by using a variable-density programme; alternatively, the whole extract can be analysed by coupling the extraction module to the analytical device from the beginning of extraction. This mode is especially effective when the analytes are highly soluble and the matrix readily penetrable.

Which choice (dynamic or static) is the better is a controversial subject. While advocates of the static mode claim that a longer contact between sample and solvent swells the matrix and facilitates penetration of the extractant in its interstices, thereby increasing the recovery, defenders of the dynamic mode argue that continuous exposure of the analyte to the pure (clean) solvent favours displacement of the analyte partitioning equilibrium to the mobile phase. The better choice obviously depends on the particular case.

A variant of the dynamic SFE mode, the *recirculation mode*, involves pumping the supercritical fluid repeatedly through the sample as shown in Fig. 4.27C. After some time, the circuit contents or part of them are propelled to the collection vessel to evaporate the fluid and concentrate the analytes; alternatively, the line is unloaded into an on-line coupled separation and/or detection system. This mode is particularly well suited to recovery and solubility equilibrium studies. Two potential problems with which it may be confronted are contamination of the extractor and loss of analytes on passage through the pump. The recovery can be continuously monitored by inserting an appropriate detector for the analytes (Fig. 4.28). The extracting fluid is recirculated through the extraction chamber, with an on-line coupled UV-visible monitor and a sampling valve that takes an aliquot of the extracted analyte from the recirculation

Recirculation mode

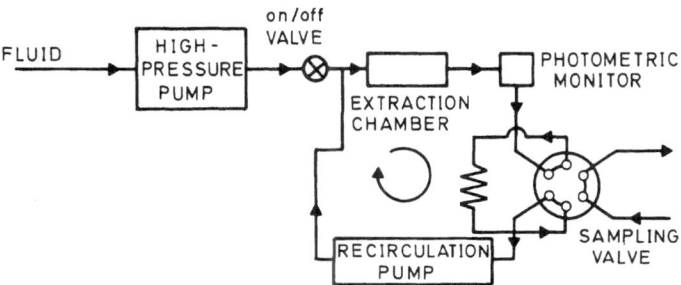

Fig. 4.28. Recirculation extraction system with UV monitoring of the process

loop and injects it into an external stream. An additional valve not shown in Fig. 4.28 flushes (floods) the sampling valve and extracting fluid lines before it is switched to its extraction position [35]. This set-up can be used in solubility studies provided the compound to be extracted is of a medium to high solubility and possesses a chromophore absorbing in the UV-visible region. A variant of this basic design affords rapid, highly repro-ducible recovery of variously soluble compounds with and without chromophores, as well as transfer of variable aliquots of the closed loop contents to a high performance chromatograph in those cases where a subsequent separation is required. Chromophores are measured by UV-visible spectrophotometry, whereas non-absorbing substances are sensed with a refractive index detector [36].

4.5   Off-Line Coupled SF Extraction/Detection

As noted in Chap. 1, supercritical fluid extraction is a sample extraction/ preparation technique (similar to liquid solvent extraction), so the SF extractor need not be coupled to a chromatograph or detector. In fact, most SFE studies carried out so far involved collection in some of the above-described modes (bubbling through liquids or analyte deposition by adsorption or cooling). The treatment to which the extract is subsequently subjected for identification or quantification depends on its complexity and the type of information required. Thus, the analyte may need cleaning, separation, etc., prior to introduction into the detector, which conditions the intervening extraction/detection treatment it is to be given. Extracted analytes can be determined by a variety of measuring options including chromatographic, spectroscopic, gravimetric and radiochemical techniques, mainly.

The off-line mode is to be preferred when a deep knowledge of the features of the extraction process concerned is required as it allows optimization of such experimental variables as the pressure, temperature, SF polarity and flow-rate, extractor volume and dimensions, extraction time and sample size.

Gravimetry is the oldest technique used for off-line detection in super-critical fluid extractions. Its utter simplicity is a result of the absolute nature of gravimetric measurements, which avoid the need and save the time taken to run a calibration graph. Both direct and indirect gravimetries have been used for quantitation following extraction (Fig. 4.29). The most serious source of error in the direct mode (weighing of the extracted analytes) is incomplete collection, which makes the collector system a key piece provided extraction is complete. In the indirect mode, based on weighing before and after extraction, the sample weight loss not ascribable

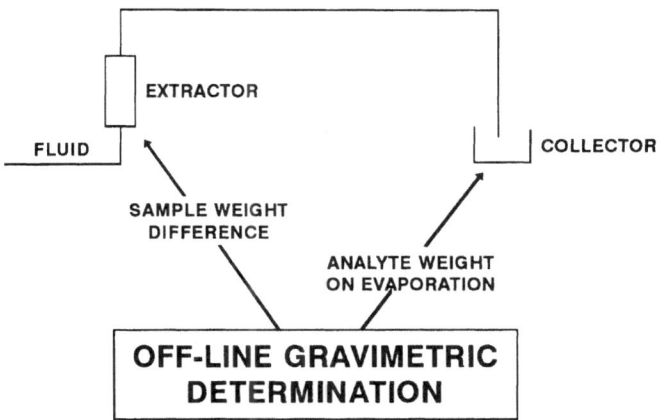

Fig. 4.29. Ways of performing gravimetric determinations following SFE

to the analytes (e.g. moisture removed by sweeping) is quantified as part of the analytes.

Ultraviolet monitoring requires the presence of a chromophore in the extracted analytes. Using CO_2 as extractant in this context is highly advantageous as the fluid is transparent to light up to around 190 nm [37,38]. The amount of information provided by a UV-visible detector is intermediate between those supplied by a mass spectrometer (MS) and a flame ionization detector (FID).

Since SFE usually entails separating the matrix from more than one analyte, the subsequent quantification often requires such analytes to be isolated from one another – particularly if they are very similar – which is usually done chromatographically. Choosing gas or liquid chromatography for this purpose relies exclusively on the features of the analytes and the properties of the collection solvent. This chromatographic step is usually performed in the on-line mode in SFC.

Gas chromatography is the most frequently used technique for separation/determination of volatile or readily evaporated organics following extraction with an SF. Schantz and Chesler extract polychlorinated biphenyls (PCBs) from sediments and polynuclear aromatic hydrocarbons (PAHs) from transformer oils using supercritical CO_2. The extract is trapped in a reversed-phased silica cartridge from which the analytes are removed with dichloromethane and, after partial evaporation of the solvent, quantified by GC. The results thus obtained are comparable to those provided by Soxhlet extraction. Under the extraction conditions used (a 0.93 g/l density), the SFE of high-molecular weight PAHs is more efficient than their Soxhlet extraction {e.g. 18% and 30% higher for benzo(ghi)perylene and indeno [1,2,3-cd]pyrene, respectively}. Also, the SFE takes only 30–60 min compared to 16 h for the Soxhlet extraction. More polar compounds require using a cosolvent (e.g. methanol) to facilitate SF extraction [39].

Hawthorne and Miller developed an off-line SFE/GC method using supercritical N_2O containing 5% methanol as modifier for the extraction of PAHs from environmental solid matrices. The extracted analytes are collected by inserting the restrictor outlet into a vial containing dichloromethane. In this way, quantitative recoveries of PAHs from urban dust and deuterated PAHs from a river sediment and ash can readily be obtained in about 30 min. Recoveries of deuterated PAHs are significantly higher than those achieved after 4 h sonication or 8 h Soxhlet extraction with benzene or dichloromethane [40].

Desorption of PAHs from various matrices has also been investigated by using off-line coupled SFE and GC. Various adsorbents including XAD-2, polyurethane, Spherocarb and Tenax were used in conjunction with a customized extractor working at pressures and temperatures above 400 bar and 200 °C, respectively [41]. Extraction effluents were cryogenically collected in a sealed vessel cooled by immersion in a liquid nitrogen bath. While the extraction chamber was kept at the required temperature with the aid of a GC oven, the restrictor was electrically heated to avoid freeze-plugging during expansion of the fluid in the collection vessel – especially when SC CO_2 was used as extractant. The extraction results for XAD-2 resin as the matrix obtained by Soxhlet extraction with dichloromethane and SFE with various fluids are listed in Table 4.7. Soxhlet

Table 4.7. Comparison of the extraction of PAHs from an XAD-2 matrix by using a Soxhlet system and various SFs

Analyte	Recovery (%)						
	Soxhlet alone[a]	CO_2[b]		Isobutane[d]		CO_2-CH_3OH[e]	
		SFE	Soxhlet[c]	SFE	Soxhlet[c]	SFE	Soxhlet[c]
Chrysene	79	84	0	86	5	88	3
Benzanthrene	86	98	0	110	0	88	0
1-Nitropyrene	93	81	0	70	0	98	0
Dibenzo[a,i]carbazole	88	54	22	96	6	97	0
Coronene	81	46	63	93	9	90	5
Decacyclene	85	6	88				

[a] CH_2Cl_2 for 16 h.
[b] CO_2 at 125 °C and 400 bar, around 200 ml of liquid.
[c] Extraction with CH_2Cl_2 of the same sample following SFE.
[d] Isobutane at 150 °C and 185 bar, around 300 ml of liquid.
[e] 20 mol% CH_3OH in CO_2 at 130 °C and 400 bar, around 210 ml of liquid.

extraction provided high recoveries for all the substances studied, and
so did SFE for low-molecular weight compounds; however, the latter
provided gradually decreasing recoveries with increase in the MW of
the substances. This effect can be ascribed to the low solubility of high-
molecular weight compounds in supercritical fluids and the relatively high
pressures and temperatures used. Using supercritical isobutane or modified
CO_2 was found to improve on the extraction efficiency of pure CO_2. The
SF extractions typically took 30–45 min, whereas the Soxhlet extractions
required 16 h for completion.

Wong et al. [42] compared the performance of extraction techniques   **Performance**
(SFE vs solid–liquid extraction with ethyl acetate as extractant) and   **comparisons**
detection systems (GC with nitrogen-phosphorus and electron capture
detectors vs ELISA) by using soil samples containing 4-nitrophenol
and parathion. The most salient advantage of SFE over conventional
extraction with ethyl acetate is the ease with which the sample can
be prepared prior to GC analysis. The supercritical extraction of 4-
nitrophenol and parathion in soils takes an overall 20–25 min and the
extract collected by bubbling through methanol can be directly injected
into the chromatograph with no preconcentration. Extraction (leaching) of
the same compounds with ethyl acetate takes 20 min in addition to the
time required for separation of undissolved sample prior to filtration;
moreover, the extracts usually require concentration to 2–3 ml prior to
GC analysis.

Figure 4.30 compares the performance of GC plus nitrogen–phosphorus
or electron capture detection with that of ELISA for 4-nitrophenol in the
form of a plot of analyte extracted (in picograms) against the normalized
peak area and the percent of control. The steep slope of the GC curves
shows this chromatographic technique to be more precise. Conversely, the
ELISA curve results in a broader working range (from 10 pg to 1 µg). The
ELISA method is also more sensitive than its GC counterpart (the GC/
ECD detection limit is 10 ppb, whereas that of the ELISA method is

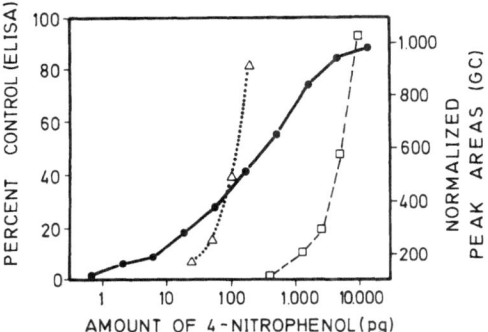

Fig. 4.30. Comparison of the
standard curves for 4-nitrophenol
obtained by ELISA (●) and with
an electron capture detector (△)
and a nitrogen–phosphorus detector (□). (Reproduced with permission of the American Chemical
Society)

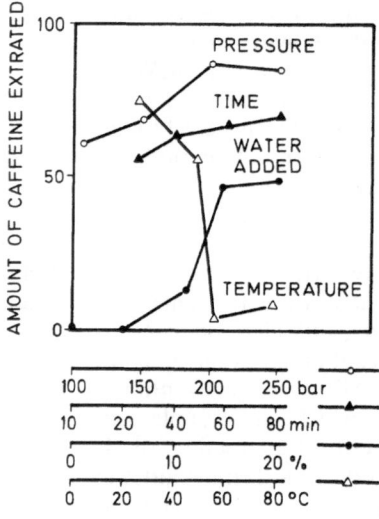

Fig. **4.31.** Influence of different variables on the extraction of caffeine from roasted coffee beans. (Reproduced with permission of Elsevier Science)

0.2–1 ppb, both in terms of dissolved substance). However, the most salient advantage of the ELISA method over the chromatographic method is its sample throughput. ELISA allows at least 15 trays (over 90 samples) to be processed in two 8-h sessions, each tray involving running a standard curve, analysing 6 samples at two concentrations and determining samples and standards in duplicate. The chromatographic determination is slower as the samples and standards must be injected in duplicate and each chromatogram takes 10–15 m to record, so only 10–15 samples and the required number of standards can be analysed in an 8-h session.

Off-line analyses of SF extracts by high performance liquid chromatography (HPLC) are very commonplace. Sugiyama et al. studied the effect of various parameters on the amount of caffeine extracted from coffee beans by using a closed, recirculating SFE system. The trapping column used was packed with active carbon and trapped analytes were eluted with 55:45 v/v methanol/water, an appropriate aliquot being injected into the HPLC system. Figure 4.31 shows the influence of several variables on the extraction yield. The amount of caffeine extracted is represented as a percentage in drinking coffee, i.e. as the proportion of caffeine in coffee. The recovery increases with pressure and the extraction time and decreases sharply with increase in the temperature. Above 60 °C, caffeine is very sparsely extracted owing to its decreased solubility in light carbon dioxide. The recovery also increases with the moisture content, which is consistent with the claims that water is essential to mass transfer with CO_2 as supercritical fluid. Such conditions as 200 bar, 48 °C, a 20% water content in the SF and an extraction time of 60 min ensure high recoveries of caffeine [43].

Vitamin K_1 can be extracted from infant foods by using SC CO_2 at 544 bar and 60 °C. Quantitative extraction is achieved in only 15 min, after which the SF is depressurized and the vitamin K_1 extracted is trapped in a small tube packed with silica and subsequently eluted with a mixture of dichloromethane and acetone. After the solvent is separated, the residue is dissolved in the eluent and analysed by reversed-phase HPLC with electrochemical detection [44]. The minimum detectable amount is 80 pg and the linear dynamic range encompasses at least five orders of magnitude. The recoveries, between 94 and 96%, have an rsd of around 6–7%.

Morphine and quinine extracted from plants with SC CO_2 can also be quantified off-line by HPLC. The analytes are extracted by heating a sealed extractor filled with a certain amount of dry ice. The results thus obtained compare favourably (higher recoveries and shorter extraction times) with those provided by extraction with subcritical ethanol or tetra-hydrofuran and Soxhlet extraction. Solid–liquid extraction of blood samples spiked with 200 µg morphine/ml was compared with SFE of the drug in terms of recovery. The mean recovery achieved in 10 determinations including SFE was 96.7% (rsd = 3.2%), whereas that provided by solid-liquid extraction was 92.2% (rsd = 4.0%). However, the time required for SFE of serum samples was rather long owing to the aqueous nature of the sample, which required lyophilization for 12 h [45].

4.6 On-Line Coupled SF Extraction/Detection

The on-line mode integrates the extraction process and the measuring step, in addition, usually, to a separation (most often chromatographic) step.

On-line methods are more sensitive since the whole extracted material   **Sensitivity** is transferred to the analyser, which also minimizes potential errors through avoidance of some manipulation steps. In addition to increased sensitivity and ease of automation, direct coupling of SFE to other techniques, particularly chromatographic ones, involves extra benefits derived from analyses where the species are not exposed to atmospheric oxygen, light or high temperatures.

4.6.1 Types of Interfaces Used

Effective coupling of SFE to a detection technique rests on availability of an appropriate interface between the two. An SF extractor can be coupled

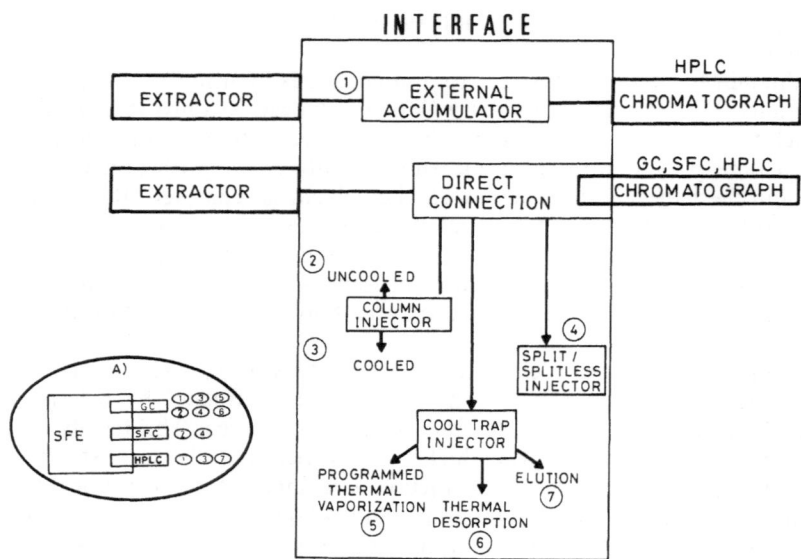

Fig. 4.32. Types of interfaces most commonly used in coupled SFE/GC systems. *A* most frequently used interfaces for each type of chromatograph (GC, SFC, HPLC)

to a chromatograph in two general ways, viz. (a) by using accumulators outside the chromatograph, and (b) by collecting the analytes directly in the chromatographic column or a retention interface. Figure 4.32 shows a scheme of the most commonly used extractor/chromatograph interfaces, which are described in detail below.

External accumulator   The function of *external accumulator* can be served by a 6-way valve such as that depicted in Fig. 4.33. The extract is trapped in the valve loop by virtue of the cooling effect of the SF depressurization. The trap can optionally be packed with a sorbent. After trapping, the eluent sweeps the retained compounds to an HPLC instrument or the loop is heated to introduce the extract into the column of a gas chromatograph. In this latter case, the system is applicable to volatile substances.

When the extracted analytes are to be retained directly in the chromatographic column or retention interface, introduction into either can be achieved in several ways, namely: by injection into the column, whether direct (SFC and GC) or with the aid of a cooling system (GC and HPLC); by sample split/splitless injection (SFE and GC); by means of a programmed temperature vaporizer (GC); and by injection into a cold trap and subsequent thermal desorption (GC) or elution (HPLC).

Direct injection   *Direct injection* into a chromatographic column after SFE (Fig. 4.34A) is performed via a T-piece in order to avoid plugging. The restrictor tip is

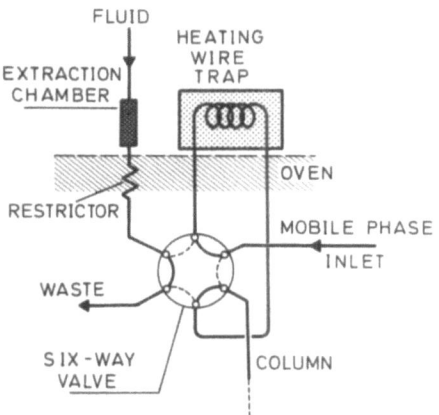

Fig. 4.33. Interface for on-line coupling of SFE to HPLC and GC based on accumulation and elution of the SF extract in the loop of an injection valve. (Reproduced with permission of Elsevier Science)

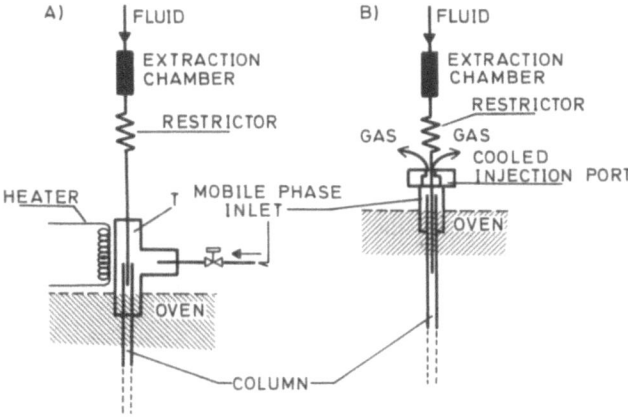

Fig. 4.34A,B. Interfaces for on-line coupling of SFE to HPLC and GC based on direct retention in the chromatographic column. **A** With, **B** Without cooling. (Reproduced with permission of Elsevier Science)

positioned above the retention zone and the carrier gas is fed coaxially with the restrictor. The whole extractant circulates through the column, into which all the extracted compounds are introduced, focussed in the retention zone. During extraction, the carrier gas flow is halted to avoid a back-pressure effect from the expanded SF. This type of interface has the disadvantages that the inner diameter of the separation column cannot cope with high flow-rates of extracting fluid and that some of the com-

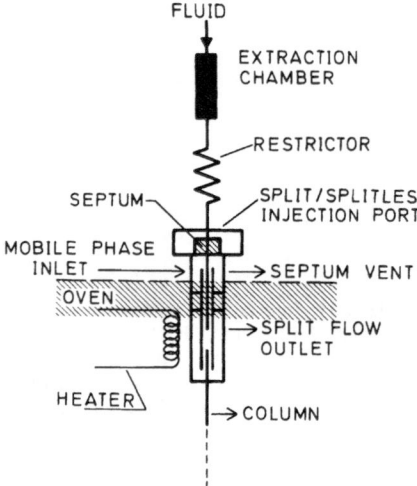

FLUID

EXTRACTION
CHAMBER

RESTRICTOR

SEPTUM

SPLIT/SPLITLESS
INJECTION PORT

MOBILE PHASE
INLET ⟶ →SEPTUM VENT

OVEN

→SPLIT FLOW
OUTLET

HEATER

→COLUMN

Fig. 4.35. Interface for on-line coupling of SFE to GC based on a conventional split/splitless injector. (Reproduced with permission of Elsevier Science Publishers)

ponents transferred to the column may contaminate or overload it. In addition, the maximum possible flow-rate of gaseous SF is limited by the presence of the capillary column, which makes use of a micro-extractor advisable. When large extraction cells are to be used, the SF from the extraction system should be introduced into the chromatograph in an intermittent fashion similar to that of the cut-off technique used in multi-dimensional chroma-tography. One such system was found to be able to deal with compounds of a higher molecular weight than thermal desorption analysis following GC [46].

Cooled column injection systems are similar to the previous ones with the difference that the depressurized extractant is removed from the system at the interface itself, so it does not go through the column. The interface in question can be a conventional cooled column injector. Figure 4.34B illustrates the functioning on one such interface in which the restrictor is inserted into the chromatographic column, which is cooled by cryogenic focussing. Extracted components are retained by the stationary liquid phase whereas volatile components are removed with the solvent.

Split/splitless sample injection systems, which are typically used with capillary columns, require no alteration for use as interfaces between an SF extractor and a gas or supercritical fluid chromatograph. Figure 4.35 shows a scheme where the injector is heated during extraction to avoid plugging while the column is cooled to "focus" the extracted components. The restrictor is inserted through the septum during extraction and sub-sequently removed. In this case, the sample is always split, the split ratio depending on the extractant flow-rate: the higher the extraction pressure

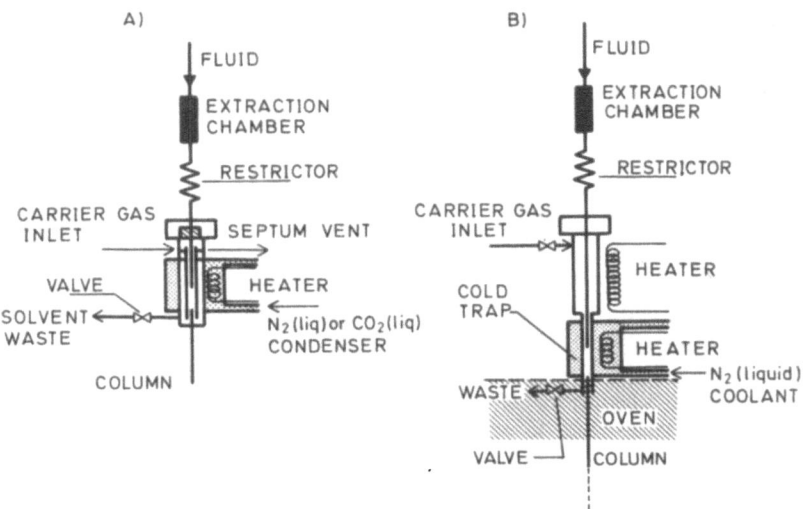

Fig. 4.36A,B. Interfaces for on-line coupling of SFE to GC and HPLC based on a cold trap (liquid N_2O or CO_2). **A** With programmed-temperature vaporization. **B** With thermal desorption. (Reproduced with permission of Elsevier Science)

is, the higher is the split ratio. This type of interface is very easy to use; however, injection is rather sluggish, so it may result in non-uniform introduction of substances of a relatively high-molecular weight into very large cells.

Programmed-temperature vaporization injectors [47] can be used unaltered (Fig. 4.36A) as interfaces for split and splitless injection, with the added advantage of on-line preconcentration by solvent removal, which allows use of high-flow rates to reduce the extraction time. As in the previous systems, the restrictor is inserted through the septum during extraction and later removed from it. The extractant and high-vapour pressure substances are released via a vent line. The injector is kept below room temperature in order to trap the extracted components. At the end of extraction, the restrictor is removed from the septum, the solvent vent line is shut and the injector temperature programme is started, so all the trapped sample components are introduced into the column for separation/determination. Those substances with boiling points at least 250 °C higher than the trapping temperature can be retained quantitatively. Because the restrictor is not heated, it may plug. There is no limit to the usable extraction chamber volume. Solutes of up to 50 carbon atoms can be transferred quantitatively across this type of interface.

The *cold trap/thermal desorption injectors* used as SFE/detection interfaces are modifications of the conventional device [48] in which plugging is avoided by placing the restrictor in a hot zone including an adsorbent-packed trap (Fig. 4.36B). During extraction, the extractant is driven to waste via a vent valve whereas extracted components are trapped on the surface of the cold trap located in the hot zone. As with the above-described vaporizer, after extraction is complete, the purge line is closed and the carrier gas channel is open while the cold trap is rapidly heated; in this way, trapped components are transferred quantitatively to the column. Cryogenically trapped analytes in SFE/HPLC can be recovered with the aid or an appropriate solvent (mobile phase).

Maximum sensitivity As in conventional injections, retaining the extract in the column results in the maximum possible sensitivity since the extracted analytes are transferred quantitatively to the chromatographic column; however, the system does not operate properly when the sample contains high concentrations of other extractable components such as water or fat, which also deposit on the column. In addition, with capillary columns, introduction of the whole gaseous effluent limits the extract flow-rate and hence increases the extraction time. The detector stability can also be affected as a result – particularly with MS detectors. This shortcoming can be circumvented by using a CO_2 vent system [49]. Coupled SFE/split GC systems offer some advantages for highly concentrated samples or matrices containing significant amounts of other extractable substances (e.g. water) that may interfere with the subsequent trapping and GC analysis.

Microextractor In 1977 Stahl [50] developed a micro-extractor that can be considered a precedent of coupled SFE/chromatographic detection. The device was used with different supercritical fluids (CO_2, N_2O, NH_3, ethylene, ethane, propane, propene, trifluoromethane and difluoro-chloromethane) to extract small amounts of samples. On emergence from the extraction chamber, the extract impinged directly on a chromatographic plate. In this way, he investigated the behaviour of various samples and established several rules correlating their properties with those of the SF used [51].

4.6.2 Coupled SFE/Gas Chromatography

Even though gas chromatography can only be applied to thermally stable compounds, it has been widely used in conjunction with SFE; this joint use, however, involves special considerations as regards the features of the interface between the two techniques. Thus, the solvent in SFE is usually a liquefied gas (CO_2, N_2O, SF_6) that returns to the gaseous state on depressurization so the extractant is a gas by the time it reaches the interface of a coupled SFE/GC system. Consequently, the extractant volume is expanded by a factor of about 1000 (1 ml of liquid CO_2 expands to about 560 ml of gas). In addition, SF extraction requires a few minutes for completion when a micro-extractor is used; as a result, a large amount of gaseous solvent must be isolated from the solutes and removed at the interface.

Coupling an SFE and a GC system is fairly simple inasmuch as depressurization and conversion of the SF into a gas both take place under conditions typical of GC.

The solvent power of pure CO_2 limits its action to non-polar or scarcely polar compounds, so substances dissolved in SC CO_2 are usually eligible for GC analysis. Such is not the case, however, with highly polar compounds, which frequently tend to decompose at the working temperatures used in GC.

In addition, the amount of solute that is transferred to the column can be diminished by formation of a frozen CO_2 plug in its head. This can be avoided by using a hot, split or splitless injector as interface or a cooled thermal desorption injector. In this way, extracted analytes can be trapped while the SF is released into the atmosphere as a gas. This operational mode allows the analysis of wet samples, in contrast to coupled SFE/ column GC, where the presence of water may plug the restrictor or column with ice deposits.

The ability to obtain quantitative resnlts by SFE/split GC was studied with a new marine sediment certified material (SRM 194) from NIST which contains certified concentrations of several PAHs based on Soxhlet extractions with methylene chloride and n-hexane/acetone for 16 h. This material is particularly difficult to extract with SFs as it contains 2% w/w elemental sulphur and 4% water. Repeated attempts at analysing it by off-line methods with collestion in a few millilitres of a liquid solvent [40,52,53] and injection into the column of a combined SFE/GC system [54,55] proved unsuccessful since elemental sulphur, present as S_8, was extracted from the sediment and deposited in the restrictor in large enough amounts to plug it within a few seconds of extraction. By contrast, SFE/split GC extractions are scarcely prone to plugging, probably because the SF is depressurized at the hot injector port. SFE/GC/MS analysis of this material provided a tall chromatographic peak for S_8. Selective ion monitoring allows PAHs to be quantified with no interference from the S_8 peak [54]. **Sulphur plugging**

Coupled SFE/split GC also allows wet samples to be analysed, in contrast to column injection, where the water present may plug the restrictor or chromatographic column on freezing during extraction. Figure 4.37 shows the results of the SFE/GC/MS analyses of the contaminated sediment described in the previous paragraph, containing 20% moisture by weight. The chromatogram obtained by using the selective ion current mode at $m/z = 57$ showed most components to be n-alkanes. Other significant speies reflected in the chromatogram included several alkyl-

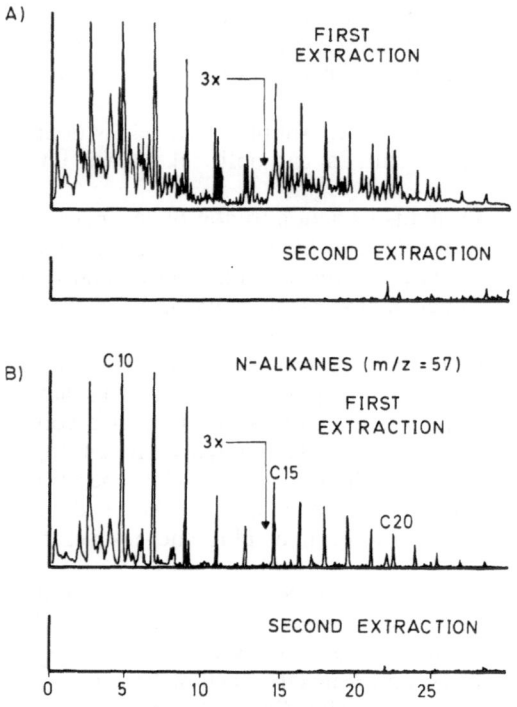

Fig. 4.37A,B. SFE/GC/MS analysis of a wet sample of petroleum-contaminated sediment. The sediment chromatograms were obtained after 10-min extraction with CO_2 at 380 nm and 50 °C by using a universal **A** and a selective mass spectrometer **B**. (Reproduced with permission of Elsevier Science)

benzene and alkylnaphthalene isomers. As can be inferred from the absence of peaks in the chromatograms obtained in a second extraction, 10 min extraction was long enough for all the components to be extracted quantitatively [56].

The presence of modifiers is occasionally incompatible with an on-line SFE/GC coupling. Such is the case with the recovery of pyrene vapour deposited on silica gel after extraction with methanol- or toluene-modified SC CO_2, which results in the loss of all the pyrene extracted despite the retention device used between the injection port and the column. Using unmodified CO_2 results in no significant loss of extracted pyrene [57].

Restrictor diameter   The restrictor diameter is central to SFE/GC applications, not only when injection is performed directly into the column, the diameter of which is dictated by the flow-rate of gas resulting from depressurized SF. The efficiency of the other types of interface is also conditioned by the

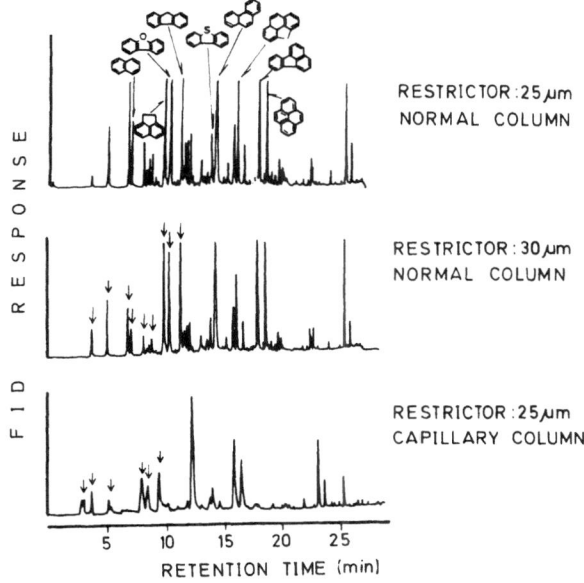

RESTRICTOR : 25 μm
NORMAL COLUMN

RESTRICTOR : 30 μm
NORMAL COLUMN

RESTRICTOR : 25 μm
CAPILLARY COLUMN

RETENTION TIME (min)

Fig. 4.38. Coupled SFE/GC/FID system for the analysis of river sediments containing petroleum as contaminant. Influence of the restrictor inner diameter and type of column used. For details, see text. (Reproduced with permission of Elsevier Science)

restrictor inner diameter. Figure 4.38 shows the influence of such a dimension on the efficiency of a coupled SFE/GC/FID system furnished with a small-bore column (lower chromatogram). The restrictor of 25 μm i.d. provides ill-defined peaks for many species. While that of 20 μm i.d. gives rise to better shaped peaks, obtaining peaks similar to those provided by a 25-μm i.d. restrictor and a wide-bore column requires using a restrictor of only 15 μm i.d. [23].

The detection systems most commonly used in this hyphenated technical approach are mass spectrometric, flame ionization and electron capture detectors. In every case, the most suitable interface is that which allows the extractant to be removed without passage through the column and hence, the detector. High flow-rates of gaseous CO_2 interfere with flame ionization detection and call for ignition prior to analysis. Conversely, electron capture detectors are not affected by the presence of large amounts of CO_2: the baseline remains stable even after the column has been flooded with carrier gas. On the other hand, the ECD provides a relatively high N_2O signal which requires removing the SF thoroughly

Commonly used detectors

after extraction by flooding the column with carrier gas for 2–3 min while the collection system is kept at the retention temperature.

Repro ducibility   The reproducibility of coupled SFE/GC systems was studied by comparing the performance of a split or splitless interface and conventional injection into a chromatograph of the same mixture of analytes (clay contaminants). Table 4.8 shows the results provided by the two procedures, which, as can be seen, are very similar in terms of rsd. Figure 4.39 shows the results of a study of the contribution to the overall error of the process of each individual unit and the overall error resulting from the coupling – this last was slightly greater than the sum of the individual contributions of the units, though closer to that arising from the extraction.

4.6.3 Coupled SFE/SFC

Supercritical fluid chromatography (SFC) has traditionally been regarded as the solution to those problems with which neither GC nor HPLC can cope. Even though this is not strictly true, SFC does solve some problems that cannot be successfully addressed by the other chromatographies. One obvious advantage of SFE in this respect is that it makes rather a suitable means for introducing samples into SFC systems. Because the solvent used to inject the sample is the same as the mobile phase, the primary requisite for effective coupling of two techniques (viz. compatibility between the output of the first system and input of the second) is met. One additional asset of on-line coupled SFE/SFC over SFE/HPLC and SFE/GC is the low probability of sample constituents that are insoluble in the mobile phase being introduced into the column.

Compatibility   Figure 4.40 shows the compatibility of some of the more common SFs
with detectors   with UV-visible and flame ionization detectors on a scale running from 10 (full compatibility) to 0 (full incompatibility).

In addition to the above-described basic types of interface for SFE/chromatographs couplings, there are several special designs for coupled SFE/SFC including the thermal-modulation interface developed by Mitra and Wilson [58]. The modulator is constructed by externally coating the topmost 15 cm of a fused silica SFC column of 0.1 mm i.d. with conductive paint to enable rapid heating by means of electric current pulses. The column inlet and outlet are connected to the extraction cell and an FID, respectively. Because the modulator possesses a small thermal mass, it can be heated and cooled very rapidly. The mobile phase containing the sample is introduced into the column through the modulator, the sample components being adsorbed in the stationary phase packed within. On application of an electric pulse, adsorbed components are desorbed as a "concentration pulse", acting as an injection within the current. The amplitude of the modulator signal is proportional to the concentration of

Table 4.8. Comparison of the precision achieved in the analysis of clay contaminants by using SFE/GC and conventional split GC

Contaminant	rsd (%)		Concentration (ng/ml)
	SFE–GC	Split GC	
2-Chlorophenol	1.8	2.0	50
Naphthalene	2.1	4.6	200
1-Chloronaphthalene	5.6	8.1	60
Hexachlorobenzene	5.8	7.8	50
Phenanthrene	4.0	3.8	300
Pyrene	4.2	5.6	200
Banzo[a]pyrene	5.5	6.4	20

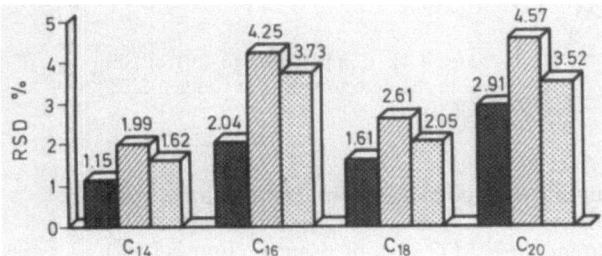

Fig. 4.39. Precision achieved in the extraction step (cross-hatched zone), the chromatographic determination (dot-hatched zone) and the SFE/GC determination of *n*-alkanes (line-hatched zone) in a clayey soil reference sample. (Reproduced with permission of Isco)

the analytes flowing through it provided the modulator is not saturated. The pulse duration is 1 s and the power supply voltage ranges between 15 and 20 V. This set-up was used for the analysis of semi-volatile organic compounds.

The solventless injector developed by Oudsema and Poole [59], which consists of a modified 6-port injection valve mounted on top of a chromatograph, a stainless steel pre-column and a wide-bore tee, has been used as an interface connecting the extraction cell at its outlet end to the injector pre-column by a 15-cm length of 15 μm i.d. fused silica capillary tubing [60].

Solventless injector

The potential of coupled SFE/SFC has been assessed by a number of authors as it solves one of the major problems faced by SFC in bioanalyses. Many pretreatments for biological samples use polar solvents, which are highly detrimental to the phase systems employed in SFC. This problem

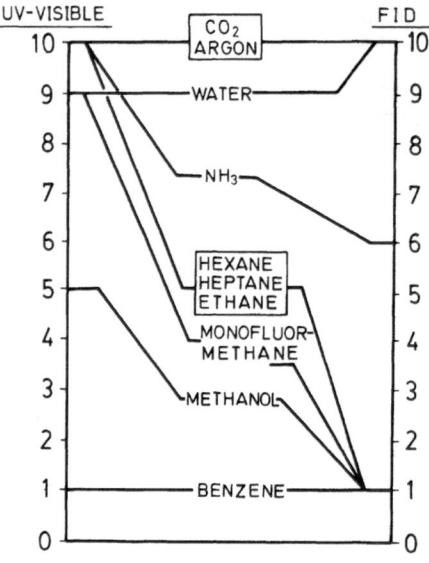

Fig. 4.40. Compatibility (0–10) of SFs with the detectors used in coupled SFE/GC systems

can readily be solved by coupling the chromatographic technique with SFE.

Proper functioning of a coupled SFE/SFC system requires that several prerequisites be met, namely:

a) The extraction chamber volume must be suited to the sample size afforded by SFC.
b) The pressure drop in the SF during transfer of the extract to the chromatograph must be minimal.
c) The chromatograph must be pressurized and equilibrated at the pressure to be used in the chromatographic determination before the extract is introduced into it.

Even though injection valves are seldom used in this context, some coupled SFE/SFC systems involve retention in the loop of one such valve [61].

All three types of ordinary SFC columns (open and packed capillaries, and normal-bore columns) are employed in SFE/SFC systems.

These coupled systems usually include switching valves that allow the extracted sample to be collected in a cooled accumulator while the same pump delivers the SF acting as the mobile phase to the chromatograph and the extracts are focussed to a trapping interface located prior to the chromatographic column, which is cooled below a preset temperature. After extraction and collection, the valves are actuated and the trap

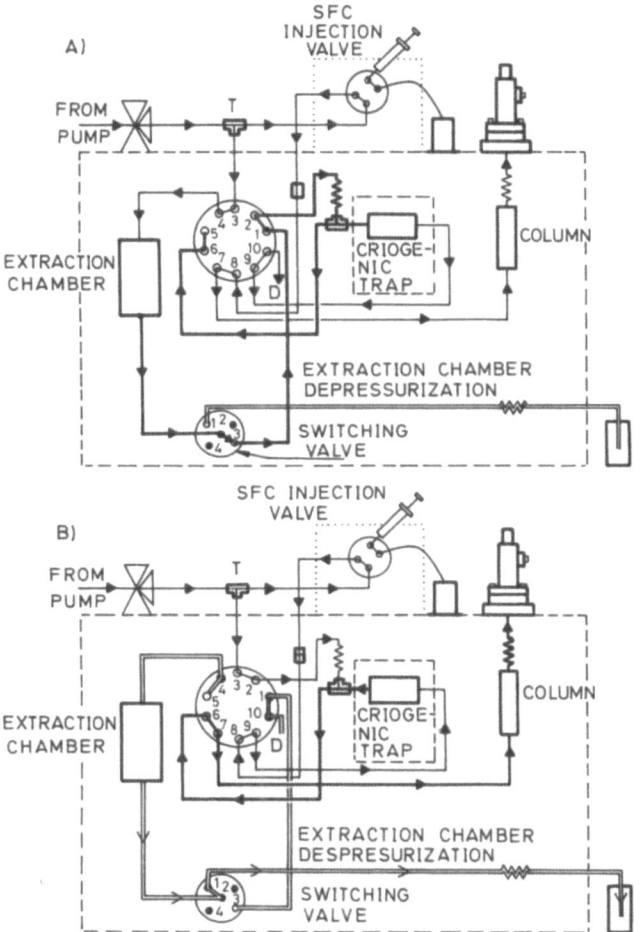

Fig. 4.41A,B. Hyphenated SFE/SFC system from Suprex. **A** Extraction position. **B** Transfer–determination position. For details, see text. (Reproduced with permission of Suprex)

temperature raised by means of the chromatographic oven itself or an external heating system. In this way, the SF transfers the extracted substances from the trap to the analytical column. Figure 4.41 shows one such system (Suprex) including three different valves (10-way/2-position, 5-way/4-position and 4-way/2-position, respectively), in addition to a tee of zero void volume. One of the tee outlets leads the supercritical mobile phase to the injcetion valve for use in conventional SFC alone. The other tee outlet leads to the 10-way valve, via which the analytes go through the restrictor onto the cryogenic trap, where they are retained. At this point, the expanded SF usually becomes a gas that is released into the atmosphere through the 10-way valve. After extraction has been completed, the pump is equilibrated for the chromatographic step and the 10- and 5-way valves are switched simultaneously (Fig. 4.41B). The mobile phase passes through the tee, the injection valve and the 10-way valve into the cryogenic trap, which is heated instantaneously at the programmed injection temperature. Subsequently, the mobile phase drives the extracted components back to the 10-way valve for introduction into the chromatographic column [62].

Use of second pump

Many SFE/GC applications entail using a second pump to pressurize the extraction chamber, the first pump being solely used to effect the chromatographic separation. These systems pose fewer technical problems and are more flexible: because they function independently of the extractor and chromatograph, they enable dynamic and static operation.

Open-loop dynamic system

Thiebaut et al. [63] use an open-loop dynamic system for the direct extraction of liquid samples in a coupled SFC system. Aqueous samples are directly injected into a supercritical CO_2 stream and extracted in a coil of adequate length. Water and SC CO_2 are immiscible, so they must be separated prior to detection; thus, the SF containing the extract is isolated from the water by means of a phase separator, trapped in a loop and then diverted to the chromatographic column by means of a switching valve. The analytes thus separated and determined so far are substances of medium polarity (phenol and 4-chlorophenol) in aqueous and urine samples; the extraction efficiency achieved exceeds 95%, with 8% repeatability (as rsd) for the overall system and 4% for each individual step.

Capillaries

The SFE technique has also been used in conjunction with capillary SFC. PAHs in coal can be separated and identified by using an open-loop system where the supercritical extract is expanded with the aid of a frit restrictor located in the sample cavity of a cooled micro-injector, the analytes being deposited by condensation while CO_2 is wasted through a vent valve. Subsequently, the loop contents are connected on-line to the mobile phase of the capillary chromatograph. Extracted analytes are detected by off-line FTIR spectroscopy after collection on a KBr disc and evaporation of the solvent [64].

The performance of the SFE/SFC association has been studied in comparative experiments involving various spectrometric detectors and SFs. The extraction/separation/detection of diphenylamine, naphthol and caffeine using CO_2 and N_2O as the mobile phase was found to give rise to no baseline drift arising from the pressure programme – the two SFs were previously compared in this respect by using a UV detector under isothermal and isobaric conditions, and found to provide similar results.

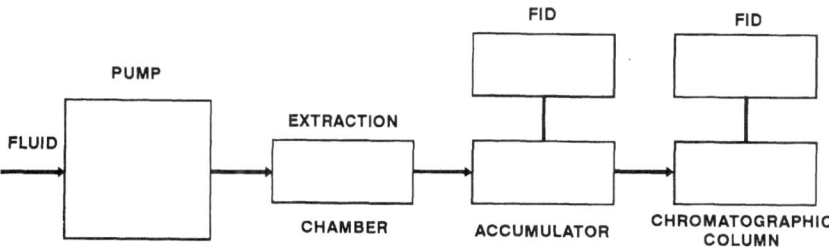

Fig. 4.42. Coupled SFE/SFC system with a dual detection system for the overall and individual analysis of a coating fibre extract

Using and FID resulted in a baseline rise as a result of the pressure programme when SC N_2O was used; nevertheless, this SF can be used in conjunction with open and packed capillary columns since the flow-rate of mobile phase through them is low enough for baseline drifts not to interfere with detection.

One other commercially available coupled SFE/SFC system is the Computer Chemical Systems instrument composed of an extractor, accumulation module and chromatograph, which was used to monitor coating fibre manufacture, either to determine all the fibre components or the raw materials, for which no satisfactory chromatographic quality control methods were available. The typical constituents of coating fibres are oils, waxes and oil mixtures (solvents, vegetable oil triglycerides and alkylethoxy surfactants), the determination of which by GC and HPLC is unfeasible owing to the formation of aqueous emulsions and the absence of chromophoric groups, respectively. Figure 4.42 shows the set-up used to develop a method for such a determination. Two FIDs located at the accumulator outlet and the column allow the overall and individual determination, respectively, of the extracted analytes. Table 4.9 testifies to the good results obtained [65].

4.6.4 Coupled SFE/HPLC

The SFE/HPLC coupling is a logical extension of the above-described hyphenated techniques aimed at addressing the analysis of extracts inaccessible to GC and SFC owing to their high polarity, molecular weight or thermal lability.

Interfacing an SFE to a liquid chromatograph is the most complex operation involved in SFE hyphenated techniques as a result of the difficulty involved in coupling a preparative technique that produces a gas at

Table 4.9. Results obtained in the analysis of coat fibres by on-line coupled SFE/SFC

Sample no.	Weight (mg)	Analyte	Area (%)	Oil/wax ratio
1	20	Waxes	32.63	2.1
		Oils	67.36	
2	20	Waxes	31.00	2.2
		Oils	69.00	
3	30	Waxes	33.26	2.0
		Oils	66.23	
4	30	Waxes	33.47	2.0
		Oils	66.52	
5	40	Waxes	32.08	2.1
		Oils	67.91	
6	40	Waxes	32.94	2.0
		Oils	67.05	

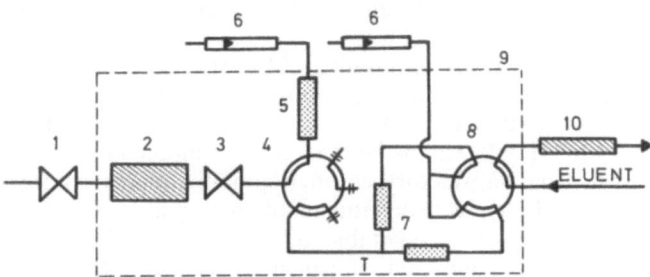

Fig. 4.43. Coupled SFE/HPLC systems based on two six-way valves. (Reproduced with permission of Elsevier Science)

the interface with a chromatographic technique that handles a liquid mobile phase. For this reason, the SFE/HPLC association has developed to a fairly limited extent notwithstanding the fact that the earliest application was reported in 1983 [66].

Figure 4.43 shows an operational scheme for a typical SFE/HPLC system. Elements 1 and 3 in the figure are pressure regulators, while 2 denotes the extraction chamber. The two techniques are coupled via two slightly modified, serially arranged high-pressure 6-way valves (4,8). The first of these valves (4) is used to detach or load the sample loop; one of its ways is connected to the waste column (5). The flow-rate of the separated SF is measured by means of a rotameter (6) and output of the second way is split with the aid of a tee into two capillaries that lead to two narrow-

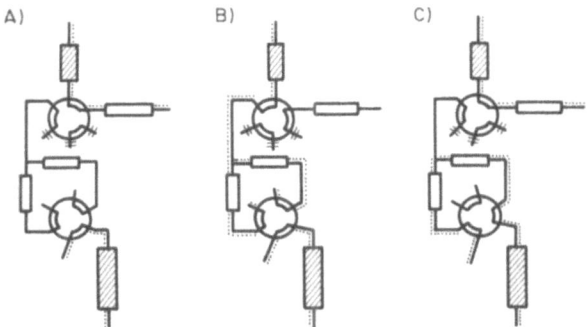

Fig. 4.44A–C. Operational diagram of the coupled system shown in Fig. 4.43. For details, see text. (Reproduced with permission of Elsevier Science)

bore columns (7). The second valve (8) functions as the injection system, the loop of which is formed by the two segments connected to the columns (7). The third segment of the valve is used as eluent inlet and outlet to the HPLC column (10). The three steps of the process (A, extraction; B, separation and extract collection; and C: pre-column elution and injection into the analytical column) are illustrated in Fig. 4.44.

Direct injection of SC CO_2 extracts into a reversed-phase HPLC system gives rise to peaks that elute near the column void volume and long baselines of up to 20 min. These perturbations are ascribed to the CO_2 co-injected with the analytes. Various attempts at avoiding or minimizing this problem such as using an inverted tee or a planar membrane separator proved the latter to the most effective choice.

4.6.5 Comparison of SFE Hyphenated Techniques with GC, SFC and HPLC

Supercritical fluid extraction has so far been most frequently used on-line with GC in its two modes (packed and capillary columns) than it has with SFC and, expecially, HPLC, probably because most of the analytes that are extracted by SFE can be separated by GC or SFC. The choice of the chromatographic technique to be coupled on-line with SFE is dictated by the analytes to be determined. If the analytes can be analysed by capillary GC, the SFE/GC association is to be preferred to the SFE/SFC coupling simply because capillary GC provides more expeditious extractions and is simpler to implement and more solidly established. However, the types of sample to which coupled SFE/GC can be applied is logically determined by the features of gas chromatography. Even though the SF can dissolve

Choice dependent on analytes

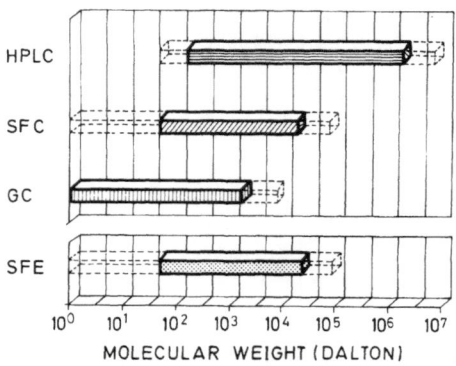

Fig. 4.45. Overlap regions of molecular weights of SF extractable species that can be determined by the three column chromatographies. (Reproduced with permission of Elsevier Science)

analytes of high boiling points, GC cannot cope with low-volatility and/or thermally unstable compounds. Determining SF extracts by HPLC offers one inherent advantage over GC since most coextracted species contain no chromophoric groups, whereas most organic substances are sensed by the flame ionization detector. Figure 4.45 shows the molecular weight ranges afforded by the three chromatographies and SFE [67].

High temperature capillary columns

One other important consideration is that the solubility of a compound in an SF cannot exceed that in the liquefied fluid; consequently, a substance that can hardly be dissolved in the liquid will also be scarcely soluble in the SC solvent. In addition, high-molecular weight and polar compounds are sparsely soluble in polar (CO_2) and slightly polar SFs (N_2O). The limitation imposed by GC can be overcome by using capillary columns at high temperatures and highly stable stationary phases. Compounds of up tp 100 carbon atoms can be determined by using these technically superior systems; however, they do not cover the whole range of substances afforded by SFE.

4.6.6 Other Hyphenated Techniques

Several combinations of SF/chromatography and a high-capacity (MS, TFIR) detector were described in the preceding sections. One such combination including an SF extractor, a multi-dimensional liquid column chromatographic system and a gas chromatograph furnished with a UV-visible and a flame ionization detector, respectively, was used for the determination of insecticides in pasture [68]. Figure 4.46 depicts the overall set-up (A) and the SFE/chromatographic system interface (B). Switching valves play a major role in the coupled system: during extraction/

A)

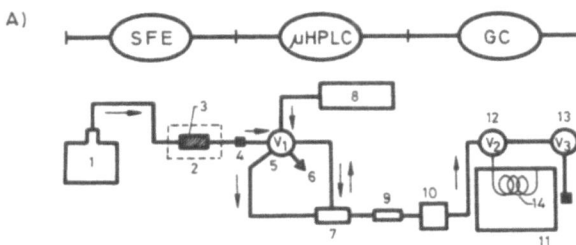

B)

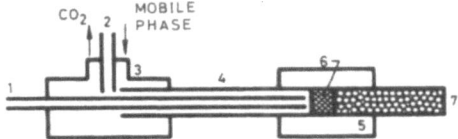

Fig. 4.46A,B. Multi-dimensional SFE/HPLC/GC system. **A** Instrumental set-up (1, extractor pump; 2, extractor heating system; 3, extraction system; 4, filter; 5, switching valve (V_1); 6, waste; 7, impact interface; 8, micro-HPLC pump; 9, micro-HPLC column; 10, UV detector; 11, GC furnace; 12, switching valve (V_2); 13, on-off valve (V_3); 14, GC capillary column). **B** Impact interface (1, linear restrictor; 2, HPLC mobile phase inlet/CO_2 waste; 3, low-void volume T-piece; 4, impact tube; 5, low-void volume junction; 6, impact zone; 7, micro-HPLC column). (Reproduced with permission of the American Chemical Society)

collection, V_1 connects the extraction chamber with the impact interface collection unit while the chromatographic column is shut by means of V_3 in order to avoid passage of CO_2 and hence flushing of the eluent from the column packing and the associated baseline drifts. Valve V_2 opens or shuts passage of the effluent from the liquid column to the GC. Table 4.10 shows the results obtained; as can be seen, this hyphenated alternative offers all the advantages of the multi-dimensional system including high selectivity, sensitivity and precision, in addition to greatly simplified sample handling and a high automatability. Samples usually requiring repeated manual clean-up can be analysed with minimal manipulation with one of these coupled systems, which thus diminish the risk of contamination and analyte losses. As can be seen in Table 4.10, sample sizes as small as 5 mg can be processed, which minimizes the use of organic solvents and the personal and environmental hazards involved in their handling.

In 1989, Frei et al. developed an original liquid-segmented SF extraction set-up that is schematically depicted in Fig. 4.47. It consists of an SC CO_2 line propelled by a syringe pump into which liquid samples are injected. A phase separator located after the extraction system allows

Liquid-segmented SF extraction

Table 4.10. Recovery achieved in the determination of chlorpyrifos in grass (160 ng/g) by on-line coupled SFE/LC/GC

Sample weight (mg)	Concentration found (ng/g)	% Recovery[a]
5.08	140	87.5
5.09	168	105
5.28	158	98.7
5.31	161	101
4.70	188	118
5.01	150	93.8
5.71	142	88.8
4.64	140	87.5
Mean		97.5
sd		10.5
rsd (%)		10.8

[a] Recoveries were calculated on the basis of 160 ng/g concentrations obtained by using off-line SFE and a conventional sample pretreatment procedure.

passage of the SF to the UV detector or the injection valve of an SF chromatograph. Notwithstanding the promising results obtained in preliminary experiments, the system has not been further developed or applied [69].

Continuous analytical technique   On-line coupling of SF extraction to a continuous analytical technique such as flow injection analysis requires special analyte trapping and elution systems, development of which has been successfully addressed by Brewster et al. [70] using a membrane phase separator in conjunction with a high-pressure sampling valve and a short capillary restrictor. The overall set-up, shown in Fig. 4.48, consists of a high-pressure pump and switching valve (V_1), in addition to the extractor sampling valve (V_2, 10-µl loop), a fused silica capillary restrictor (15 cm × 50 µm i.d.), a membrane separator, a low-pressure 6-way valve (V_3) and a diode array spectrophotometer. All tubing used is of stainless steel, while connections are made of fluorinated polymer. The void volume between the detector and the injection valve (including V_3) is 300 µl. Extraction is performed in the recirculation mode, during which extract samples are taken at regular intervals in order to monitor the process. When equilibrium is reached, V_3 is switched to connect the phase separator with the detector and V_2 is actuated. After 60 s (the time required for the extractant to expand through the restrictor and phase separator), V_1 is switched and the solvent pumped through the sampling valve to drive the sample to the separator and onto the detector. Calibration is done by replacing the high-pressure injection valve with the sampling valve and injecting aliquots of standard through it.

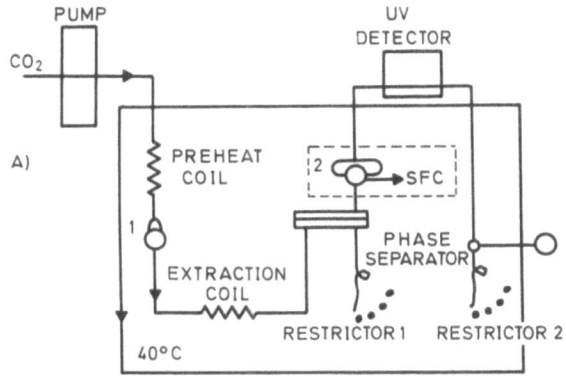

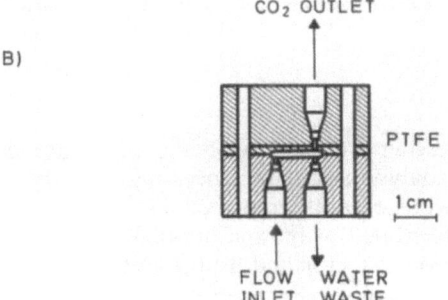

Fig. 4.47A,B. Segmented-SF extraction system for aqueous samples. **A** Scheme of the experimental set-up (1, sample injection valve; 2, chromatographic injection valve). **B** Cross-sectional view of the phase separator). (Reproduced with permission of Elsevier Science)

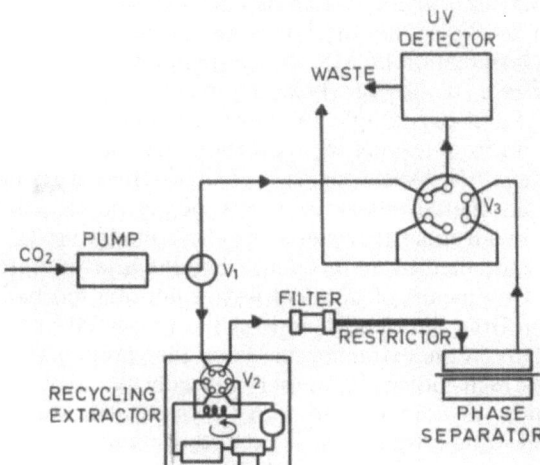

Fig. 4.48. Coupled SFE/FIA system including a phase separator and three valves. For details, see text

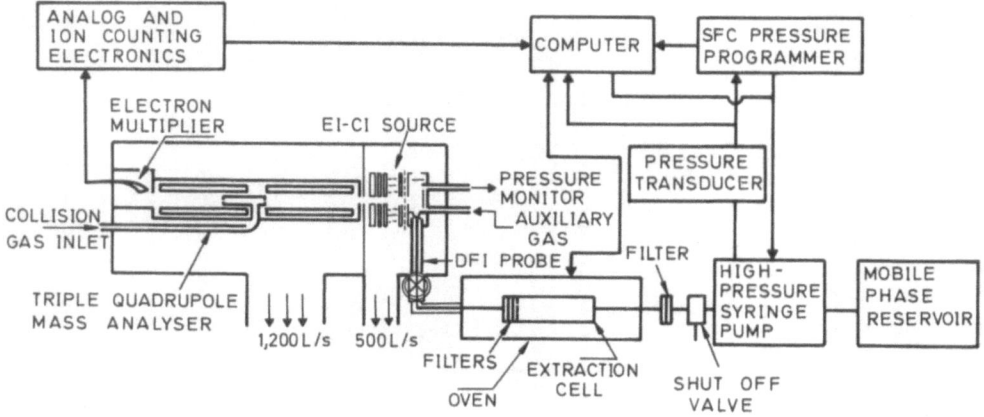

Fig. 4.49. Coupled SFE/MS system. For details, see text. (Reproduced with permission of the American Chemical Society)

A combined sample processing system involving SF extraction, adsorption/desorption, on-line analysis and data processing, and the preliminary results it provided, were recently reported [71]. The system consists of an SF extractor furnished with an automatic adsorption/desorption unit, a flame ionization detector and a fibre optic detector (a transmittance/absorbance configuration).

Direct introduction of fluids   A previously developed interface for direct introduction of fluids in coupled SFC/MS [72] was used to connect an MS detector on-line with an SF extractor as shown in Fig. 4.49, thereby accomplishing micro-scale extraction and direct determination by MS. A high-pressure syringe pump is used to propel the SF to a stainless steel extraction cell of 680 µl volume accommodating up to 0.5 g of dry sample. A series of filters of decreasing pore size prevent the solid from leaving the extraction chamber and hence plugging of the restrictor (a fused silica capillary of 35–50 cm × 8 µm i.d.), which is connected to another capillary of 1 m × 50 µm i.d. for direct injection into the three-quadrupole spectrometer. The temperature of the extraction chamber is maintained by a GC oven, while the injection probe is thermally conditioned by means of an electrical heater-assisted hot-air system. A pressure ramp from 1 to 5 bar/min over the range 100–300 bar is used to ensure selectivity in the extraction based on the variations in the solubility of solutes of a high molecular weight with density. The results obtained by determining mycotoxins in wheat samples showed the SFE/MS to offer the following advantages: (a) reduced sample manipulation as a result of avoiding discrete extraction, purification and/or derivatization; (b) ease of analysis and interpretation; and (c) compatibility of SFE with a variety of chemical ionization reagents which enable selective, sensitive determinations of compound families without the interference of coextracted substances, which yield high background signals [73].

When matrix effects cannot be avoided by selective ionization, tandem MS may help to increase the selectivity and, also frequently, the sensitivity [74].

Two on-line coupled SFE/Fourier Transform Infrared Spectrometry (FTIR) approaches were recently reported. One, developed by a Japanese group, uses a syringe pump connected to an FTIR spectrometer to monitor the extraction dynamics of stearic acid methyl ester, linoleic acid methyl ester and DL-α-tocopherol [75]. The other, developed by Kirschner and Taylor, involves SFE/SFC equipment in which the conventional chromatographic column is replaced with a 1 m × 0.0020 in i.d. transfer line of high-quality stainless steel. The line is joined via a zero-dead volume piece to a 50-μm fused silica transfer line leading to the high-pressure IR flow-cell. Adequate back-pressure is maintained by a 60-μm i.d. deactivated fused silica tapered restrictor attached to the flow-cell exit transfer line. The restrictor tip is housed within an FID heating block to reduce plugging of the exit orifice. The system was originally perfected for the extraction and quantification of n-tetracosane from Celite and subsequently applied to more complex samples including fibre/fabric finishing oils with comparable success [76].

An IR detection system lying between off-line and on-line systems was previously described in Sect. 4.3.5.2 [33].

4.7 Comparison of the Off-Line and On-Line Modes

Supercritical fluid extraction not directly coupled to detection or separation/detection is inherently simpler to implement as it only requires the extraction step to be optimized and the extract can be analysed by any suitable method; on the other hand, the off-line mode requires interpretation and optimization of each unit separately, in addition to appropriate coupling. In combining an SF extractor on-line with a detection system, a compromise must be made between their individual optimal operation conditions. On-line extracted/determined analytes cannot be used for subsequent analysis by a different method. Also, the coupling interface may pose problems not faced by the off-line mode.

The chief advantage of SFE coupled on-line to chromatography is avoidance of human involvement between the extraction step and chromatographic analysis (automation), and the ability to achieve maximum sensitivity through quantitative transfer of the extracted analytes to the chromatographic column, which allows smaller sampler sizes to be used and results in enhanced precision.

While the choice in every case is dictated by the particular problem addressed, some general guidelines can be given in this respect:

– The off-line mode is to be preferred for developing new applications, while the on-line mode is better suited to routine analyses. The latter, however, is only recommended when the sample matrix from which the solutes are to be extracted is invariable in terms of chemical composition, size, particle size, etc.

Optimization – Optimization procedures for the off-line mode are more versatile than those for the on-line mode, so they enable optimal, uncompromised functioning of each individual unit.

Flexible – Off-line applications are more flexible as regards the addition of modifiers, changes in the flow-rate or pressure, etc.; in addition, they are less markedly affected by the sample properties (nature of the matrix and/or analyte concentrations) and allow multi-analysis of the extract.

Figure 4.50 compares schematically the most salient advantages and disadvantages of both modes by way of summary.

Automatability The automatability of the off-line mode can be increased by using an autosampler to collect the extracts (from the same or different samples) and introduce them into the chromatograph (Fig. 4.51). The autosampler can also be regarded as a highly useful "interface" as it lessens many of the disadvantages of the off-line mode.

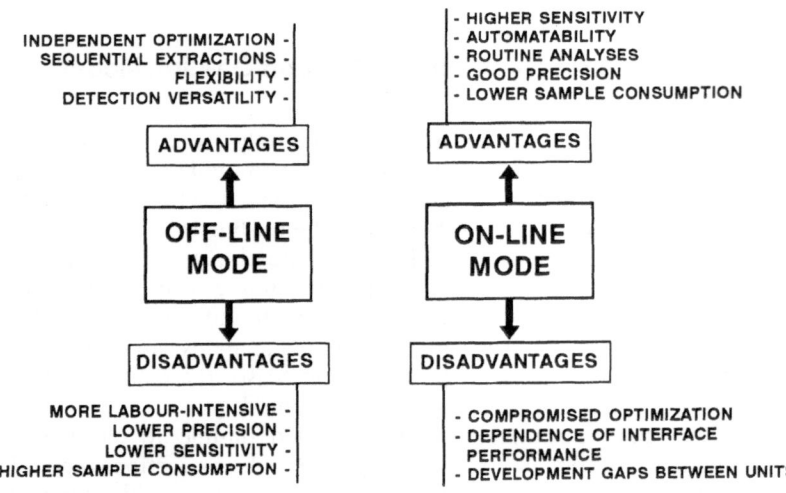

Fig. 4.50. Advantages and disadvantages of the on-line and off-line SFE mode

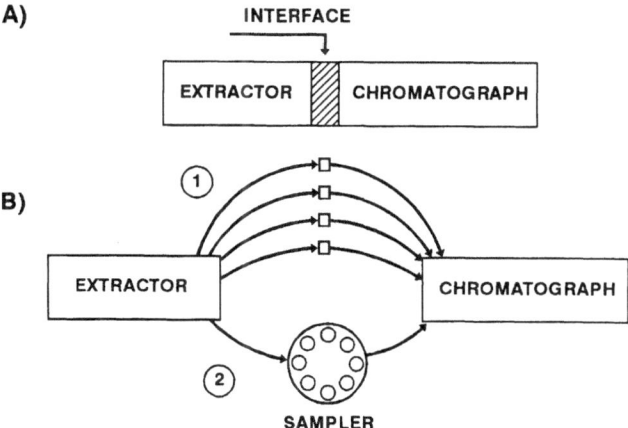

Fig. 4.51A,B. Ways of connecting an SF extractor to a chromatograph. **A** On-line or interface mode. **B** Off-line mode, with (1) individual and (2) joint transport of extracts the chromatograph

Table 4.11. Comparison of various features of analyses for PAHs in urban dust by using off-line SFE and on-line coupled conventional SFE/GC

Parameter	Conventional extraction	SFE	
		Off-line	On-line with GC
Sample size	1000 mg	20 mg	2 mg
Liquid solvent volume	450 ml	3 ml	0 ml
Extraction time	48 h	1 h	15 min
Extract concentration time	3 h	0–10 min	0 min
Extraction/concentration time	16 h	20 min	20 min
Minimum time for analysis of a whole sample	3 days	2 h	1 h

Table 4.11 compares practical aspects of off-line SFE and on-line coupled SFE/conventional liquid extraction in the determination of PAHs in a standard of urban air particulates. For each sample, off-line analysis and two different SFE/GC associations provide results that are highly consistent with the certified concentrations obtained with conventional extraction methods involving liquid solvents (between 2 and 7 µg/g, depending on the PAH). However, the time needed to prepare the sample (from reception to delivery of the final results) and the amount of sample required are reduced by over two or ders of magnitude.

4.8 Commercially Available Supercritical Fluid Extractors

SFE systems have reached an unprecedented state of development in the context of analytical instrumentation. As a rule, manufacturers are cautious in introducing new products. Such novelties are usually endorsed by much research (i.e. a host of literature references), which, to some extent, guarantees profitability of the investment incurred. Commercialization of SF extractors has departed from this trend since, following scarce, unsystematic laboratory research and application, manufacturers started to launch the earliest SFE systems; at present, at least seven firms, most of international prestige [Computer Chemical Systems (CCS Instrument Systems, Inc.), Carlo Erba–Fisons, Dionex–Lee Scientific Division, Hewlett–Packard, Isco Corporation, Jasco Ltd and Suprex, in alphabetic order] market a wide range of SFE apparatuses of quite similar performance. Gilson Medical Electronics recently marketed one such apparatus as well.

CCS (Avondale, PA) is one of the pioneers in the commercialization of instruments and apparatuses for analytical-scale SFs; in fact, this firm launched its first fully automated SF chromatograph in 1986 and its earliest extractor with a built-in FID and coupled SFE/SFC system in 1987. One of the latest extraction modules are those of the "on-site environmental analyzers–Champ" range, specially designed for field work (usually under **Field work** hazardous atmospheres). These are portable apparatuses that can work at up to 6000 psi and 250 °C and use extraction chambers of 0.5, 5 and 10 ml. The laboratory extractor currently offered by this firm, the Model 3100-100 Benchscale SFE System, can be used either for SFE or as a reactor for processing SFs. Its specifications include working pressures and temperatures of up to 5000 psi and 200 °C, respectively, for processing. It can manage six extraction chambers simultaneously, in addition to a restrictor heating system that allows the pressure and flow-rate to be regulated independently and extracts to be collected in aerated, detachable containers.

Analytical Carlo Erba–Fisons (Milan, Italy) makes the "SFE Analyzer 3000 **and preparative** Series", which can be used for analytical and preparative purposes and is **work** supplied in two basic configurations that differ in the extraction unit: with one or up to six cells for the sequential extraction of six samples; with a single syringe pump or several in a tandem arrangement; and with a wet or dry collection system. The cell capacity ranges between 0.5 and 40 ml. The extractor can readily be coupled to a gas or SF chromatograph.

Dionex-Lee Scientific Division (Salt Lake City, UT) introduced its SFE-**Unlimited** 703 model in 1991. The system features unlimited pumping capacity (with **pumping** one or two pumps) and a chamber located in an oven that is operational **capacity** between room temperature and 150 °C, and eight extraction cells arranged in parallel for simultaneous extraction of up to eight samples. The cell volume ranges between 0.5 and 32 ml. The specially designed restrictor can be heated up to 250 °C. Other special features include the ability to measure the flow-rate of each extraction cell (both the individual and the overall value are shown in a display) and a maximum working flow-rate of 18 ml/min.

The Hewlett–Packard model 7680A SFE was launched in 1990. It consists of a modulecontaining a dual piston pump, the extraction chamber – which includes the restrictor, the analyte trap, flushing solvent reservoirs, a mini-pump for delivery of flushing solvents and a fraction collection tray. The extractor is controlled via a PC workstation runningsoftware for monitoring the extraction in real time in addition to other applications. Up to three different fluid cylinders (e.g. CO_2 and two different premixed CO_2–modifier fluids) can be fitted to the extraction module, fluids being selectable during the process. The analytes are collected in a cryogenic trap and eluted with the solvents from the two containers. The maximum flow-rate of compressed fluid is 4 ml/min.

ISCO (Lincoln, NE) offers two modular extractors for laboratory and field work. The laboratory unit (the SFE System 1200) was introduced in 1990 and the field unit in 1991. Both include dual-cell extraction modules and a convenient mechanism for loading and coupling to the extractor. The propulsion system, a syringe pump, can be of two types: (a) 100-ml capacity, maximum pressure and flow-rate 10000 psi and 25 ml/min, respectively; (b) 260-ml capacity, maximum pressure and flow-rate 7500 psi and 90 ml/min. The extracted material is collected by a solvent contained in encapsulated tubes. The latest extractor marketed by ISCO is an automated model furnished with a 24-position carousel which can load 24 samples in disposable bar-coded cartridges. A programmable wash unit eliminates carry-over between extractions. The apparatus can be operated at up to 10000 psi, 150 °C and a flow-rate of 40 ml/min.

The SF extractor from Jasco Ltd (Japan) is an off-line system that allows collection of up to ten fractions and continuous monitoring of the extraction process by means of a UV detector.

Suprex Corporation (Pittsburgh, PA), launched it SFE-50 model, originally used in the off-line mode, in 1988. One salient feature of this extractor is flexibility: it allows fractions to be collected unattended by means of a series of multi-way valves. For example, by using an 8-line manifold (4 lines connected to 4 extraction cells) and a 12-way, 2-position valve connected to 8 restrictors, two fractions of each of the 4 extraction cells can be collected with no intervention of the operator (see Sect. 4.3.3.6). Suprex later introduced two additional systems: the Prepmaster, which includes a dual piston pump and can be used in field work; and the MPS-225 model, which is based on a syringe pump and is primarily intended for on-line coupling to SF chromatographs. The latest model from this firm is an automatic extractor (the Autoprep 44) that allows

Unattended fraction extraction

Table 4.12. Specifications of selected commercially available detectors

Feature	DIONEX–LEE SCIENTIFIC SFE-703	HEWLETT–PACKARD 7680A	ISCO SFE SYSTEM 1200	SUPREX SFE-50
Propulsion system				
– Pressure range	Piston pump <10000 psi	Piston pump 77–383 bar (1100–5.560 psi)	Syringe pump <7500 psi	Syringe pump <7500 psi
– Temperature range	Room temperature	5°C	Room temperature	Room temperature
– Flow-rate range	1.0–10.0 ml/min (liquid)	0.15–4.0 ml/min (liquid)	1.0 µl/min–90 ml/min (liquid)	1.0 µl/min–9.0 ml/min
– Pump capacity	Continuous	Continuous	266 ml	240 ml
– Accuary	±1% (15–40°C, 100–600 atm)	±2% (340 bar thermostated pump head)	±2%	±2%
– Loading and bleeding time	NA^a	NA	2.5 min	3 min
– Operational mode	Pressure	Density/flow-rate	Pressure	Pressure, density
Extraction fluid				
– Fluid choices	CO_2, CO_2 + modifiers	CO_2, CO_2 + modifiers	CO_2, CO_2 + modifiers	CO_2, CO_2 + modifiers
– Modifier separate action	Possible	No	Possible	Possible
Extraction chamber				
– Temperature range	35–150°C	40–120°C	<150°C	<200°C
– Maximum programmable extraction time	900 min	300 min per extraction step	999 min per step	600 min
– Maximum equilibration time	NA	6 min per extraction step	Selectable	NA
– Number of extraction chambers	8, in parallel	1	2, in parallel	4, in series or parallel

Extraction chamber				
– Type	Stainless steel	Stainless steel	Stainless steel, PEEK[b]	
– Volume	3.5–10.5 ml (340 atm) 10–32 ml (680 atm)	1.5 ml, 7.0 ml	0.5 ml, 2.5 ml, 10 ml	0.15–50 ml
– Placement	Horizontal	Vertical	Vertical	Vertical or horizontal
– Dimensions	5 cm × 0.94 cm i.d. (3.5-ml chamber)	8.3 cm × 1.1 cm i.d (7-ml chamber)	6 cm × 0.75 cm i.d. (2.5-ml chamber)	3.8 cm × 1 cm i.d. (3-ml chamber)
– Maximum pressure	680 atm	580 atm	680 atm	500 atm
– Sealing	PEEK	PEEK	PEEK	PEEK
– Frit (pore size)	0.5 and 2.0 μm	2.0 μm	2.0 μm	0.5 and 2.0 μm
– Fitting	Requires nuts	Manual control	Manual control	Requires nuts
– Leak sensor	No	Closed-loop control	No	No
Depressurization device				
– Restrictor dimensions	Special design	Variable resistor with electronically controlled pressure	25 cm × 50 μm i.d., 375 μm e.d fused silica tube	60 cm × 50 μm i.d., 375 μm e.d fused silica tube
– Restrictor temperature range	<250°C		<150°C	<150°C
Collection system				
– Type	Glass vials	Adsorbent trap	Glass vials	Glass vials
– Temperature	2°C	Variable heating/cooling (−30 to 80°C)	Ambient	Ambient

[a] NA = not applicable.
[b] Atomic.

Table 4.13. Features of selected commercially available extractors

Assets	Other features
DIONEX–LEE SCIENTIFIC SFE-307	
– High throughput. Up to 8 parallel extractions.	– Does not allow static extractions.
– The dual piston pump provides a continuous CO_2 flow.	– Noisy pump.
– Working temperatures up to 130 °C.	– No pressure or temperature programming during extraction.
– Special restrictor design can be heated automatically up to 150 °C.	– The chambers require special tools for assembly and connection.
– Dual-vial collection chamber enables effective entrapment of volatiles. The vials can be cooled to 2 °C during extraction.	– The collecting vials are difficult to assemble (the assembly process is labour-intensive and contamination-prone).
– The CO_2 flow-rate is recorded in ml/min depressurized gas as total volume in each vial.	
HEWLETT–PACKARD 7680A	
– The dual-piston valve system enables continuous circulation of CO_2 or any fluid from the three tanks at the back of the extractor.	– Only a single extraction at a time is possible.
	– Very slow (equilibration with the initial conditions takes a long time).
– The variable-output restrictor enables precise control of the flow-rate up to 4 ml/min.	– Extraction conditions of other extractors are difficult to adjust since it is not pressure and temperature, but density and temperature than can be controlled.
– The pressure and flow-rate can be adjusted separately.	
– Easy to mount extraction vials (no tools required).	– Collecting and eluting the extracted analytes calls for thorough optimization (the trap temperature and packing, solvent type and volume, flow-rate and number of required washings must be specified).
– Static and dynamic operation.	
– Fully automated system. Interface graphs allow any step of the extraction process to be accessed. The real, updated extraction time can be read out.	– Requires CO_2 to cool the pump head, restrictor output the trap.
	– Changing from the 1.5-ml to the 7-ml extraction chamber requires altering the hardware.
– Leaks can be detected by means of a close-loop control system.	
– Robot-compatible.	
– Allows collection of several fractions by altering the pressure/temperature conditions.	
ISCO 1200 SYSTEM	
– Ease of chamber changeover (no tools required).	– Limited syringe capacity: 260 ml.
– Wide availability of reusable and disposable PEEK cartridges.	– The system includes eight needle valves that are rather sensitive to pressure and heat, and prone to leak.
	– Only works in the isothermal mode.
– Dual extraction chambers with independent valves for parallel extraction of samples.	– The extraction must be stopped manually by actuating the needle valve.
– Static and dynamic operation.	
– Working temperatures up to 150 °C.	
– Pump loading takes 2–3 min. The syringe pump requires no cooling with CO_2 while loading.	
– Clogged restrictors can readily be replaced.	

Table 4.13. *Continued*

Assets	Other features
– Two syringe pumps allow up to five extraction modules to be used simultaneously (i.e. up to ten samples to be processed). – The modifier can be added dynamically by combining two syringe pumps (one for SF supply and the other for modifier supply).	
SUPREX SFE-50 SYSTEM – Operational flexibility: the multi-way valve allows automatic collection of fractions. The extractor can be operated under a pressure or density programme. The temperature can also be programmed. – Dynamic and static operation. – The microprocessor can store up to 25 programmes. – The modifier solvent valve allows modifiers to be directly added to the syringe pump. – The syringe pump requires no CO_2 cooling while loading. – Can be modifier to perform up to four simultaneous or sequential extractions.	– Limited syringe capacity: 250 ml. – The pump takes around 12 min to load. – The "autopump" feature is only partly useful – the pump should allow automatic filling as well. – Extraction chamber assembly and connection require tools. – Extraction chambers of more than 3 ml are difficult to accommodate in the furnace when multi-way valves are used. – The rotors of multi-way valves tend to leak above 125 °C. – Replacing the restrictors is labour-intensive.

up to 44 samples to be processed unattended. It also enables analysis of urgent samples and the use of randomly placed extraction cells of 0.5–10 ml capacity. The system includes six disassemblable trays with seven sample positions each that can be used to implement up to ten methods per tray, including static and dynamic extraction. The restrictor flow-rate is controlled by a computer over the range 1–7 ml/min. The cryogenic collection module included can operate from −50 to 40 °C and collect up to 16 fractions per sample. The trap is heated ballistically and the solvent used to wash and recover the sample can be passed at a programmable flow-rate of up to 10 ml/min, thereby increasing the sample desorption efficiency and minimizing carry-over between samples. The solvent can be purged with nirogen.

Recently, López Avila et al. compared the performance of four ex- **Practical** tractors from as many firms (viz. the Dionex–Lee Scientific SFE-703 **comparisons** model, the Hewlett–Packard 7680A model, The Isco SFE System 1200 and the Suprex SFE-50 model) in order to test their applicability to environmental problems and provide the manufactures involved with unbiased information on the advantages and disadvantages of their respective models [77]. Table 4.12 compiles the features of the extractors tested, while Table 4.13 lists both advantages of interest to environmental analysis (left column) and "other features" that may be advantageous orundesirable depending on the particular purpose, so they should be considered cautiously.

References

1. Jinno K (1992) Hyphenated techniques in supercritical fluid chromatography and extraction. Elsevier, Amsterdam
2. Westwood SA (1993) Supercritical fluid extraction and its use in chromatographic sample preparation. CRC, Florida
3. Pariente GL, Pentoney SL Jr, Griffiths PR, Shafer KH (1987) Anal. Chem. 59:80
4. Pawliszyn J (1990) J. High Resol. Chromatogr. 13:199
5. Mauldin RF, Vienneau JM, Wehry EL, Mamanton G (1990) Talanta 37:1031
6. Hawthorne SB, Miller DJ, Walker DD, Whittington DE, Moore BL (1991) J. Chromatogr. 541:185
7. Schmidt S, Blomberg L, Wannnman T (1989) Chromatographia 28:400
8. Furton KG, Rein J (1991) Chromatographia 31:297
9. Hedrick J, Taylor LT (1989) Anal. Chem. 61:1986
10. Hedrick JL, Taylor LT (1990) J. High Resol. Chromatogr. 61:1986
11. Ong CP, Ong HM, Yau Li SF, Lee HK (1990) J. Microcol. Sep. 2:69
12. Lohleit M, Bachmann K (1990) J. Chromatogr. 505:227
13. Hawthorne SB, Krieger MS, Miller DJ (1988) Anal. Chem. 60:472
14. Hawthorne SB, Krieger MS, Miller DJ (1989) Anal. Chem. 61:736
15. Jahn KR, Wenclawiak B (1988) Chromatographia 26:315
16. Lohleit M, Hillmann R, Bachmann K (1991) Fresenius J. Anal. Chem. 339:470
17. Fjeldsted JC, Kong RC, Lee ML (1983) J. Chromatogr. 279:449
18. Chester TL, Innis DP, Owens GD (1985) Anal. Chem. 57:2243
19. Guthrie EJ, Schwartz HE (1986) J. Chromatogr. Sci. 24:236
20. Smith RD, Udseth HR (1983) Anal. Chem. 55:2266
21. Cortes H, Pfeiffer CD, Richter BE, Stevens TS (1987) U.S. Patent #4, 793:920
22. Hawthorne SB, Miller DJ, Krieger MS (1988) Fresenius Z. Anal. Chem. 330:211
23. Hawthorne SB, Miller DJ (1987) J. Chromatogr. 403:63
24. Dionex Corporation (1993) Elements of supercritical fluid extraction. Salt Lake City
25. Mulcabey LJ, Taylor LT (1992) Anal. Chem. 64:981
26. Thomson CA, Chesney DJ (1991) J. Chromatogr. 543:187
27. Hedrick JL, Taylor LT (1990) J. High Resol. Chromatogr. 13:312
28. Vannoort RW, Chevert JP, Lingeman H, De Jong GJ, Brinkman UAT (1990) J. Chromatogr. 505:45
29. Nielen MWF, Sanderson JT, Frei RW, Brinkman UHT (1989) J. Chromatogr. 474:388
30. Schäfer K, Baumann W (1989) Fresenius Z. Anal. Chem. 332:884
31. Ashraf-Klorassani M, Taylor LT, Zimmerman. P (1990) Anal. Chem. 62:1177
32. Wright BW, Wright CW, Gale RW, Smith RD (1987) Anal. Chem. 59:38
33. Pariente GL, Pentoney SL Jr, Griffiths PR, Shafer KM (1987) Anal. Chem. 59:808
34. Pyle SM, Setty MM (1991) Talanta 38:1125
35. Nair B, Hube JW III (1988) LC-GC 6:1071
36. Maxwell RJ, Hampson JW, Cygnarowicz-Provost M (1991) LC-GC 9:788
37. Wright BW, Fulton JL, Kopriva AJ, Smith RD (1988) In: Charpentier BA, Sevenants MR (eds) Supercritical fluid extraction and chromatography: techniques and application (ACS Symposium Series. No 366). American Chemical Society, Washington, DC, 44

38. Kiebman SA, Levy EJ, Lurcott S, O'Neill S, Guthrie J, Ryan T,Yocklovich S (1989) J. Chromatogr. 27:118
39. Schantz MM, Chesler SN (1986) J. Chromatogr. 363:397
40. Hawthorne SB, Miller DJ (1986) J. Chromatogr. 24:258
41. Wright BW, Wright CW, Gale RW, Smith RD (1987) Anal. Chem. 59:38
42. Wong JM, Li OX, Hammock BD, Seiber JN (1991) J. Agric. Food Chem. 39:1802
43. Sugiyama K, Saito M, Hondo T, Senda M (1985) J. Chromatogr. 332:107
44. Schneiderman MA, Sharma AK, Mahanama KRR, Locke DC (1988) J. Assoc. Off. Anal. Chem. 71:815
45. Ndiomu DP, Simpson CF (1988) Anal. Chim. Acta. 213:237
46. Lohleit M, Hillmann R, Baschmann K (1991) Z. Anal. Chem. 339:470
47. Lavy RJ, Storozynsky E, Raney RM (1991) HRC 14:661
48. Nielen MWF, Sanderson JT, Frei RW, Brinkman UATh (1989) J. Chromatogr. 474:388
49. Nielen MWF, Sanderson JT, Frei RW, Brinkman UATh (1989) J. Chromatogr. 474:388
50. Stahl E (1977) J. Chromatogr. 42:15
51. Stahl E, Schilz W, Schultz E, Willing E (1978) Angew. Chem. Int. Ed. Engl. 17:731
52. Hawthorne SB, Miller DJ (1987) Anal. Chem. 59:1705
53. Raymer JH, Pillizzari ED (1987) Anal. Chem. 59:1043
54. Hawthorne SB, Miller DJ (1987) J. Chromatogr. 403:63
55. Hawthorne SB, Miller DJ, Krieger MS (1989) J. Chromatogr. Sci. 27:347
56. Hawthorne SB, Miller DJ, Langenfeld JJ (1990) J. Chromatogr. Sci. 28:2
57. Mauldin RF, Vienneau JM, Wehry EL, Mamanton G (1990) Talanta. 37:1031
58. Mitra S, Wilson NK (1990) J. Chromatogr. Sci. 28:182
59. Oudsema JW, Poole CF (1992) J. High Resol. Chromatogr. 15:185
60. Oudsema JW, Poole CF (1992) Fresenius J. Anal. Chem. 344:426
61. Sugiyama K, Saito M, Hondo T, Senda M (1985) J. Chromatogr. 332:107
62. Suprex Catalogue and User's Manual
63. Thiebaut D, Cherver JP, Vannoort RW, De Jong GJ, Brinkman YATh, Frei RW (1989) J. Chromatogr. 477:151
64. Raynor MW, Davies IL, Clifford AA, Williams A, Chalmers JW, Cook BW (1988) J. High Resol. Chromatogr. 11:766
65. Yocklovich SG, Watkins JC, Levy EJ (1990) J. Chromatogr. 505:273
66. Unger KK, Roumeliotis P (1983) J. Chromatogr. 282:519
67. Jinno K (1992) Hyphenated techniques in supercritical fluid chromatography and extraction. Elsevier, Amsterdam p. 257
68. Cortes J, Green LS, Campbell RM (1991) Anal. Chem. 63:2719
69. Thirbaut D, Chervet JP, Vannoort RW, De Jong GJ, Brinkman UATh, Frei RW (1989) J. Chromatogr. 477:151
70. Brewster JD, Maxwell RJ, Hampson JW, private communication
71. Liebman SA, Levy EJ, Lurcott S, O'Neill S, Guthrie J, Ryan T Yocklovich S (1989) J. Chromatogr. Sci. 27:118
72. Fjeldsted JC, Lee ML (1982) Anal. Chem. 54:1883
73. Kalinoski HT, Udseth HR, Wright BW, Smith RD (1986) Anal. Chem. 58:2421
74. Smith RD, Udseth HR, Hazlett RN (1985) Fuel 64:810
75. Ikushima Y, Saito N, Hatakeda K, Ito S, Arai M, Arai K (1992) Ind. Eng. Chem. Res. 31:574
76. Kirschner CH, Taylor LT (1993) Anal. Chem. 65:78
77. Lopez-Avila V, Dodhiwala NS, Benedicto J (1991) Evaluation of four supercritical fluid extraction systems for extracting organics from environmental samples. Short course on SFE, New Orleans

5 Analytical Applications of Supercritical Fluid Extraction

5.1 Introduction

The high throughput and versatility of SFE usually leads novices to undertake short, isolated studies rather than systematic development of specific applications, which sooner or later results in the appearance of unexpected problems. Supercritical fluid extraction is a recent, still fairly unexplored technique that must be consolidated with the rigour exercised in developing other analytical methodologies.

In addressing SFE applications for the first time, novices should bear in mind the essential difference between this and other separation techniques. Newcomers should therefore become acquainted with the underlying technology and behaviour of samples in SFs by first searching available literature for the SF to be used in relation to the analytes to be separated or similar substances. Occasionally, the literature on its sister technique, supercritical fluid chromatography, provides information of interest in this respect.

The two chief factors to take into account in addressing an analytical SF extraction process are (a) the analytes to be separated and determined, which make the object of the analysis, and (b) the matrix that contains them. Both the analytes and the matrix, as well as the association between the two can vary widely. The solubility of the analytes in the SF and potential interactions between the matrix and SF are two other relevant considerations.

The next step in developing an SFE application involves determining the extractability of standard solutions of the analytes in straightforward, non-adsorptive matrices such as filter paper, diatomaceous earth or sand. Special care should be exercised to avoid any residual solvent remaining in the matrix after the analyte standard is added so as not to alter the SF properties.

These preliminary extractions provide information on the behaviour of the analytes free from any matrix effects, which are usually rather complicated. If the analytes are not soluble enough over a reasonable SF density range, nor in modified fluids, then SFE is inadvisable for the intended purpose.

Use of simulated sample After the most suitable solvent and working conditions have been chosen, a simulated sample is subjected to SFE in order to identify any coextracted interferences and/or detect any matrix–solute interactions that might delay extraction of the analytes. Matrix effects can usually be overcome by altering the system operating conditions and/or using a modifier.

Finally, the analyte is extracted from the real sample. There are few affordable reference standards for the wide variety of matrices available, and even fewer containing variable amounts of analytes, which occasionally poses virtually unsurmountable problems.

Supercritical fluid extraction is useful for analyte concentrations of a few parts per million to a few parts per billion. If the sample in question contains high concentrations of readily soluble solutes, then a proportionally high extractant flow-rate must be used in order to avoid plugging the restrictor. In this situation, lowering the flow-rate to compensate for the analyte solubility may lead to incomplete extractions. Also, when high preconcentration factors are required (e.g. with analytes at the parts-per-trillion level), one should take into account that coextracted species are concentrated as a result and SF impurities have a greater effect on selectivity as they are gathered in the collection system. **Analyte concentration range**

One frequent mistake in choosing modifiers is relying on previous experience with conventional extraction. There are a host of references to the use of methanol as modifier for CO_2, yet this is no reason to use it in SFE as well since matrix-solvent–analyte interactions are highly influential on the extraction process and must be taken into account in choosing an appropriate SF modifier.

Supercritical fluid post-extraction strategies used after the analyte is collected are analogous to those commonly employed for identification and/or quantification of analytes. If collection is performed on-line with extraction, then the requisites discussed in Sect. 4.6 must be met.

5.2 Variables Affecting Extraction Quality

Analytical laboratory work must be conducted according to pre-established guidelines based on several antagonistic parameters that can be represented at the apices of a tetrahedron (Fig. 5.1), namely: throughput, accuracy, economy and personnel safety/comfort. As a rule, the aim is to expend as short a time as possible in developing the process while achieving a reasonably high accuracy in the results at a moderate cost and with adequate personnel safety. Placing special emphasis in any of these quality parameters introduces an imbalance that can be graphically represented by a distorted tetrahedron where one apex or edge moves away from the others, to their detriment. For example, making a process more expeditious usually results in decreased accuracy, possibly increased costs

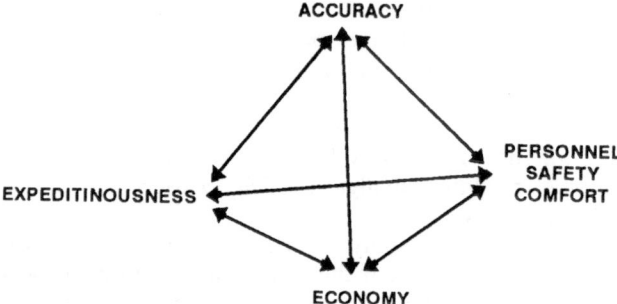

Fig. 5.1. Tetrahedron representing the properties of the analytical process and their relationships

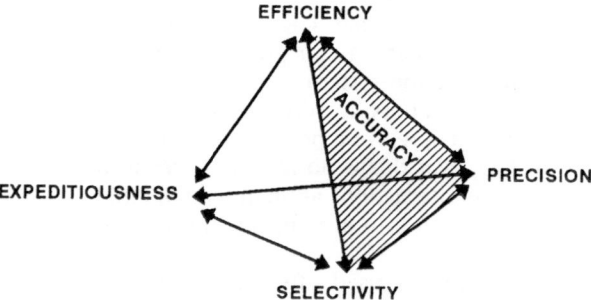

Fig. 5.2. SFE quality parameters and their relationships

and certainly diminished personnel safety and comfort. Likewise, increased accuracy is achieved at the expense of a lower throughput, higher costs and a more laborious process development; on the other hand, reducing costs is detrimental to accuracy and the analyst's safety and comfort. Finally, increased staff safety and comfort call for increased costs and frequently results in diminished throughput and accuracy.

Developing an SFE process entails obtaining data in the form of several parameters that define the goodness of the extraction process and must be optimized for appropriate performance. Such parameters are throughput, efficiency, precision and selectivity. The relationship between the last three determines a fifth, closely related parameter: accuracy. Because of their opposing relationships, accomplishing a balance between them requires a compromise to be made, as shown graphically by the tetrahedron in Fig. 5.2. Shifting the balance to one or more parameters involves detracting from the others.

The wide variety of variables that affect SFE can readily be altered for specific purposes (e.g. achieving rapid, selective, efficient or precise extraction) by optimizing the process for one or more parameters.

The most influential variables on extraction quality parameters are as follows:

a) The supercritical fluid used as extractant, which can be chosen from among a large variety of options, used at variable pressures and temperatures (i.e. densities), as well as flow-rates, and modified with suitable agents in order to alter its polarity, all of which determine its solvent properties.
b) The extraction time, which, once the extractant is chosen, allows such parameters as throughput and efficiency to be altered at will.
c) The particle size, amount of sample, pore size and unknown material thickness, on which efficient extraction in a reasonably short time relies heavily. The size of heterogeneous samples can have a decisive effect on reproducibility.
d) The nature of the matrix, which influences the selectivity, quantitativeness and analyte separation rate achieved in the extraction process through matrix–analyte interactions.
e) The volume and dimensions of the extraction chamber and hence the time required to achieve a preset efficiency.

The kinetics of extraction is ultimately the key to performance in SFE; it is determined by the above-mentioned variables and conditions quality parameters, all of which depend on it to a greater or lesser extent. In fact, most of the solute is extracted over a fairly short time at the beginning of the process, after which the extraction rate falls sharply; thus, accomplishing 99% extraction typically takes ten times longer than extracting 50% of the analyte. This often calls for a compromise between efficiency and throughput to be made.

The typical steps ivolved in an extraction process (separation of the analyte from the matrix surface, dissolution of the analyte in the fluid and transfer of dissolved substances to the bulk fluid) call for a thorough optimization study of the variables affecting each in order to achieve the best possible quality parameters in accordance with the pursued objective.

The simplex method makes an excellent choice for optimization of SFE **Simplex method** as it allows the response function to be adjusted in order to favour one or more quality parameters (or all at once). However, the large number of variables involved, most of which are unrelated, make it difficult to apply. In any case, one can always choose those variables that have the greater effects or are more closely related. Table 5.1 shows the results obtained in a study aimed at minimizing the extraction time and maximizing the extraction efficiency of oils from rubber by using the sample size and

Table 5.1. Use of the simplex method for optimizing several parameters affecting SFE recovery

Point	Sample size (g)	Extraction time (min)	Extraction T (°C)	Recovery (% w/w)
1	1.5	120	75	21.6
2	0.5	60	75	22.4
3	1.5	90	125	20.6
4	0.5	120	125	26.0
Ideal	0.5	120	75	29.5

extraction time and temperature as variables, and the percent recovery as response function [1].

5.2.1 Nature and Composition of the Extractant

The type of SF used is central to the development of extraction and should be chosen in accordance with the properties of the analytes to be extracted without disregarding the nature of the sample matrix. Figure 5.3 is highly illustrative of the strong influence of the supercritical fluid. It shows the results obtained in the extraction of soyabean sediments with SC CO_2 and N_2O under identical conditions; as can be seen, N_2O resulted in more efficient extraction in a shorter time and with substantially lower consumption [2].

The numerical results given in Table 5.2 were obtained in the SFE/GC/MS analysis of PAHs in a marine reference material (sediments) by using supercritical CO_2 and N_2O as extractant at 400 atm and 50 °C for 10 min, samples (30 mg) being analysed in triplicate [3]. Both extractants provided results that were consistent with the certified values for PAHs with molecular weights of 178–202. On the other hand, PAHs of higher MWs (228,

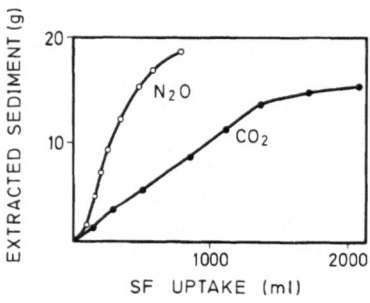

Fig. 5.3. Influence of the nature of the SF and its consumption on the extraction at 150 atm and 50 °C of soyabean sediments

Table 5.2. Influence of the type of SF used on the extraction–determination of PAHs in marine sediments by coupled SFE/GC/MS

PAH	MW	Concentration (μg/g) ± sd		
		MCR[a]	SC CO₂	SC N₂O
Phenanthrene	178	0.58 ± 0.06	0.59 ± 0.15	0.54 ± 0.08
Anthracene	178	0.20 ± 0.04	0.20 ± 0.05	0.22 ± 0.04
Fluoranthene	202	1.22 ± 0.24	1.37 ± 0.26	1.45 ± 0.14
Pyrene	202	1.08 ± 0.20	1.29 ± 0.20	1.31 ± 0.16
Benz[*a*]anthracene	228	0.55 ± 0.08	0.38 ± 0.01	0.60 ± 0.08
Chrysene	228	0.67[b]	0.46 ± 0.07	0.69 ± 0.12
Benzo[*b*] and benzo[*k*]fluoranthene	252	1.22[c]	0.74 ± 0.09	1.38 ± 0.15
Indeno[1,2,3-*cd*]pyrene	276	0.57 ± 0.04	0.12 ± 0.03	0.56 ± 0.08
Benzo[*ghi*]perylene	276	0.52 ± 0.08	0.15 ± 0.05	0.56 ± 0.09

[a] NIST certified reference material.
[b] Uncertified value.
[c] The two compounds are chromatographically indistinguishable.

252 and 276) were recovered by only 70, 60 and 25% with CO₂ in 10 min, even though recovery could be made quantitative by extending the extraction time. By contrast, SC N₂O provided quantitative recovery of all the PAHs assayed within 10 min.

The presence of a modifier can significantly alter the solvent power of an SF; in fact, modifiers can be used in a variety of types and proportions to suit the SF polarity to that of the analyte, as well as to favour displacement of the analytes from the matrix active sites. As can be seen in Fig. 5.4, on constancy of all other variables, the nature of the modifier (Fig. 5.4A) and its proportion (Figs 5.4A and B) can decisively influence the extraction process [4]. A modifier usually acts by reducing the extraction time and hence SF consumption. Quantitative extraction of benzocaine from pharmaceutical preparations (tablets) over a given time interval requires using SC CO₂ with a density of 0.65 g/ml; adding a mere 150 μl of methanol to the sample lowers the required density to only 0.35 g/ml. While the need for a cosolvent and its nature should be checked in every specific application, the properties of the matrix and analyte involved are final in choosing a modifier. For example, the extraction of benzocaine from tablets is inhibited by the presence of water, which, however, substantially enhances the extraction of caffeine from coffee beans [5].

Modifiers alter solvent power

5.2.2 Pressure

The SF pressure in the extraction chamber has a decisive effect on efficiency through the SFE density, as discussed in Sect. 3.6.1. The bar

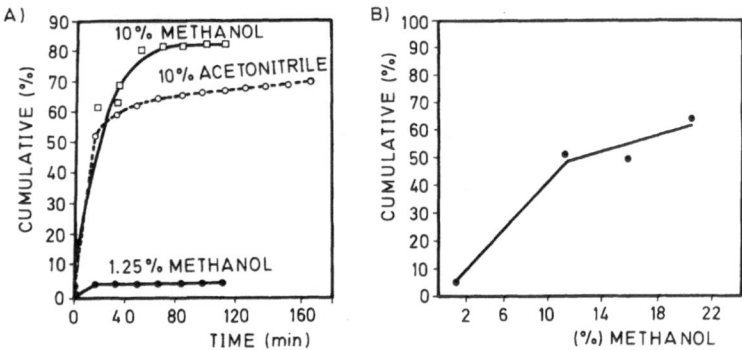

Fig. 5.4A,B. Influence of the type (**A**) and proportion (**A,B**) of modifier used on the extraction at 340 bar and 100 °C of diuron from soils. (Reproduced with permission of Elsevier Science)

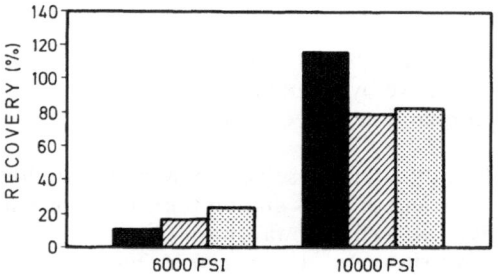

Fig. 5.5. Influence of the SF pressure on the extraction of PAHs, HC and PCBs from oyster tissue

graph in Fig. 5.5 illustrates the effect of pressure on the extraction of various environmental organic pollutants (PAHs, CHs and PCBs) in oyster tissue; as can be seen, the useful pressure range of a given extractor can be crucial to its performance. The behaviour of an extraction system as a function of pressure is quite similar whatever the SF used as extractant (Fig. 5.6A) [6]; this must therefore be taken into account economy-wise, particularly in routine analyses. In fact, at a given SF uptake, the extraction efficiency can be doubled or trebled by using an appropriate pressure (Fig. 5.6B) [7]. However, the pressure influences other quality parameters including selectivity, on which it can have a marked effect (see Fig. 5.7, which illustrates the extraction of PAHs adsorbed on deactivated glass beads with CO_2 at several pressures, with separation–determination by capillary GC [7]).

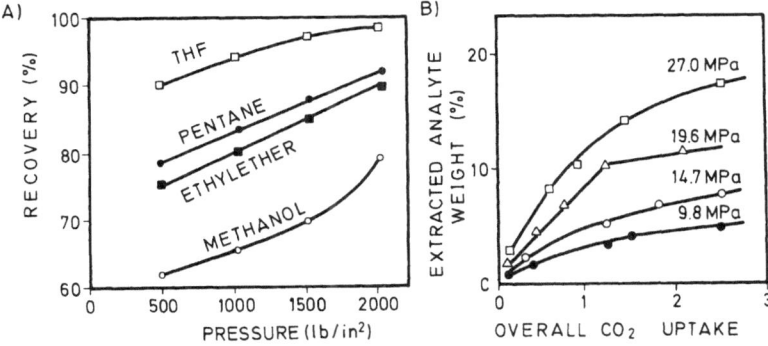

Fig. 5.6A,B. Influence of pressure on SFE. **A** Effect of the type of SF used on the extraction of phenanthrene from pitch at a constant temperature of 240 °C. **B** Joint influence of pressure and the CO_2 uptake. (Reproduced with permission of the Royal Society of Chemistry and Elsevier Science)

5.2.3 Temperature

Changing the temperature of the extraction chamber has a triple effect as it alters the analyte volatility, extraction kinetics and SF density. These effects can be of the same or opposite sign since, while the analyte volatility and extraction kinetics are usually favoured by increased temperatures, the resulting decreased density gives rise to varying effects of the previous two factors on the extraction process.

The temperature within the extraction chamber determines both the efficiency achieved over a given extraction interval and that reached at equilibrium (Fig. 5.8A). Operational costs in terms of SF consumption can be more than halved by raising the temperature by 20 °C (Fig. 5.8B) irrespective of the extractant used (Fig. 5.8C). The most drastic conditions one can use do not always provide the best results, as can be seen in Fig. 5.8D for the extraction of mebeverine alcohol from plasma samples adsorbed in C_{18} cartridges. Again, the properties of the matrix and analyte are clearly seen to allow for optimal extraction under fairly mild conditions.

5.2.4 Flow-Rate

As can be seen in Fig. 5.9A, the percent recovery of diuron from soils increases on doubling the SF flow-rate from 2.5 to 5.0 ml/min, while a further increase to 7.5 ml/min has a dramatic adverse effect on the efficiency. This trend can be ascribed to the pressure drop along the

Adverse effect of pressure drop

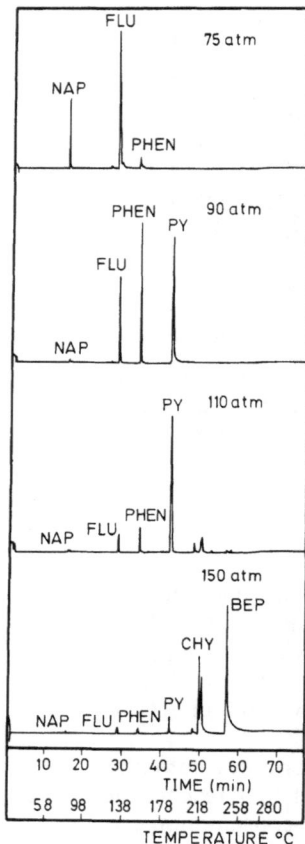

TEMPERATURE °C

Fig. 5.7. Influence of pressure on selectivity in the extraction of PAHs retained on deactivated glass. The determination was carried out off-line with gas chromatography and flame ionization detection. NAP = naphthalene; FLU = fluorene; PHEN = phenanthrene; PY = pyrene; CHY = chrysene; BEP = benzo(e)pyrene. (Reproduced with permission of the American Chemical Society)

extraction cell, which varies (increases) markedly with the flow-rate (Fig. 5.9B). At a constant temperature of 75 °C, such a drop is of 100, 400 and 1000 psi at 2.5, 5.0 and 7.5 ml/min, respectively. Maintaining the increase in efficiency with increase in the flow-rate requires minimizing pressure drops in the extraction cell, which in turn calls for a deep knowledge of their behaviour.

5.2.5 Extraction Time

The time during which extraction is allowed to proceed is final to achieving adequate efficiency, but also has a strong effect on selectivity. The SF solvent power and the variability of analyte–matrix interactions results in

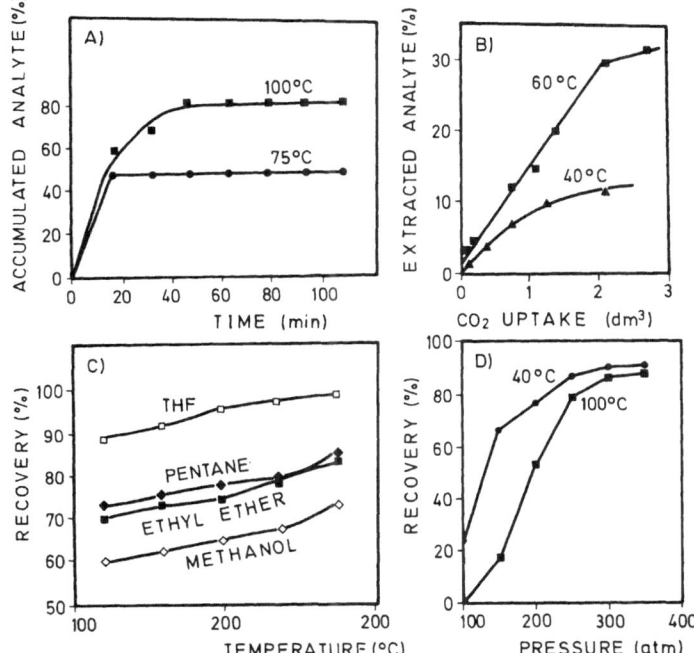

Fig. 5.8A–D. Influence of temperature on SFE. **A** Kinetics of extraction of diuron from soils at two different temperatures and a constant pressure of 340 bar using 9:1 CO_2/methanol as SF. **B** Variation of the amount of tristearin extracted at a constant pressure of 19.6 MPa and two different temperatures with the CO_2 uptake. **C** Variation of the amount of phenanthrene extracted from pitch with various SFs as a function of temperature. **D** Variation of the amount of MEBOH extracted from plasma retained in C_{18} cartridges by using 95:5 CO_2/methanol as a function of pressure and temperature. (Reproduced with permission of Elsevier Science)

widely variable temporal limits for SFE, which obviously influence the throughput and analytical costs involved. Table 5.3 shows a broad picture of the influence of the extraction time on the recovery of 4-nitrobiphenyl, 2-nitrofluorene and fluoranthene retained in XAD-4 resin using SC CO_2 containing various modifiers; the figure reflects the influence of three major variables of the extraction process, namely: the nature and proportion of modifier, the extraction time and the analyte properties [10].

Dynamic extractions require considering one further temporal factor: the equilibration time or interval preceding extraction proper over which the sample is in contact with the SF under the extraction conditions.

The extraction efficiency is considerably affected by the equilibration time, as shown in Table 5.4 for 1,4-naphthoquinone supported on filter

Equilibration time

A)

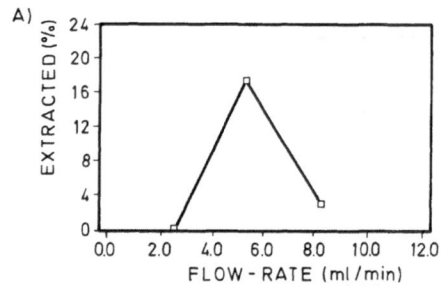

B)

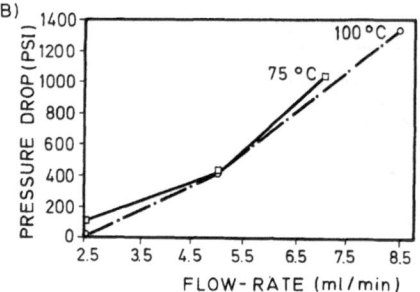

Fig. 5.9A,B. Influence of the flow-rate on SFE. **A** Extraction of diuron from soils with 9:1 CO_2/methanol at 350 bar and 75 °C. **B** Variation of the pressure drop along the extraction tube as a function of the flow-rate at two different temperatures. (Reproduced with permission of the American Chemical Society)

paper. Under given working conditions, the percent analyte recovery is roughly doubled on changing the equilibration time from 5 to 20 min [11].

5.2.6 Sample and Analyte Properties

The nature of the sample as regards both the matrix and the analyte affect the efficiency of the extraction process and, obviously, its throughput, but also its precision and, naturally, its selectivity.

Analytes added to solid supports may behave very similarly towards extraction. Such is the case with PCBs supported on Tenax, Spherosil, Florisil and RP-18, all of which provide very similar recoveries by extraction with SC CO_2 at 20 MPa and 50 °C for 16 min (Table 5.5). The working conditions were strong enough for any differences in the behaviour of the analytes arising from their MWs to be cancelled [12]. In other instances, the nature of the solid support is decisive, as with PAHs adsorbed on glass wool, filter paper or active carbon. The last was found to retain the analytes very strongly (Fig. 5.10) since extraction with SC CO_2 at 300 atm for 15 min resulted in only 20% recovery over the temperature range assayed (40–100 °C); on the other hand, filter paper and glass wool only hindered or prevented quantitative extraction of naphthalene [13].

Table 5.3. Influence of the nature and proportion of modifier, extraction time and type of analyte on the SFE of aromatic compounds retained in XAD-4 resin

Extraction time (min)	CO_2	CO_2 modifier				
		6% methane[a]	6% acetone[a]	6% EtOAc	6% n-hexane	12% n-hexane
15	b	0.9	b	5.5	12.5	11.9
30	0.1	1.5	b	15.1	37.3	37.3
60	0.4	59.6	25.1	36.0	78.6	74.5
90	8.7	78.6	74.2	76.4	81.5	76.4
120	49.6	84.5	70.6	55.7	92.1	74.7
180	83.9	84.6	70.2	78.4	73.2	56.1
Remaining[c] in XAD-4	1.5	b	0.9	b	0.2	b
15	d	0.4	d	4.0	d	1.2
30	0.1	0.6	0.2	7.6	1.1	20.2
60	0.1	3.8	40.4	7.5	18.0	40.3
90	0.2	25.2	59.4	41.7	42.7	67.2
120	0.5	65.3	62.6	48.1	70.9	71.2
180	11.5	79.7	63.9	64.6	65.0	59.3
Remaining[c] in XAD-4	52.5	0.6	d	0.9	1.0	0.1
15	e	e	e	e	e	e
30	e	e	e	e	e	e
60	e	e	e	e	e	e
90	e	e	4.7	3.1	12.7	25.9
120	e	e	31.7	20.7	27.0	66.2
180	e	e	49.1	34.8	63.0	61.1
210	e	e	48.6	42.3	ND[f]	ND[f]
Remaining[c] in XAD-4	e	e	6.3	29	11.3	e

[a] 50 mg added.
[b] Below the detection limit (0.09%).
[c] Conventional extraction with ethyl acetate.
[d] Below the detection limit (0.06%).
[e] Below the detection limit (4.5%).
[f] Not determined.

Based on the previous example, some adsorbents can be used to **Adsorbent** selectively immobilize certain interfering species present in a complex **effects** liquid sample prior to the analytical determination, thereby enhancing the overall selectivity of the analytical process. Such adsorbents as Celite, sodium sulphate, magnesium sulphate, Florisil, alumina and polyurethane foams have been used in on-line coupled extraction applications in order to remove interferents and water from various matrices. Analytes that are bonded irreversibly to adsorbents can give rise to spurious results. The

Table 5.4. Effect of the equilibration time on the extraction of 1,4-naphthoquinone on filter paper

Equilibration time[a]	Recovery (%)
5	49
10	71
15	88
20	90

[a] CO_2 at 8000 psi and 60 °C.

Table 5.5. Recovery of PCBs retained on different supports

PCB	Tenax	Spherosil XOA 200	Florisil	RP-18
28	99 ± 2.3	101 ± 14.9	93 ± 13.0	95 ± 8.5
52	98 ± 4.5	100 ± 9.7	102 ± 6.1	97 ± 6.3
101	100 ± 1.1	97 ± 10.4	94 ± 12.1	97 ± 7.7
138	93 ± 11.9	103 ± 8.6	97 ± 9.6	101 ± 1.3
153	98 ± 3.8	105 ± 13.2	94 ± 14.3	102 ± 5.5
180	99 ± 2.3	99 ± 5.5	101 ± 8.7	94 ± 10.8

adsorbent can be used by mixing it with an appropriate amount of sample prior to introduction into a single extraction chamber or in two serially arranged chambers holding the sample and adsorbent prior to transferal to a GC. One typical application of the latter arrangement is the characterization of clean orange juice. In order to cancel the effect of water, which is present in a high proportion in the juice, an extraction cell furnished with an adsorbent molecular sieve is placed between the sample extraction chamber and the GC. In this way, orange juice can be characterized quantitatively since such compounds as limonene, terpenes and sesquiterpenes, which interfere with the separation/determination, are isolated or their appearance in the chromatogram delayed.

A comprehensive study of the influence of the sample matrix on SFE was carried out for the determination of PAHs in urban dust (SRM NBS 1649), fly ash and river sediments using various supercritical fluids (C_2H_6, CO_2 and N_2O) in the presence and absence of methanol as modifier. The results, shown in Fig. 5.11, proved modified N_2O to be the most efficient extractant, followed by modified CO_2 and unmodified N_2O. As far as the matrix is concerned, the reference material was the most readily extractable. On the other hand, the precision achieved was similar in all instances [14].

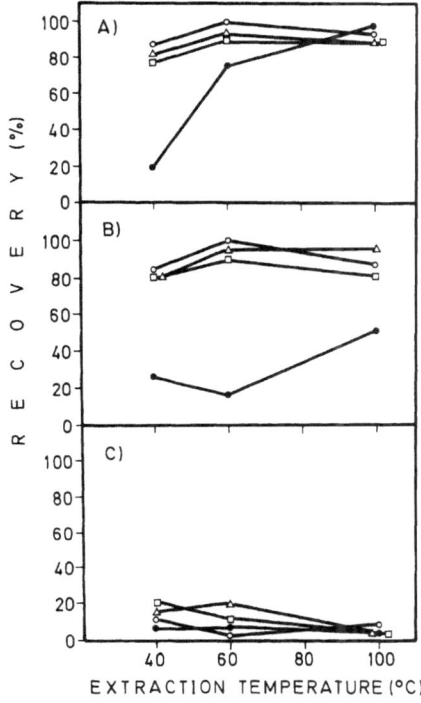

Fig. 5.10A–C. Influence of the support used on the recovery of PAHs retained in (**A**) glass wool, (**B**) filter paper and (**C**) active carbon. (●) Naphthalene; (○) anthracene; (▲) pyrene; (□) 1-nitropyrene

The sample homogeneity is particularly influential on the reproducibility of SFE when small samples are involved. The effect was studied by extracting a polar drug (an aromatic carboxylic acid) from animal feed. Mixing the drug is crystal form with the feed resulted in recoveries between 50 and 85%, with an rsd of 20.6%. On the other hand, a slurry of the drug dissolved in methylene chloride with dried, ground, homogenized feed gave rise to recoveries of around 85% with an rsd of only 2% [15].

Sample homogeneity

The effect of particle size on extraction quality parameters (particularly efficiency and throughput) has been widely studied. The effect arises from the fact that the SF is first brought into contact with the sample surface, from which the analytes are separated, subsequently, the extraction rate falls as a result of the analytes within the matrix having to travel increasingly longer distances to reach its surface and be incorporated into the bulk SF. This behaviour, predicted by Bartle et al. [16], was checked [17] by using various additives adsorbed on PVC and is shown in Fig. 5.12 for the stepwise extraction of the diphenyl ester of carbonic acid from polybutylene terephthalate sheets of variable thickness. The efficiency,

Particle size and extraction quality

A)

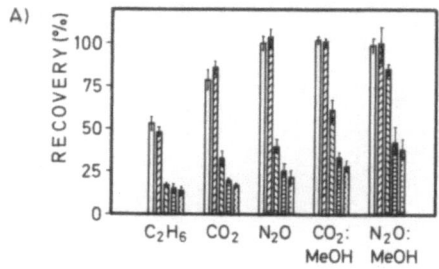

B)

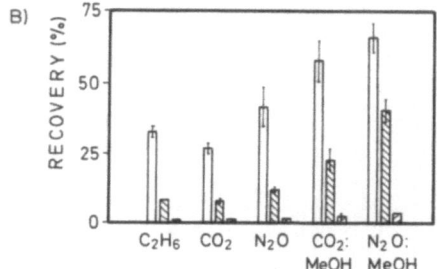

C)

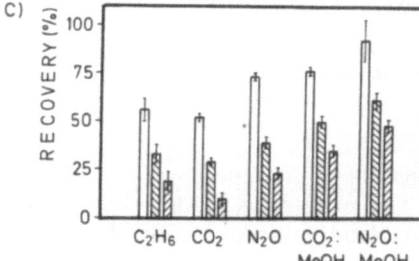

Fig. 5.11A–C. Influence of the sample matrix on the extraction of PAHs with various SFs. **A** Urban dust certified reference material [analytes: ☐ fluoranthene, ◺ benzo[*a*]anthracene, ◿ indeno[1,2,3-*cd*-pyrene, ▤ benzo[*ghi*]perylene]. ▦ **B** Fly ash (analytes: d_{10}-phenanthrene, d_{10}-pyrene, d_{12}-perylene). **C** River sediments (analytes: d_{10}-phenanthrene, d_{10}-pyrene, d_{12}-perylene). (Reproduced with permission of the American Chemical Society)

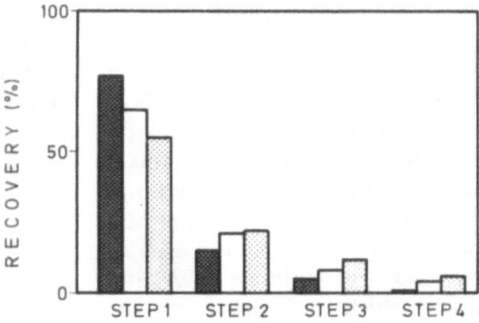

Fig. 5.12. Influence of particle size on the recovery of carbonic acid diphenyl ester from polybutylene terephthalate. (▦) 25 µm. (☐) 50 µm. (▦) 100 µm. (1–4) Sequential extractions. (Reproduced with permission of Pergamon)

Table 5.6. Influence of the cell geometry on the supercritical extraction of PCBs supported on various matrices in terms of percent recovery and standard deviation

PCB	Percent recovery (standard deviation)			
	C_{18}		Florisil	
	1:1	1:11	1:1	1:11
2-Chlorobiphenyl	29.6 (0.5)	29.1 (0.1)	40.4 (1.5)	41.5 (1.6)
2,3-Dichlorobiphenyl	20.2 (0.3)	15.2 (0.7)	39.3 (0.9)	40.1 (1.1)
2,4,5-Trichlorobiphenyl	18.7 (1.2)	12.9 (0.9)	39.2 (0.5)	40.9 (0.7)
2,2',4,4'-Tetrachlorobiphenyl	13.9 (0.2)	11.2 (1.4)	38.9 (1.2)	41.1 (1.1)
2,2',3',4,6-Pentachlorobiphenyl	16.8 (0.5)	9.5 (0.3)	41.4 (1.0)	42.8 (0.4)
2,2',4,4',5,6'-Hexachlorobiphenyl	10.4 (0.7)	6.6 (0.1)	40.2 (0.8)	41.1 (0.6)

Extraction time = 21 min.

Table 5.7. Influence of the analyte concentration on the SFE of PAHs supported on XAD-4 resin

PAH Added (μg)	Recovery (%)		
	4-Nitrobiphenyl	2-Nitrofluorene	Fluoranthene
50	76.7 ± 8.8	87.0 ± 2.5	70.8 ± 3.3
5	88.5 ± 12.7	92.3 ± 8.3	60.6 ± 5.7

Extraction conditions: 6000 psi, 180 min, 88:12 CO_2/n-hexane.

referred to the first extraction, was 22% higher for a 25-μm thick than for a 100-μm thick support [18].

The effect of the analytes in terms of their nature and concentration, and the nature of the sample, whether real or spiked, must be borne in mind in designing SFE applications. When the analytes to be extracted from a given matrix differ in their extractive properties, the working conditions to be used will be a compromise favouring the quality parameters of interest. Table 5.6 lists the average recoveries of several PCBs adsorbed on C_{18} and Florisil achieved by using cells of the same volume and different dimensions [19].

The analyte concentration can affect the extraction efficiency if it exceeds or is close to the SF saturation limit. On the other hand, non-quantitativeness due to increased analyte concentrations may not be the result of extraction proper, but of saturation of the liquid or solid sorbent used for collection. By way of example, Table 5.7 shows the efficiency lost on increasing the PAH concentration added to an XAD-4 resin at two different concentrations [10]. .

Comparative studies on real vs synthetic samples of semi-volatile organic compounds on the EPA list revealed that, for sufficiently soluble analytes, recoveries of added analytes were excellent, whereas those from real samples under the same working conditions were considerably lower since analytes in environmental matrices are much more strongly bound to them. For example, most PAHs are readily soluble in pure SC CO_2; however, their quantitative extraction from real samples requires the use of a modifier or a supercritical fluid interacting with the matrix active sites in a more efficient way [20].

Conditions for quantitative results   Accomplishing quantitative results in the extraction of native analytes thus calls for a balanced combination of physical, chemical and physico–chemical factors. Thus: (a) the analyte should be readily soluble in the SF; (b) the SF should be able to overcome matrix–analyte interactions, thereby favouring displacement of the analytes from the matrix active sites; and (c) the kinetics of analyte desorption should be fast enough for extraction to be accomplished in a reasonably short time [21].

5.2.7 Collection Systems

Unsatisfactory results in the overall analytical process may arise not from extraction proper but from improper collection of extracted analytes. Such is the case with the extraction/separation/determination of oils from citrus fruit by SFE/SFC/FID using a cryogenic trap for collection. A trap temperature of $-10\,°C$ resulted in substantial analyte losses relative to $-65\,°C$ (Fig. 5.13) [23].

5.2.8 In-situ Derivatization

Implementing a chemical reaction in the extraction cell can be aimed at one or more of the following goals: (a) extending application of SFE to polar or even ionic analytes; (b) facilitating the subsequent separation/determination step (e.g. by enhancing chromatographic resolution or the detector response); and (c) improving extraction quality parameters.

In situ derivatization of polar analytes such as fatty acids in *Escherichia coli* in order to improve their extraction and facilitate their subsequent separation by GC was recently reported. The derivatization/extraction is performed in two steps. Chemical derivatization is first carried out in a static fashion and is followed by extraction of the derivatized analytes under a dynamic regime. After separation and collection in 3 ml of methanol or methylene chloride, the reaction products (methyl esters) are

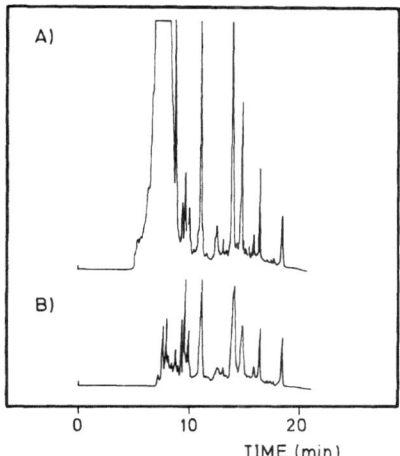

Fig. 5.13A,B. Effect of the trap temperature on the collection efficiency of citrus fruit oil for SFC determination with a flame ionization detector. **A** −65°C. **B** −10°C. (Reproduced with permission of Preston Publications)

injected into a capillary GC equipped with a flame ionization and an electron capture detector. The result thus obtained were compared with those achieved by manual extraction with liquid solvents, concentration of the organic phase, drying in an N_2 stream and derivatization of the analytes prior to injection into a GC. As can be seen from Table 5.8, the two sets of results were quite consistent and the precision, as rsd, quite similar [23]. A comparison of SF extraction with and without in situ derivatization revealed the conversion reaction to facilitate extraction of polar species and provide extracts suitable for analysis by capillary GC with no additional sample preparation. Figure 5.14 shows the enhanced efficiency achieved in the extraction of tocopherols from soyabean sludge on esterification with ethanol [2].

2,5-Dichlorophenoxyacetic acid has been derivatized in situ by silanization and methyl esterification, and subjected to ion-pairing and ionic displacement in order to enhance its recovery from soils. A comparison of the CO_2 extraction results obtained for the different derivatives and with those provided by conventional Soxhlet extraction revealed methyl esterification and ionic displacement during SFE to be the most effective for quantitative extraction of the acid [24].

The two chief goals of in situ derivatization (viz. enhanced extraction and separation/determination) can be achieved to a greater extent by choosing the most suitable reaction for the purpose. Thus, the determination of sulphonated aliphatic (SA) and aromatic surfactants (AS) in sewage sludge was carried out with quantitative (>90%) extraction of the analytes from the sludge as their tetrabutylammonium ion-pairs, which formed butyl esters also quantitatively at the injection port of a gas

Table 5.8. Effect of in situ derivatization/SFE of fatty acids from phospholipids. Comparison with the conventional method

Fatty acid	Concentration (µg/dry g)	
	Conventional method	Derivatization/SFE
Tetradecanoic acid	5.2 ± 0.1	5.9 ± 0.6
Pentadecanoic acid	0.6 ± 0.0	0.6 ± 0.1
cis-Evadex-9-enoic acid	0.7 ± 0.1	1.1 ± 0.1
Hexadecanoic acid	40.3 ± 1.0	45.1 ± 4.4
Hexylcyclopropanecetanoic acid	23.5 ± 0.7	25.0 ± 2.1
cis-Octadec-11-enoic acid	1.9 ± 0.1	2.3 ± 0.1
Octadecanoic acid	0.5 ± 0.1	0.6 ± 0.1
2-Hexylcyclopropanedecanoic acid	15.0 ± 0.4	14.4 ± 1.0

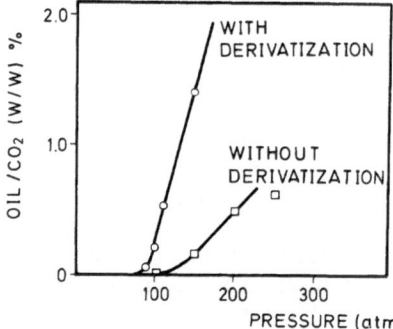

Fig. 5.14. Influence of in situ esterification on the extraction of tocopherols from soyabean sludge with SC CO_2

chromatograph and were determined by GC/MS with no class fractionation of the extracts [25].

In any case, the best example of how SFE can be extended to polar **Metal ion** compounds is the extraction of metal ions. Direct extraction of metal ions **extraction** by SC CO_2 is known to be highly inefficient because of the charge to be neutralized and the weak solute–solvent interactions involved. However, if the metal ions are bound to organic ligands, their solubility in SC CO_2 can be significantly increased. Recently, Laintz et al. showed metal–diethyl dithiocarbamate (DDC) complexes to be sparsely soluble in supercritical CO_2 [26,27]; however, if fluorine is replaced with hydrogen in the ligand, as in bis(trifluoroethyl)dithiocarbamate (FDDC), the resulting metal–FDDC complexes exhibit significantly high solubilities in the SF. Based on these results, Laintz et al. developed an experimental approach to the extraction of Cu(II) from aqueous solutions and silica surfaces using SC CO_2 containing LiFCCD as extractant. The SF is passed through solid

LiFDDC contained in a 10-ml high-pressure stainless steel cylinder which functions as the ligand extraction vessel. The fluid phase containing the dissolved FDDC is subsequently introduced into a high-pressure stainless steel cell holding either the aqueous Cu(II) solution or solid $Cu(NO_3)$ adsorbed on silica. The extraction cell is fitted with 1.27–cm quartz windows. Mixing of the two phases is accomplished by stirring with a Teflon-coated stir bar driven by a magnetic motor. The system portion that includes the view cell and stirrer motor is accommodated in a UV-visible spectrophotometer equipped with a data station to monitor the formation of $Cu(FDDC)_2$ in the supercritical CO_2 phase [28]. The results obtained in the study open up interesting prospects for supercritical fluid extraction of ionic species.

5.2.9 Other Factors

That are some factors extraneous to the extractor that affect the features of the overall analytical process by acting on extraction proper or a subsequent step.

Thus, ultrasonication increases the extraction efficiency. This type of energy can be directly applied to the sample – whether as such or with the modifier – or be introduced during extraction (e.g. in the equilibration and/or extraction step). The favourable effects of ultrasounds on extraction are ascribed to the following phenomena:

Ultrasonication

a) The formation of sinusoidal waves that give rise to marked density and pressure changes which favour mass transfer.
b) Density waves probably induce solute convection from within the material pores as they are much smaller in the matrix than in the bulk fluid.
c) The SF acoustic flux is believed to diminish the resistance to external mass transfer.
d) Ultrasounds probably increase the desorption rate via a local heating effect.
e) Porous materials undergo changes in their internal structure that facilitate extraction.

Ultrasounds increase the rate of extraction of chrysene from various adsorbents and that of caffeine from coffee beans. This source of external energy appears to be especially useful with small extraction cells, which preclude use of conventional stirring procedures.

Such typically analytical parameters as sensitivity and selectivity can be improved in the steps following extraction, whether by isolating the analytes from coextracted species or by removing the latter. Occasionally, the detector itself provides the required selectivity and/or sensitivity. Such

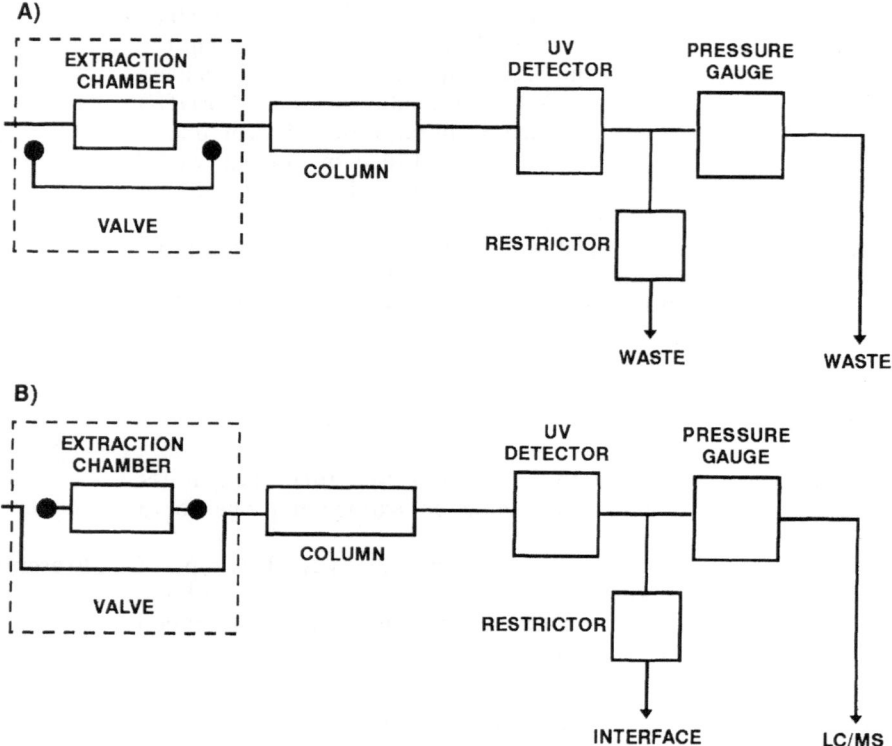

Fig. 5.15A,B. On-line clean-up of drug extracts from animal tissues prior to introduction into a mass spectrometer. **A** Extraction step. **B** Measuring step. (Reproduced with permission of Elsevier Science)

is the case with the SF extraction of animal tissues for isolation of drugs, which results in the simultaneous withdrawal of non-polar endogenous compounds. Using a mass spectrometer coupled on-line with the SF extractor is out of the question as the detector would be rapidly contaminated. The problem can be avoided by using coupled SFE/SFC in order to enhance the cleanliness of the flow reaching the spectrometer taking advantage of the fact that most of the drugs are non-polar, whereas most of the coextracted material is polar. A polar amino column is used for separation. During extraction with unmodified CO_2, extracted drugs are strongly retained at the column head, whereas most of the extracted endogenous substances are rapidly eluted and wasted. The more polar drugs and endogenous substances are eluted by adding a modifier to the mobile phase. Figure 5.15 shows a scheme of the coupled system used.

The UV detector employed prior to the switching valve (SV) is intended to sense species in the fluid in order to drive them to waste or the MS depending on their retention times. [29].

5.2.10 Quality Parameters

Notwithstanding the many factors affecting SFE quality parameters (Fig. 2) whose influence is known and was discussed in the previous sections, there are still some gaps in the knowlege of this separation technique owing to the lack of systematic research on all possible influential factors; as a result, the unsatisfactory or unexpected performance of some SFE systems cannot be justified in clear terms.

This lack of knowledge on each and every factor has given rise to widely disparate precision results which range from fully acceptable in terms of recovery and reproducibility in the analysis of mebeverine alcohol (MEBOH) added to plasma samples (Table 5.9) or PCBs on various sorbents (Table 5.6) to rsd values as high as 50–70% for PAHs in reference soil samples (Table 5.10). However, intermediate precisions are the norm, with medium precision and disperse values for some analytes. This reveals the existence of unknown – and hence uncontrolled – influential factors, as reflected in the recoveries of PAHs added to soils (see Table 5.11). Taking into account that the imprecision of an overall analytical process is the summation of those of all the steps involved, it is extremely difficult to allocate a given percent error to extraction irrespective of the other steps that allow quantification and hence calculation of the mean deviation and/or the actual analyte concentration in order to determine the extraction precision and/or accuracy. By way of example, Table 5.12 lists the rsd

Intermediate precision

Table 5.9. Precision and recovery achieved in the extraction/determination of mebeverine alcohol (MEBOH) in plasma

Extraction	MEBOH added (ng/ml)	MEBOH found [(ng/ml) ($\pm$ rsd)][a]	% Recovery [mean (% rsd)]
SLE[c]/SFE[b]	10.0	10.1 $\pm$ 0.1	101.0 (1.0)
	50.0	47.8 $\pm$ 0.1	95.6 (0.2)
	250.0	227.7 $\pm$ 2.0	91.1 (0.9)
SLE[c]	10.0	9.87 $\pm$ 0.12	98.7 (1.2)
	50.0	49.4 $\pm$ 1.8	98.8 (3.6)
	250.0	236.5 $\pm$ 2.3	94.6 (1.0)

[a] Errors are expressed as standard deviations obtained in triplicate measurements.
[b] SFE working conditions: 95:5 CO_2/methanol, 350 atm, 40 °C, 0.3 ml/min, 10 min.
[c] Conventional solid–liquid extraction.

Table 5.10. Efficiency and precision achieved in the SFE of 15 soil pollutants (MRC) in two parallel extraction experiments involving 7 aliquots each

Analyte[a]	Certified value	Experiment 1[b]		Experiment 2[b]	
		Recovery[b] (%)	rsd (%)	Recovery[c] (%)	rsd (%)
Naphthalene	32.4 ± 8.2	40.1	37.4	50.3	12.6
2-Methylnaphthalene	62.1 ± 11.5	56.1	23.7	69.5	16.0
Acenaphthylene	19.1 ± 4.4	53.1	23.9	56.4	25.0
Acenaphthene	632 ± 105	80.5	17.2	87.1	18.9
Dibenzofuran	307 ± 49	52.9	16.9	90.7	18.3
Fluorene	492 ± 78	71.4	18.3	77.3	22.4
Phenanthrene	1618 ± 348	98.2	17.9	90.8	28.8
Anthracene	422 ± 49	65.7	23.6	68.0	30.0
Fluoranthene	1280 ± 220	79.3	26.0	83.0	40.5
Pyrene	1033 ± 289	77.2	20.8	67.1	41.2
Benzo[a]anthracene	252 ± 38	61.8	34.8	58.6	52.7
Chrysene	297 ± 26	60.0	34.7	53.5	53.1
Benzo[b+k]fluoranthene	152 ± 22	47.2	45.3	38.7	65.4
Benzo[a]pyrene	97.2 ± 17.1	30.1	60.8	22.6	73.0
Pentachlorophenol	965 ± 374	137	38.3	71.7	60.0

[a] Soil sample size (MCR): 1 g.
[b] Seven simultaneous extractions with the same extractor.
[c] Relative to certified values.

Table 5.11. Recovery of PAHs added to soils

PAH	Concentration (ppm)	Recovery (%)	rsd (%)
Naphthalene	20	97.2	18.8
Fluorene	20	96.4	4.5
Phenanthrene	20	98.3	6.5
Pyrene	110	99.0	3.4
Chrysene	62	93.7	5.1
Benzo[b]fluoranthene	70	86.7	4.5
Benzo[a]pyrene	43	82.7	4.7
Dibenz[a,h]anthracene	22	84.1	6.5
Coronene	40	84.2	8.8

values obtained in the determination of n-alkanes. The concentration for each alkane was calculated on the basis of SFE/GC/FID determination and split injection of a standard containing n-alkanes in the rage C_6-C_{30}. The data listed in the third column are based on the peak areas obtained by triplicate SFE/GC/FID analysis of 1.30-g samples. Finally, the data in

Table 5.12. Comparison of the precision provided by SFE/GC/FID and conventional extraction plus split GC in the determination of *n*-alkanes in a polluted sediment

n-Alkane	Concentration in sediment (ng/g)	rsd (%)	
		SFE/GC/FID	Conventional extraction
Dodecane (C_{12})	80	1.7	5.6
Tridecane (C_{13})	370	11	10
Tetradecane (C_{14})	820	11	20
Hexadecane (C_{16})	2100	12	11
Octadecane (C_{18})	1400	7.4	8.5
Eicosane (C_{20})	860	1.7	11
Docosane (C_{22})	450	2.6	4.6
Tetracosane (C_{24})	310	3.9	7.4
Octacosane (C_{26})	230	5.3	5.0
Tricosane (C_{30})	140	6.2	2.6

the right-most column were calculated by considering the peak areas obtained in three conventional, split injections of an analyte extract in methylene chloride (manual extraction). Therefore, taking into account the erratic dispersion in the rsd value for each experimental series, it is rather difficult to draw reliable conclusions as regards the precision of the extraction step [31]. Table 5.5 supports this assertion as well.

With the above-mentioned reservations as regards still obscure influential factors, extraction *throughput* and *efficiency* are determined by two types of factors related to experimental variables and the sample/analyte couple. Thus, the experimental variables include (a) the nature of the SF (and modifier) used; (b) the temperature and pressure, which govern the SF density; (c) the SF flow-rate and overall volume used; (d) the extraction time; (e) the extraction cell volume and dimensions; (f) the extraction cell void volume; (g) the sample size to cell volume ratio; and (h) the analyte collection efficiency. On the other hand, the sample- and analyte-related variables that influence throughput and efficiency in the extraction process are as follows: (a) the analyte solubility; (b) the strength with which analytes are retained at the matrix active sites; (c) the SF ability to compete with the analytes for such sites; and (d) the location of the analyte (inside the matrix or at its surface).

Throughput and efficiency

In developing an SFE method, it is often useful to divide the overall process into three steps in order to determine its efficiency. This entails answering three basic questions, namely:

a) How do the analytes partition between the matrix and SF?
b) Once in the SF, are the analytes flushed quantitatively from the extraction cell?

c) Are the analytes quantitatively collected on depressurization?

These three questions can to some extent be answered individually and dictate jointly the optimal procedure to be used for sample pretreatment prior to the SFE proper.

Selectivity   *Selectivity* is the most flexible quality parameter of SFE relative to other separation techniques including liquid–liquid extraction and, especially, those starting with a solid sample. Virtually all of the internal and external parameters of the extraction process can be altered to enhance the overall selectivity of the analytical process. Thus, as regards internal factors:

a) The type and composition of the SF may be chosen to enable or prevent the extraction of some substances in terms of their molecular weight. Also, the need for a modifier is imposed by the polar or non-polar character of the species to be extracted.
b) The analyte solubility can be adjusted via the density, which in turn can be changed via the pressure and temperature and need not be taken to extreme values (see Table 5.3).
c) The equilibration and extraction time are central to selectivity as they determine which species are effectively removed from the matrix (see Tables 5.4 and 5.5, and Fig. 5.11).
d) In situ derivatization may have a great impact on the extraction selectivity [23] since the analyte properties can be altered by choosing an appropriate (bio)chemical reaction in order to establish significant differences with other sample components.

The external factors that can be adjusted to enhance the selectivity of the overall analytical process are smaller in number, though not in significance – probably because their effects are better known. Essentially, there are two such factors, viz. (a) implementation of a separation step after extraction by means of an on-line or off-line coupled apparatus or instrument (usually a column chromatograph), and (b) use of a detector endowing the determinative step with enhanced selectivity to offset any deficiencies of the preliminary steps in this respect.

Selective extraction of a single compound from a complex matrix is hardly achieved unless the analyte in question is rather different from the other sample components. Most often, the sample includes several compound families, i.e. suites of compounds that can be grouped on the basis of their chemical properties. Such is the case with complex environmental samples, which may contain various families of organic substances including PAHs, PCBs, alkanes, etc. There are two different approaches to resolving complex samples (class-selective extractions). One is only used when a single compound family is to be extracted, so optimization is aimed at isolating the series of interest and avoiding extraction of the other components as far as possible. The process is usually conducted in a single step, as in the extraction of PAHs shown in Fig. 5.7, where the selectivity is a function of the SF pressure [7]; that of various organic compounds by use of different sorbents [13]; and that of aromatic amines in the presence of primary and secondary aliphatic amines based on the nature of the SF (CO_2 rather than N_2O, which extracts all of them) [33]. This approach is also applied to the extraction of analytes from a matrix also soluble in the SF, as in the selective extraction of chlorinated organic pesticides in fat using CO_2 at a pressure near that causing the matrix to be dissolved as well (120 atm and 40 °C).

The second approach is used when different compound families need be extracted separately, which is accomplished by employing a sequential procedure described in the following section.

5.3 Sequential Extractions

One of the most salient features of SFs is that their solvent power can be adjusted by altering their density through one of the two variables on which it depends: pressure and temperature. By adjusting these two parameters, using a modifier to change the solvent polarity and choosing an appropriate extraction time, a given sample can be subjected to sequential extractions under the most suitable conditions for each step in order to extract a given compound family. In this way, a series of analytes can be extracted from the same sample by raising the SF solvent power in pressure steps (density changes) or increasing the modifier concentration by using a two-pump system with programmable pressure and solvent delivery.

For example, SC CO_2 fails to extract dioxins from ash; however, it is used to clean up the sample as it removes weakly adsorbed organic matter. Subsequently, 2% methanol is added to the CO_2 to effect extraction of the dioxins [34].

Changing the polarity of a solvent (e.g. pure or methanol-modified CO_2) allows the same sample to be extracted for non-polar (n-alkanes from C_{11} to C_{14}) and polar compounds such as nicotin and dicyclohexylamine in a second extraction in the presence of the modifier [35]. **Changing solvent polarity**

Variations in the solvent power of SFs resulting from density changes have been used both to separate families of compounds of different polarity and to fractionate compounds of a given group according to molecular weight. For example, caffeine can be selectively extracted from coffee beans with SC CO_2 at 40 °C and a density of 0.7 ng/ml, as shown by the band at 101 cm^{-1} in the IR spectrum of the extract (Fig. 5.16A) [10]. Raising the density to 0.8 or 0.9 ng/ml provides an extract that exhibits an IR band at 1745 cm^{-1}, typical of a lipid (Fig. 5.16B).

A)

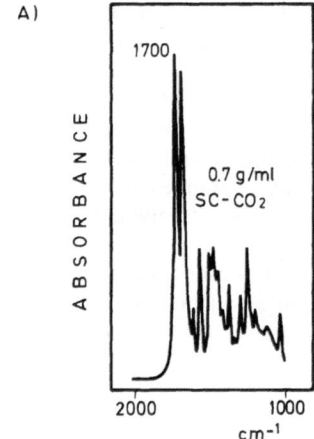

B)

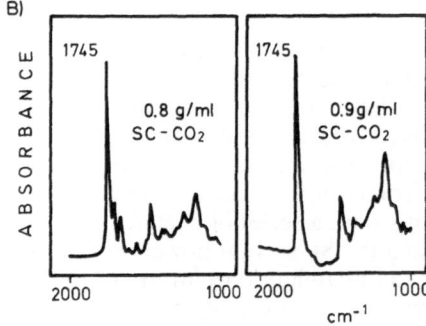

Fig. 5.16A,B. Influence of the SF density on selectivity in the extraction of caffeine from coffee beans. IR spectrum of the extract obtained at an SC CO_2 density of (A) 0.7, (B) 0.8 and 0.9 ng/ml. (Reproduced with permission of the American Chemical Society)

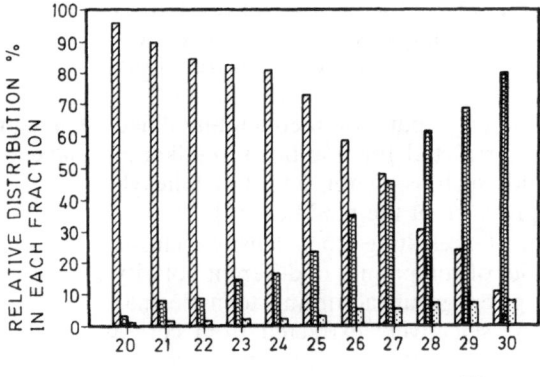

Fig. 5.17. Influence of pressure changes on selectivity in the extraction of *n*-alkanes added to previously extracted aerosol particles. (Reproduced with permission of Springer)

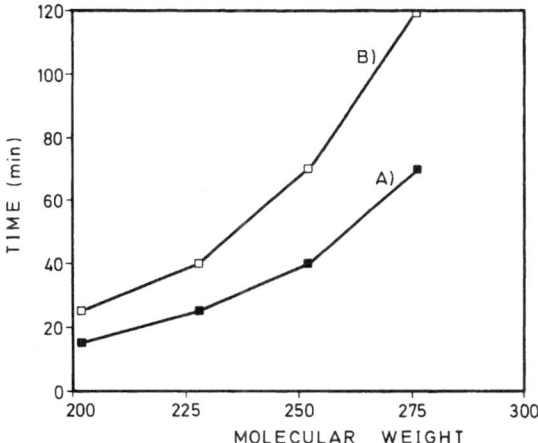

Fig. 5.18. Influence of the molecular weight of PAHs on the time needed to achieve a preset recovery: (A) 50%. (B) 80%

Compounds of a given group in environmental samples can be fractionated by using the sampling–extraction unit described in Sect. 4.3.3.5 (Fig. 4.11). Figure 5.17 illustrates the separation of *n*-alkanes in terms of their MW at three different pressures (9, 15 and 22 MPa). As can be seen, the recovery of the heavier compounds increases as the density is raised [36].

The dependence of selectivity on the analyte molecular weight is reflected in Fig. 5.18, which shows the time needed to achieve 50% and 80% recovery for various PAHs added to a silica sample as a function of their molecular weight [37]. A more comprehensive study performed on 80 PAHs at three different extraction times (3, 5 and 7 h) provided a broad view of their behaviour and the feasibility of their sequential extraction [32].

By simultaneously adjusting the pressure and extraction time one can enhance the selectivity of sequential extractions. Thus, the extraction of *n*-alkanes, PAHs and heteroatomic PAHs adsorbed on polyurethane foam under two sets of working conditions (80 atm for 5 min and 380 atm for 20 min) provided acceptable recoveries of *n*-alkanes in the first extraction which decreased with increase in the molecular weight (Table 5.13). Extraction of PAHs under these conditions was very limited and decreased with increasing MW [38]. Because extraction was quantitative under extreme conditions, selective extraction could be achieved by subjecting the same sample to two extraction steps, as can be seen from Table 5.14 for *n*-alkanes and PAHs extracted from urban dust (1 mg of sample) in

Table 5.13. Supercritical CO_2 extraction of mixtures of *n*-alkanes, PAHs and heteroatomic PAHs in polyurethane foam under two different sets of working conditions

Analyte	Molecular weight	$x \pm sd$ A	B
***n*-Alkanes**			
Dodecane (C_{12})	170	99 ± 6	97 ± 1
Hexadecane (C_{16})	226	102 ± 7	94 ± 5
Eicosane (C_{20})	282	102 ± 5	87 ± 10
Tetracosane (C_{24})	338	102 ± 3	77 ± 6
PAHs			
Naphthalene	128	100 ± 2	52 ± 11
Phenanthrene	178	100 ± 2	9 ± 7
Pyrene	202	103 ± 5	3 ± 4
Chrysene	228	100 ± 7	1 ± 1
Perylene	252	97 ± 6	<1
Heteroatomic PAHs			
Acetophenone	120	97 ± 4	84 ± 8
Quinoline	129	106 ± 6	50 ± 15
Dibenzofuran	168	100 ± 4	37 ± 7
Fluorenone	180	103 ± 6	14 ± 3
Dibenzothiophene	184	101 ± 2	10 ± 3
Carbazole	167	83 ± 5	1 ± 1
Benzo[*b*]naphtho[2,3-*d*]furan	218	100 ± 6	4 ± 3

A: 380 atm for 20 min.
B: 800 atm for 5 min.

two steps: one at 75 atm for 5 min and the other at 300 atm for 15 min. By coupling the separation process to GC/MS these compound families can be fractionated and quantified in less than 2 h, which shows the significant contribution of SFE to expeditiousness in the analytical process – it accelerates this usually lengthy, tedious solid sample treatment step [39].

Extractive fractionation   Extractive fractionation of dissolved analytes adsorbed on a solid support under the dynamic extraction regime with programmed SF density and/or polarity changes is quite similar to supercritical fluid chromatography – particularly if the extractor is connected on-line to a suitable detector. This approach has been used to fractionate families of alkanes, alkenes and aromatic compounds in gasolines by extracting–eluting each family on a silica column and transferring each group of compounds to a capillary GC for isolation of individual components [40,41].

Plastic manufacture   Plastic manufacture entails characterizing additives and oligomers at different stages along the industrial process. Coupled SFE/SFC provides an expeditious, convenient means for selective extraction of both types of compounds and their quantification with sequential extraction in two steps:

Table 5.14. Sequential extraction of *n*-alkanes and PAHs from urban dust into SC NO$_2$ by using coupled SFE/GC/MS

Analyte	Percentage in each fraction	
	Fraction 1 (75 atm)	Fraction 2 (300 atm)
***n*-Alkanes**		
C$_{22}$	95	5
C$_{23}$	94	6
C$_{24}$	96	4
C$_{25}$	92	8
C$_{26}$	89	11
PAHs		
Phenanthrene	38	62
Fluoranthene	36	64
Pyrene	34	66
Benzo[*a*]anthracene	21	79
Benzo[*a*]pyrene	2	98
Indeno[1,2,3-*cd*]pyrene	ND	>95
Benzo[*ghi*]perylene	ND	>95

ND = not detected.

the former (2000 psi, 50 °C, 30 min) extracts additives, whereas the latter, under more drastic pressure conditions (6000 psi, 50 °C, 15 min) extracts oligomers [42].

Selective extraction through fine-tuned working conditions can be achieved in a number of still unexplored ways, so research into this topic is a fairly open field that offers very interesting prospects. Figure 5.19 shows the solubility profile for the extraction of a hypothetical natural product containing various compound families. The three volatility/polarity "cross-sections" in the graph are difficult or even impossible to achieve with liquid solvents.

5.4 General Applications of SFE

While developments in analytical-scale supercritical fluid extraction have seldom been achieved through the rigorous, systematic study that establishing the foundational principles of a new technique requires, research carried out so far has indeed revealed – so instrument manufacturers must believe – the vast potential of this novel technique in the analytical process step calling for the greatest achievements. Far from exhaustive, Tables

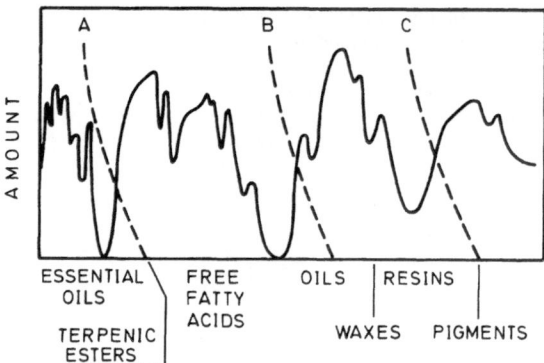

Fig. 5.19. Solubility profile for a hypothetical extraction of natural products. (A–C) Zones resolved by altering the SC CO_2 pressure. (A) 800 psi and 160 °C. (B) 1450 psi and 60 °C. (C) 4350 psi and 60 °C

5.15–5.19 provide a broad picture of selected SFE applications that illustrate the fields, matrices and analytes for which SFE has so far proved to be a useful alternative to solving real problems, and also the gaps to be filled by exploiting the full potential of this technique in order to clearly establish its analytical uses.

The following sections discuss the different types of samples and analytes addressed by SFE as well as its current scope of application, in the light of Tables 5.15–5.19.

5.4.1 Types of Samples

Application to liquid samples

Support for non-solid samples

As can be seen from Tables 5.15–5.19, the samples typically subjected to the extractive action of SFs are mostly – though not exclusively – solid; in fact, SFE has also been directly applied to some liquid samples [92,108, 110,130]. As a rule, samples are introduced in the solid state into SF extraction cells, even though the starting sample can be solid, semi-solid, liquid or gaseous. In the last three cases, the sample is supported on an appropriate material in order to assure effective attack by the SF. Solid supports are not only used for liquid, gaseous or semi-solid samples, though. Most research work conducted so far on solid samples has not involved real samples by synthetic samples prepared from a selected sorbent (a natural matrix where presence of the analytes of interest was previously excluded, or a synthetic support such as polyurethane foam, glass wool, etc.) with which a solution of the analytes is homogenized. Quantitative evaporation of the analyte solvent is compulsory since any

Table 5.15. Environmental applications of SFE

Matrix	Analyte	SF	Modifier	Temperature (°C)	Pressure	Extraction time (min)	Collection	Connection	Detection	Ref.
Ash	p-Dioxins	N_2O, CO_2	2% methanol 5% toluene		350–400 atm	40–120 (static) 10–60 (dynamic)		Off-line	GC/EM	43
Soils	Oils and fats	CO_2		60	250 atm	60	Cryog. trap	Off-line	IR	44
Soils	Pesticides	CO_2		80	7500 psi	20	Cryog. trap	Off-line	Gravimetric	45
Clayey soils	Alkanes (C_{14}–C_{20})	CO_2		90	5352 psi	27		Off-line	GC/FID	46
Water (filter)	Pesticides	CO_2		50	5000 psi	40		Off-line	GC/ECD	47
Marine sediments	PAHs	CO_2	5% methanol	100	5000 psi	120	Gryog. trap	Off-line	GC/FID	48
River sediments	PCBs	CO_2	5% methanol	90	5352 psi	40	Cryog. trap	Off-line	GC/ECD	49
Insoluble deposits	PAHs	CO_2		35	9.6 MPa	20	Cryog. trap	Off-line	Gravimetric, FTIR, GC/MS	50
Coal	Pentane, isopropylene	CO_2		280	10–100 atm		Direct introduction	On-line	MS	51
Clay	Organic pollutants	CO_2		60	400 atm	20	Injection port	On-line	GC/FID	52
Urban dust sediments	PCBs and PAHs	CO_2			345 bar	240	Cryog. trap	Off-line	GC/FID, ECD	53
Soils	Linuron, diuron and metabolites	CO_2	methanol, ethanol, acetonitrile	55–100	200–330 bar	15	Cryog. trap	Off-line	Gel chromatography	54
Tar soils	PAHs	CO_2	5% methanol	50	150–350 atm	60–420	Liquid bubbling	Off-line	GC/MS	32
Air	Volatile toxins	CO_2	6–12% methanol, acetone, ethyl acetate	50	200–6000 psi		Methanol bubbling	Off-line	GC/ECD	20

Table 5.15. *Continued*

Matrix	Analyte	SF	Modifier	Temperature (°C)	Pressure	Extraction time (min)	Collection	Connection	Detection	Ref.
Soils and plants	^{14}C residues in pesticides	Methanol		250	150 bar			Off-line	HPLC/LSCD	55
Smoke, urban dust, river sediments	PAHs, alkanes	CO_2, N_2O		45	75, 300 atm	15	Cryog. trap	On-line	GC/FID, MS	39
Soils, plants	Sulphonyl ureas, herbicides	CO_2		65		<60	Cryog. trap	On-line	SFC/UV, LSCD	56
Soils, plants, smoke particulates	Terpenes, phenols	CO_2		50	25 MPa	12	Cryog. trap	On-line	GC/FID, ECD	57
Gas-oil fuel	PAHs, alkanes	CO_2	5% ethanol	45–65	200–300 atm	30	CH_2Cl_2 bubbling	On-line	GC/MS	58
Sediments	PCBs	CO_2	2% methanol	40	200 atm	25	Cryog. trap	On-line	GC/ECD	59
Cigarette smoke, NBS standards	Nicotin, PAHs	N_2O		45	300 atm	10	Cryog. trap	On-line	GC/FID/MS, ECD	60
Marine sediments in alumina or silica gel	PAHs	NH_3		180			Dicct in detector	On-line	MS	61
Urban dust, ash, rivers sediments	PAHs	CO_2, N_2O, C_2H_6	5% methanol	45, 60	300 atm	30	CH_2Cl_2 bubbling	Off-line	GC/MS	14
Soil	Chlorinated pesticides, PCBs	CO_2				10		Off-line, on-line	GC/ECD, FID	36
Coal ash	Pyrene	CO_2, isobutane	Methanol/ toluene	250	230, 400 atm		Methanol bubbling through pump head	Off-line, on-line	GC/FID	62

Matrix	Analyte	Fluid	Modifier	Temperature (°C)	Pressure	Time (min)	Collection/trap	Mode	Detection	Ref.
Air, gas-oil fuel, cigarette smoke	PAHs, n-alkanes, PCBs	CO_2		45	380 atm	10	CH_2Cl_2 bubbling through pump head	Off-line, on-line	GC/MS	38
Tar, coal	Phenanthrene	Pentane, diethyl ether, methanol THF		120–180	500 lb/in²	15	Cryog. trap	On-line	HPLC/UV	6
Soils	Pesticides (4-nitrophenol and parathion)	CO_2	5% methanol	50	2000 psi	25		Off-line	ELISA, GC/ECD, specific thermionic detector	63
Fly ash	Benzodioxins, benzofurans	N_2O, CO_2	10% benzene	40	400 atm	30–180	n-Hexane bubbling	Off-line	GC/MS, ECD, FID	64
Refinery sludge	PNAs	CO_2		40	350 bar	4.2	Cryog. trap	Off-line	GC/MS	65
Sludge	Resin, fatty acids	CO_2		50	350 bar	15	Cryog. trap	Off-line	GC/ECD	66, 67
Soils	TCPA	CO_2		40–45	100–316 bar	40	Cryog. trap	Off-line	GC/ECD	68
Sludge	Resin, fatty acids	CO_2	1 : 1 methanol/formic acid	80	365 bar	5 (static) 10 (dynamic)	Adsorption on C_{18}	Off-line	GC/ECD	69
Soils	Hydrocarbons	CO_2		60	359 bar	10	Adsorption on C_{18}	Off-line	GC/FID	70
Soils	PCBs, DDT, toxaphene	CO_2		40	100 atm	10	Adsorption on Teflon, Viton, steel	Off-line	GC/MS	71
Soils	Dioxins	CO_2		40	2000–4000 psi	40	Adsorption on Al_2O_3	Off-line	GC/MS	34
River sediments	Herbicides	CO_2		40–80	230 bar	30	Methanol bubbling	Off-line	GC/FID	72
Clay	n-Alkanes	CO_2		80	3200 psi	16.5	n-Hexane bubbling	Off-line	GC/FID	73
Sediments	PCBs	CO_2		80	5880 psi	40	n-Hexane bubbling	Off-line	GC/ECD	73

Table 5.15. *Continued*

Matrix	Analyte	SF	Modifier	Temperature (°C)	Pressure	Extraction time (min)	Collection	Connection	Detection	Ref.
Shale	Alkanes (C_7–C_{35})	CO_2		60	4000 psi	21	n-Hexane bubbling	Off-line	GC/FID	74
Soils	Orgonachlorine pesticides	CO_2		55	5000 psi	15	n-Hexane bubbling	Off-line	GC/FID	74
Soils	PAHs	CO_2		90–120	4980 psi	120 (3 fractions)	n-Hexane bubbling	Off-line	GC/MS	74
Soils	Herbicides	CO_2	5% methanol	60	5000 psi	20	2-Propanol bubbling	Off-line	GC/NPD	74
Soils	Benzofuran	CO_2		40	300 atm	30	n-Hexane bubbling	Off-line	GC/MS	75
Soils, sediments	PCBs, DDT, PAHs	CO_2	10% methanol	60–70	150–300 atm	30–70	n-Hexane, methanol or CH_2O_2 bubbling	Off-line	GC/MS, ECD	76
Soils	DDT	CO_2	5% methanol	40–80	100 atm	5	Acetone bubbling, active carbon impaction	Off-line		77
Turf	Insecticides (chlorpyrifos)	CO_2	2% methanol	100	100–400 atm	30	Cryog. trap	On-line	GC/ECD, UV	78
River sediments	PCBs	CO_2	0.8% toluene	50	300 atm		Cryog. trap	Off-line	GC	79
Resin plug	Gas-oil exhaust	CO_2		45	300 atm	10	Cryog. trap, injection port	On-line	GC/MS	80
Quartz fibre filter	Nicotin, phenols, N-containing organics	CO_2		45	300 atm	10	Cryog. trap, injection port	On-line	GC/MS	80

Matrix	Analytes	Fluid	Modifier	Temp.	Pressure	Time	Collection	Mode	Detection	Ref.
Soil	Herbicides (linuron, diuron)	CO_2	300 µl ethanol	80–120		3–60	High-pressure valve	On-line, off-line	CFS/LSCD	81
Gas-oil-contaminated sediments	PAHs, n-alkanes	CO_2, N_2O		45	400 atm	30	Cryog. trap	On-line	GC/MS	3
Petroleum waste sludge, railroad bed soil	PCBs, PAHs	N_2O, CO_2, $CHClF_2$	Methanol	50–100	100–400 atm	30–45	Liquid bubbling	Off-line	GC/ECD	82
Soil, sand	Pesticides	CO_2	Acetone, ethyl acetate, methanol	50	250 bar	15	Liquid bubbling	Off-line	GC/FID	83
Marine paint	Organotins	CO_2	0.3% formic acid	60	125 atm	25	Pre-column	On-line	SFC/FID	84
Sewage sludge	Aromatic surfactants	CO_2		80	400 atm	5	Liquid bubbling	Off-line	GC/MS	25
Soil	2,4-D	CO_2		80	300–380 atm	30–90	n-Hexane, benzene bubbling	Off-line	ECD	124

THF = tetrahydrofuran; GC = gas chromatography; ELISA = Enzyme linked immunoassay; ECD = electron capture detector.
TCPA = 2,3,5-Trichlorophenoxyacetic acid; FID = flame ionization detector; PCBs = polychlorinated biphenyls; MS = mass spectrometry.
DDT = Dichlorodiphenyltrichloroethane; NPD = nitrogen-phosphorus detector; PAHs = polyaromatic hydrocarbons; UV = ultraviolet.
LSCD = liquid scintillation counting detector; IR = infrared 2,4-D = 2,4-dichlorophenoxyacetic acid; PNAs = polynuclear aromatics.

Table 5.16. Applications of SFE to food analysis

Matrix	Analyte	SF	Modifier	Temperature (°C)	Pressure	Extraction time (min)	Collection	Connection	Detection	Ref.
Dog feed	Fat	CO_2		80	7500 psi	20	Cryog. trap	Off-line	Gravimetric	45
Coffee	Caffeine	CO_2		35	9.6 MPa	20	CH_2Cl_2, bubbling	Off-line	FID	50
Animal feed	Methadone	CO_2		60	8000 psi	20	Adsorption on silica gel	Off-line	HPLC/ECD	11
Edible fats and oils	Peroxides	CO_2	5% ethanol	40	140 kg/cm²	7	Injection port	On-line	SFC/UV	9
Soya sediments	Tocopherol	CO_2, N_2O		50	90–350 atm		Liquid bubbling	Off-line	TLC/FID	2
Soya sediments	Tocopherol	CO_2		35–70	200–400 bar		Liquid bubbling	Off-line	HPLC/UV	85
Olive oil	Acids	CO_2		35	9.6–22 MPa	60	Adsorption in C_{18}	Off-line	GC/FID	86
Sage	Antioxidants (fractionation)	CO_2		60, 100	100–400 bar	64	Adsorption in C_{18}	Off-line	NMR	87
Edible oils	Aromas	CO_2		55	167 bar		Adsorption in C_{18}	Off-line	GC/FID	88
Animal fat	Pesticidas	CO_2		60	340 atm		Solid adsorbent	Off-line	GC/ECD	89
Coffee	Caffeine	CO_2	0.2% water	20–80	100–250 bar	20–80	Cryog. trap	On-line	GC/DAD	90
Lemon	Oils	CO_2		45	90–170 kg/cm²	30	Cryog. trap	Off-line	UV/DAD	91
Egg yolk	Cholesterol	CO_2		45	17.7 MPa	30	Cryog. trap	Off-line	SFC/FID	92
Coffee	Caffeine	CO_2	water	50	350 bar	15	Adsorption in C_{18}	Off-line	GC/ECD	93

Sample	Analyte	Fluid	Modifier	Temp	Pressure	Time	Collection	Mode	Analysis	Ref
Hops	Essential oils and bitter compounds, lipids and waxes	CO_2	6% hexane	50	92 bar	15	Adsorption in C_{18}	Off-line	GC/ECD, HPLC/UV	94
Animal feed	Drugs (aromatic carboxylic acids)	CO_2		50			Adsorption in C_{18} or steel	Off-line	HPLC/UV	15
Cereals	Aflatoxins	CO_2		60	4000 psi	10 (static) 15 (dynamic)	Isopropyl alcohol bubbling	Off-line	HPLC, TLC	74
Foods	Fat	CO_2	methanol	80	7500 psi	12	CH_2Cl_2 bubbling	Off-line	Indirect gravimetry	74
Coffee	Caffeine, acids (C_{16}–C_{18}), triglycerides	CO_2		40–60	10–15 MPa	30	1:1 Isooctane/benzene bubbling	Off-line	GC/MS, SFC, HPLC/MS	95
Spices, bubble gum, orange, pine needles, cedar wood	Aromas and fragrances	CO_2			80–300 atm	10	Cryog. trap, injection port	On-line	GC/FID GC/MS	22
Wheat	Nicotoxins	CO_2		60	100–300		Direct fluid injection	On-line	MS. GC/MS	96
Marjoram, spices	Aromas and fragrances	CO_2		45	300	10	Cryog. trap, injection port	On-line	GC/MS	80
Wheat	Herbicides (linuron, diuron)	CO_2	300 µl ethanol, 100–300 µl methanol	80–120		3–60	High-pressure valve	Off-line, on-line	SFC/LSCD	81
Wheat	Tocopherol	CO_2	ethanol	40	750 bar		Silica column	On-line	SFC/UV	97
Potatoes, almonds	Organotins	CO_2	0.3% formic acid	60	125 atm	25	Pre-column	On-line	SFC/FID	84

ECD = electrochemical detector; SFC = supercritical fluid chromatography; HPLC = high performance liquid chromatography; DAD = diode array detector; NMR = nuclear magnetic resonance.
For other acronyms, see footnote to Table 5.15.

Table 5.17. Industrial applications of SFE

Matrix	Analyte	SF	Modifier	Temperature (°C)	Pressure	Extraction time (min)	Collection	Connection	Detection	Ref.
Polymers	Monomers, oligomers	CO_2	10% methanol	60	7000 psi	20	Cryog. trap	Off-line	Gravimetric	98
Plastics	Additives	CO_2		60	350 atm		Cryog. trap	On-line	SFC/UV	99
Rubber	Additives	CO_2		70	7000 psi	30	Acetone bubbling	Off-line	Gravimetric	100
Polymeric surfactants	Oligoethers	CO_2		35	9.6 MPa	20	Cryog. trap	Off-line	SFC/FTIR, FID	50
PVC	Additives	CO_2	0–15% ethanol	45–95	35–45 MPa	20–230	Cryog. trap	Off-line	SFC/UV, direct gravimetry	17
Petroleum tar	Petroleum fractions	Toluene		320–400	29–76 bar		Cryog. trap	Off-line	Gel chromatography	54
Plastics	Low-MW compounds	CO_2		40–100	360 bar		C_{18} sorbent	Off-line	GC, HPLC, FID	101
Polyethylene	Oligomers	CO_2		70	400 atm	30	CH_2Cl_2 bubbling	Off-line	SFC/FID	102
Polymers	Additives	CO_2		<150	<415 atm		Cryog. trap	On-line	GC/MS	103
Wood pulp	Organics	CO_2		60	329 bar	33	Cryog. trap	Off-line	GC/MS	104
Propellants	Stabilizers	CO_2		50	350 bar	30	C_{18} sorbent	Off-line	GC/FID	105
Paper, wood	Anthraquinone	CO_2		65	8000 psi	20	Silica sorbent	Off-line	HPLC/ECD	106
Polystyrene	Additives	CO_2		70	7000 psi	45	CH_2Cl_2 bubbling	Off-line	GC/FID	74
Polymers	Additives	CO_2		55	200 atm	10	Injection port	On-line	GC/FID	18
Polypropylene	Irganox-1010	CO_2		55	300 bar	10	PTV	On-line	GC/FID	35
Low-density polymers	Additives	CO_2		100	450 atm	10–30	Cryog. trap	On-line	SFC/FID	107
Polymers	Additives, oligomers	CO_2		50	2000, 6000 psi	15	Cryog. trap	On-line	Serial SFC/UV/FID	42

PTV = programmed temperature vaporizer; FTIR = Fourier transform infrared.
For other acronyms, see footnotes to Tables 5.15 and 5.16.

Table 5.18. Biomedical applications of SFE

Matrix	Analyte	SF	Modifier	Temperature (°C)	Pressure	Extraction time (min)	Collection	Connection	Detection	Ref.
Plasma	Drug metabolites	CO_2	5% methanol	40	350 atm	10	2-Propanol bubbling	Off-line	GC/MS	9
Urine	Phenol	CO_2		40			Direct in detector	On-line	UV detector or SFC	108
Enzymes	Enzymes	CO_2	3–6% ethanol 0.1% water	35	200 atm	60				109
Serum	Cholesterol	CO_2		45	17.7 MPa	30	Cryog. trap	Off-line	SFC/FID	92
Animal tissues	Drug residues	CO_2				8.5	Direct	On-line	SFC/MS/MS, on-line UV	30
Plasma/XAD-2	Mitomycin C	CO_2	12% methanol	50–60	28–30 MPa		Cryog. trap	On-line	SFC/ photometric	94
Glass wool	Steroids	CO_2, $CHClF_2$		50	13.8–18 MPa	15–30		Off-line	SFC/FID	110, 111

For acronyms, see footnotes to Tables 5.15–5.17.

Table 5.19. Miscellaneous applications of SFE

Matrix	Analyte	SF	Modifier	Temperature (°C)	Pressure	Extraction time (min)	Collection	Connection	Detection	Ref.
Pharmaceuticals	Drugs	CO_2		50, 60	329, 350	5	Acetonitrile bubbling, solid adsorption	Off-line	HPLC/MS	10
Silica gel/AgNO$_3$	Triglycerides, esters, fatty acids	CO_2	Ethyl acetate	40–60	7.8–27.0 MPa			Off-line	Gravimetric, GC	8
Glass wool	Pesticides	CO_2		40	10–20 MPa		Microcolumn	Off-line	HPLC	112
Tenax, florisil, alumina, coal, C$_{18}$, fly ash	PCBs, chlorinated benzenes, dioxins	CO_2, N_2O		40	3000 psi 6000 psi	15, 90	Solid sorbent	On-line	GC/ECD	113
Water	Phenols	CO_2		40	300 bar		Bubbling, cryogenic T	Off-line	GC/FID/UV	114
Water	Pyrimidine compounds	CO_2	Methanol, tetrabutylammonium bromide	50	5000 psi	15–20	Bubbling, cryog. trap	Off-line	SFC/FID/UV	115
Motor oil	PAHs	CO_2		50	10 MPa				SFC/UV	116
Silica/C$_{18}$	PAHs	CO_2		100	4500 psi		Bubbling	On-line	UV-visible	117
Chewing tobacco	Nicotin	CO_2	1 ml ethanol/ 100 mg sample	60	8000 psi	20	Cryog. trap	Off-line	GC/FID	118
Celite, soils, silica	Amines	CO_2, N_2O		60	400 atm	20	Impact–injection port	On-line	GC/FID/UV, FTIR	33

							Direct	On-line	SFC/IR	119
Water	Phosphonate	CO_2	NaCl	75	350 atm	60				
Plastic film	Wax	CO_2		80	364 bar	30	Stainless steel ball impaction	Off-line	GC/FID	120
Dog collars	Pesticides	CO_2		45	350 bar	10	C_{18} sorbent	Off-line	GC/FID	121
Active carbon	Nitrate esters	CO_2	2% methanol	45			C_{18} sorbent, steel balls	Off-line	GC/FID	122
Tablets	Vitamers	CO_2		60	3000 psi	25	n-Hexane bubbling	Off-line	HPLC	74
Tablets	Benzocaine	CO_2		40	104 bar	20	Adsorption in C_{18}	Off-line	GC/FID	4
Glass beads	PAH	CO_2		50	80–200 atm		Cryog. trap	On-line	GC/FID	4
Silica/C_{18}	Alkanes, nicotin, DCHA	CO_2	1.6% ethanol	45	250 bar	10	PTV	On-line	GC/FID	106
Tenax	HCB, PCBs	CO_2		42	20 MPa	35	Cryog. trap	On-line	GC/UV, ECD	123
Glass wool, active carbon, filter paper	PAHs	CO_2		40–100	100–300 atm		Adsorption on glass wool, $CHCl_3$ bubbling	Off-line	SFC/FID	74
Solid–aqueous	Metal ions	CO_2	LiFDDC	35	79.3 bar	60	In situ monitoring		Photometry	28

DCHA = Dicyclohexylamine; HCB = Hexachlorobenzene; LiFDDC = bis(trifluoroethyl)dithiocarbamate–Li complex. For other acronyms, see captions to Tables 5.15–5.17.

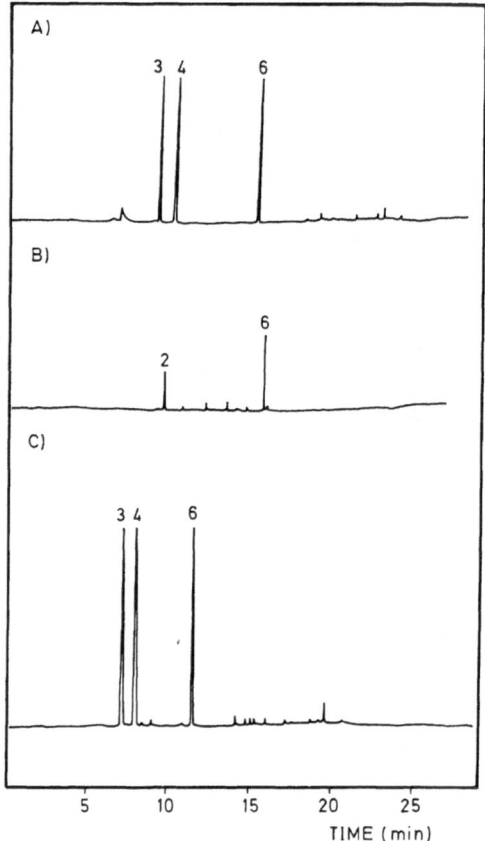

Fig. 5.20A–C. Influence of the type of support on selectivity in the extraction of amines with SC CO_2 under identical working conditions. **A** Celite. **B** Silica. **C** Soil. (Reproduced with permission of the American Chemical Society)

residues may alter the SF polarity and hence its action to an extent dependent on the fluid and solvent properties, and the amount of solvent retained.

The extensive studies performed on the different types of supports used in SFE have shown the great influence of the material employed on the extraction efficiency, selectivity and precision. By way of example, Fig. 5.20 illustrates the behaviour of three different supports (Celite, silica and soil) in the CO_2 extraction of seven amines (heptylamine, N-methylheptylamine, 2-ethylaniline, tributylamine, dodecylamine, diphenylamine and octadecylamine) under identical working conditions. While Celite and the soil allowed 2-ethylaniline, tributylamine and diphenylamine (denoted by numbers 3, 4 and 6, respectively in Fig. 5.20) to be readily desorbed, silica only permitted extraction of ethylaniline and diphenylamine [33]. As

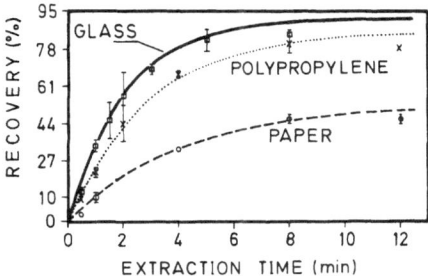

Fig. 5.21. Influence of the type of support on the recovery achieved as a function of time using SC CO_2 at 3000 psi and 60 °C

regards efficiency and precision, Fig. 5.21 shows the recoveries achieved in the extraction of glucose pentaacetate supported on various materials (glass, polypropylene and paper). As can be seen, the more polar the support was, the stronger was the binding to polar analytes and the more difficult their separation.

5.4.1.1 Solid Samples

As noted above, most research work on SFE involving *solid samples* has been conducted on synthetic, spiked samples. These provide an insight into the behaviour of the analyte and matrix that should only be taken as a guide, though. Comparative studies involving real and spiked samples have shown the former to deviate from the behaviour one would expect from the results obtained in preliminary studies from synthetic samples. As a rule, analytes bind more strongly to real samples and/or occur at greater depths within them – occasionally, extractable, unexpected contaminants present in a real sample can alter the SF properties. These factors and others of possibly unknown nature and effects, make predicting the behaviour of a real sample from results provided by a simulated sample impossible, which significantly detracts from the interest of many studies carried out so far in this context. Such studies, though, allow one to draw some general conclusions. In this respect, the three steps involved in the extraction process (desorption, diffusion and dissolution in the SF) contribute to a different extent to determining the rate of the process, which, as can be seen from Table 5.20, is a function of the nature of the extracted material. Thus, desorption is the most troublesome step with environmental samples, while diffusion poses the most serious problems with polymers, and dissolution into the SF is usually the least conflicting of all three steps.

In dealing with real samples it is customary to determine the goodness of the SFE process by comparison with the performance of Soxhlet ex-

Table 5.20. Rate-determining steps of the overall SFE of
various types of sample

Step	Sample		
	Environmental	Polymers	Foods
Desorption	+ +	–	+
Diffusion	+	+ +	+
Dissolution	–	+	+

traction, which is allowed to proceed exhaustively (20 h) in order to ensure quantitativeness and is arbitrarily assigned 100% efficiency [64].

Effect of moisture   The presence or absence of moisture in solid samples is one other major factor to be considered as its effect depends on the nature of the sample and type of analyte involved, and, to a lesser extent, on whether SFE is coupled to another separation or technique potentially affected by this experimental variable. Thus, the presence of moisture is decisive for adequate extraction of caffeine from coffee beans [90]. On the other hand, water poses a serious problem to on-line coupled SFE/GC when the analytes are desorbed directly at the column head since it also deposits in the stationary phase. If this step is unavoidable, water should be carefully removed from moist samples. Studies performed with gas-oil contaminated sediments containing 20% moisture [3] showed that air-drying of the samples at ambient temperature prior to introduction into the SFE/GC system resulted in complete loss of the more volatiles analytes (C_{19}–C_{11} alkanes), which were the most concentrated species in the wet sample. The problem posed by analyte collection coupled on-line with GC can be overcome by using split or splitless sample injection, which in addition avoids the adverse effect of sulphur present in the matrix. One other way of avoiding moisture problems is by mixing the samples with diatomaceous earth, which absorbs the water and increases the contact surface with the SF. This is the usual procedure employed with semi-solid samples (e.g. yoghurt).

Real solid samples are often better mixed with an inert material (e.g. glass wool for chewing gum samples [118], sand or glass wool for egg yolk samples [92], etc.) in order to avoid compaction and increase the contact surface so as to enhance the efficiency of the process.

5.4.1.2 Liquid Samples

Extractions of liquid (usually aqueous) samples with SFs are somewhat uncommon, even though they can be of use in those cases where the aqueous phase is incompatible with direct injection in GC and SFC. In addition, the ability to selectively extract analytes can be of special interest when the aqueous phase concerned contains a large number of interferents.

Aqueous samples can be extracted with an SF in two different ways, viz. by isolating the aqueous matrix prior to extraction in order to avoid the problems arising from the low miscibility of water with SC CO_2; and by directly introducing the aqueous sample into the extractor with the aid of a special cell.

Available procedures for removing the aqueous matrix, which implies concentrating the analytes to be separated, are quite varied and the best choice in each case depends on the properties of the analytes and interferents. Among others, such procedures include lyophilization [114], the use of adsorbents (e.g. anhydrous sodium sulphate [125] and diatomaceous earth [126]), and mixing the sample with an appropriate sorbent (sand [127], paper strips, Celite [128]) and then evaporating the solvent as described above for preparation of spiked solid samples. In any case, the most flexible approach involves using discs of sorbent material [129,130] or a pre-column packed with a material providing not only preconcentration but also enhanced selectivity in the separation. Small columns packed with a hydrophobic sorbent have proved to be highly efficient with biological samples (e.g. plasma) [9,131]. The sample is pumped through the pre-column, which is subsequently eluted with water to remove hydrophilic compounds and dried in a nitrogen stream to evaporate the polar solvent used. Finally, the pre-column itself, accommodated in the sample chamber, acts as the extraction cell in the selected SFE mode.

Biological samples

When the properties of the analytes and/or the aqueous matrix make water removal inadvisable, the liquid sample is directly subjected to the action of the SF with all the provisions involved. Thus, a special extraction cell such as that depicted in Fig. 4.19 is required. Also, because water is slightly soluble (0.3%) in supercritical CO_2 – the SF used in every methodology involving direct introduction of aqueous samples – a small amount of moisture is swept with the analytes, which must be taken into account in order to avoid potential complications after the extraction step.

Special cell for aqueous samples

The extraction mode used is crucial when aqueous samples are to be directly introduced into a chromatographic system. In the dynamic mode, where the pure fluid is continuously passed through the sample, uninterrupted extraction of the water contained in the SF (0.3%) may result in plugging of the restrictor and activation of the trapping agent or the chromatographic phase. One additional problem is faced in collecting the extract of an immiscible solvent for subsequent chromatographic analysis owing to the two phases formed. Extraction is of little interest if it takes place in this way since the analytes and part of the sample are merely moved from one container to another. On the other hand, the static mode

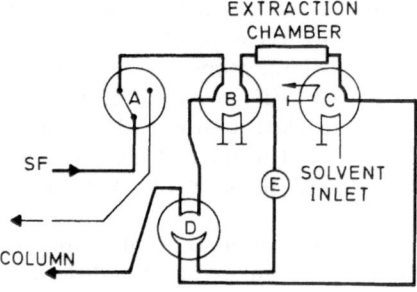

EXTRACTION
CHAMBER

SF

SOLVENT
INLET

COLUMN

Fig. 5.22. SFE recirculation systems. A–C Six-way valves, D switching vales, E recirculation pump. (Reproduced with permission of the American Chemical Society)

is also inefficient and sluggish as the pressurized cell remains quiescent until equilibrium is reached. Therefore, this type of sample is best processed by the recirculation mode, which can be implemented prior to introduction of the extract into an SFC column. A recirculation system (Fig. 5.22) typically consists of three six-way valves in addition to the recirculation pump, the extraction cell and connecting tubes, all of which are accommodated within a controlled-temperature oven. A deactivated fused silica capillary of 1 m × 1000 µm i.d. is used as interface. In this way, phosphonates were extracted from aqueous media for subsequent determination by SFC with flame ionization or UV spectroscopic detection [119], as was cholesterol in serum [92] and active principles in pharmaceuticals [132]. The segmented-flow method developed by Frei et al. for separating phenol and 4-chlorophenol from aqueous samples (Fig. 4.47) is a special case of extraction from aqueous samples with SC CO_2. Liquid segmentation of SC CO_2 provides a contact interface that ensures a high extraction efficiency [108].

5.4.1.3 Gaseous Samples

Analyte losses   Environmental analyses frequently involve gaseous samples, which is hardly surprising taking into account the large number and variety of toxic substances that are released into the atmosphere. The chief problem posed by volatile pollutants is that conventional collection and desorption methods result in analyte losses. Most studies on air pollutants concentrate on the less volatile organics, which are trapped with solid sorbents or as particulates on filters. Such adsorbents as XAD, Tenax, Chromosorb polymers and polyurethane foam are routinely used to collect compounds including PAHs, hydrocarbons, PCBs, nitrogen-, sulphur- and halide-containing substances, and pesticides, which are subsequently recovered by Soxhlet extraction or sonication. Both involve long extraction times and large solvent volumes, and give rise to incomplete extractions and/or decomposition of the sample. Extracts are subsequently vacuum-concentrated in a rotavapor or in a nitrogen stream prior to analysis – in addition, bioassays require replacing the solvent with a compatible one after extraction. Alternatively, samples can be recovered by thermal desorption and collection of the analytes at a cryogenic trap. This procedure is limited by the stability of the analyte and sorbent concerned and can also result in incomplete recovery. Desorption of the sample from a sorbent material by SFE overcomes many of the shortcomings of this step and provides enhanced selectivity if appropriate working conditions are used.

Desorption from sorbent material

SFE applications in this context encompass various types of sorbents and analytes. A comparison of the recovery of pyrene vapour deposited on chimney coal ash achieved by SFE, conventional Soxhlet extraction and ultrasound-aided extraction provided similar results, with no analyte losses on depressurization of the liquid solvent or at the head of a chromatographic column when pure CO_2 was used as extractant. However, the presence of a modifier such as methanol or toluene gave rise to the loss of all pyrene at the column head [62].

By adsorbing organic environmental pollutants such as PCBs, PAHs and n-alkanes on an appropriate material (e.g. polyurethane foam), compound families can be sequentially extracted at properly adjusted pressure and extraction times, as shown in Tables 5.13 and 5.14. This support also provides good results as collector for cigarette smoke, gas-oil exhaust products and road tar volatiles [38].

Some organic mutagens (dichloromethane, ethylene dibromide, 4-nitrophenol, 2-nitrofluorene and fluoranthene) encompassing a wide volatility range (from 349 mm Hg for dichloromethane to 0.01 mm Hg for fluoranthene) and several adsorbent materials including charcoal, Carbosieve SIII, XAD-4, Tenax Ta and Chromosorb 112 were used in a study where, following extraction and collection, the analytes were determined by a mutagenicity bioassay and GC with electron capture detection. The results showed SFE extracts from the sampling to be fit for bioassays with no additional concentration or solvent changeover. Sample losses were minimal and extracts analysed directly. Extraction of the studied compounds posed no problem except for highly volatile substances such as dichloromethane, which volatilized in the depressurized CO_2 stream – this could have been avoided by using a reliable sealing system in any case [20].

A special unit for sampling and extraction of aerosols was described in Sect. 4.3.3.5 (Fig. 4.11). The unit pumps air at a flow-rate of $1 \, m^3/h$ through quartz fibre filters. The results obtained in the selective extraction of n-alkanes adsorbed on the filters in the sampling cell itself are shown in Fig. 5.17.

5.4.1.4 Sample Size

Even though analytical-scale SFE can be applied to sample weights between 1 mg and a few hundred grams, SFE samples are typically smaller than 10 g since larger samples call for vast amounts of SF for quantitative extraction and subsequent trapping of the extracted analytes is rendered increasingly difficult by growing amounts of sample. Analytical-scale SFE usually involves fluids that are gaseous under ambient conditions; consequently, the efficiency of a given trapping procedure will depend on how the analytes are recovered on depressurization (e.g. an SC flow-rate of 1 ml/min gives rise to a gas flow-rate of around 500 ml/min after depressurization). As a rule, the lower the flow-rate is – particularly with volatile analytes – the more quantitative is collection of extracted analytes. In fact, quantitative collection of fairly volatile substances calls for SF flow-rates not higher than 1 ml/min. Taking into account that the sample leaves a void volume of 30% in the cell, passing 10 times that void volume of SF through samples of 1, 10 and 100 g takes about 3, 30 and 300 min, respectively.

Accordingly, shortening the extraction time entails using smaller samples [133], which is not always possible if sample homogeneity is to be assured or reasonable sensitivity obtained for analytes at very low concentrations.

Minimum void volume   One way of decreasing the extraction time without altering the sample size is by ensuring as low as possible a cell void volume. This was the principle behind the extraction of various organic environmental pollutants supported on polyurethane foam. The foam plugs used to adsorb atmospheric analytes were 2.5 cm long × 2.5 cm diameter and were compressed for placement in an extraction cell of 0.75 ml inner volume [38]. The extraction time can also be reduced for a given amount of sample by **Measuring surface area** increasing its surface area, as can be seen from Table 5.21, which shows the percent recoveries of various polyethylene additives obtained by using the same weight (12 mg) in the form of compact beads or particles of 30–50 mesh. Extraction of all the analytes was more efficient for the smaller particles, i.e. those with the larger specific surface area, where the analytes had to travel the shorter distances to reach the matrix surface [107].

The usable sample size is also conditioned by the analyte concentrations in the sample and the sensitivity of the determination system used. Thus, an on-line or off-line coupled chromatographic system imposes the **Maximum sample size** maximum size one can use. As a rule, quantitative transferral of extracted analytes to an on-line coupled system can only be achieved with fairly small sample sizes. Thus, extraction, separation and determination of lyophilized hamster faeces by SFE/capillary SFC/MS was feasible with 1 mg in the on-line mode but required as much as 300 mg in the off-line mode [12]. Studies involving various sample sizes and on-line coupled

Table 5.21. Influence of the sample specific surface area on the SFE efficiency of various polyethylene additives in two steps[a]

Additive	Compact pellets		Ground pellets	
	1[b]	2[b]	1[b]	2[b]
BHT/BHEB[c]	81.0	–	98.0	–
Isonox 129	68.0	18.0	98.9	1.0
Irganox 1076	72.0	20.0	96.0	3.8
Irganox 1010	50.0	34.0	94.4	4.5

[a] Sample size, 12 mg; extraction time, 30 min.
[b] Both extractions carried out under the same conditions.
[c] Both analytes coelute.

SFE/GC showed large samples and samples containing high concentrations of analytes to be unfit for collection of the extract at the head of the chromatographic column [134]. If, because of the sample heterogeneity, use of a fairly small amount is unwise, a split injection port must be used [3,57]. Therefore, unless the sample is rather heterogeneous, there is a trend towards using small sample and grain sizes in order to reduce costs and time and avoid problems derived from inefficient SF extractions, which entails using miniature extraction cells and set-ups [91]. This trend **Miniaturization** to miniaturization reflects in the design of customized mini-extraction cells of inner volumes of a few tens of a microlitre for use with samples deposited on adsorbent materials; even smaller cells (microcells of $3-4\,\mu l$ inner volume, bound by the hole in the outlet filter) are used with highly viscid or solid samples accommodated in the extractor itself [116]. Performance of these cells is improved by sonication since their minute size precludes any other form of stirring [59].

Miniaturization may increase the imprecision arising from the increasing errors made in using extremely small samples since the absolute error made in weighings (usually $\pm 0.1\,mg$) is constant whatever the sample size. A microbalance is of aid in these cases.

5.4.2 Types of Analytes

Even though SFE encompasses a wide range of non-polar, polar and even ionic compounds of a wide variety of molecular weights (see Tables 5.15–5.19), this was not so in the beginning.

In the early stages of analytical-scale SFE development (around 1986), this extraction technique was applied to relatively non-polar analytes of low molecular weights, viz. those which lent themselves readily to subsequent isolation by GC. This was chiefly the result of one of the most affordable fluids fit for use as SF extractants being CO_2, of a marked non-polar character.

The earliest SFE studies were mostly concerned with the extraction conditions (pressure and temperature, chiefly) and aimed at maximizing the solubility of the analytes of interest in each case. While this approach can be useful when the analytes account for a substantial proportion of the overall matrix (e.g. fat in meat products), maximizing solubility is secondary when the analytes are present in small or even trace amounts in the matrix. In such cases, the analytes need only be soluble enough for being removed from the extraction cell and the solubility threshold (viz. the pressure at which an analyte becomes appreciably soluble) is useful as a guide only.

The extraction rate in SFE is seemingly controlled by solubility constraints in two general cases, namely: (a) when the analytes are present at very high concentrations and may saturate the detector, and (b) when the analytes have a polar character or a high molecular weight and are thus inadequately soluble for removal from the extraction cell (i.e. when the pressure threshold is not reached). Both call for the use of a modifier, which must be carefully chosen in order to make the analyte and extractant polarity compatible. If the analyte is highly polar, then it will be scarcely soluble in unmodified CO_2 and require addition of a cosolvent such as methanol or acetone in order to raise the SF solvent power. The search for new modifiers is an area of great interest now that errors formerly made in choosing them on the basis of available experience from conventional techniques are avoided.

Use of modifier

The influence of the solvent polarity on the extraction of polar analytes was clearly shown in experiments performed with a C_{18}-modified silica gel matrix to which solutes of variable polarity were added for subsequent extraction at 250 bar and 45 °C with SC CO_2, both in pure form and mixed with methanol. Collection of the extracted analytes by means of a programmed-temperature vaporizer as interface to a capillary gas chromatograph equipped with a flame ionization detector showed pure CO_2 to extract non-polar solutes (C_{11}–C_{14}) only after 10 min extraction; on the other hand, methanol-modified CO_2, more polar than the pure SF, also extracted the more polar compounds such as nicotin and dichlorohexyl-amine in an even shorter time [35].

Matrix contaminants

In addressing the extraction of a given analyte, one should bear in mind that the extractant properties can also be altered by matrix contaminants, which may change the solvent power of the SF. One other SF modifying effect arises from the presence of water. In fact, moisture affects extractions with CO_2, not through immiscibility, but through competition for the matrix active sites. On the other hand, water can hinder extraction of non-polar compounds by forming a barrier around the matrix, thereby preventing penetration of CO_2.

The properties of extractable analytes are also largely dependent on the form in which they occur in the matrix. Thus, analytes located at or near the matrix surface can be more readily brought into contact with the extractant. On the other hand, analytes occupying crevices or interstices hinder access of the supercritical fluid and thus require the sample to be more finely divided by sonication, vigorous shaking or addition of a saturated salt solutions for biological samples, or by grinding through freezing or lyophilization otherwise. **Form of analyte in matrix**

The wide variety of samples that can be addressed in extracting analytes precludes establishment of general rules based on the nature of the analyte. By way of example, consider the different behaviour of a given analyte occurring naturally in a sample (a native analyte) or being added to it [21,113]. In the light of research carried out so far, the following generic rules can be formulated in this respect:

a) Native or endogen analytes are extracted more slowly than are externally added analytes.
b) A high analyte solubility under supercritical conditions does not assure a high extraction efficiency.
c) The best SF in terms of analyte solubility does not result in the best possible efficiency as this depends markedly on the matrix properties.

In any case, these effects must be combined with some physico–chemical factors in order to achieve adequate efficiency in the extraction of native analytes. Thus: (a) the analyte should be readily soluble in the SF; (b) the SF should be able to overcome matrix–analyte interactions; and (c) the kinetics of analyte desorption should be fast enough for the extraction to complete in a reasonably short time.

5.4.3 Scope of Application of SFE

Compared to industrial applications, analytical uses of SFE for sample preparation are fairly recent and their rapid growth is the result of major advantages (e.g. decreased time and costs, increased safety and protection from environmental contamination) over other extraction techniques.

Tables 5.15–5.19 summarize the features of selected applications in various analytical fields. Also, Fig. 5.23 compares usage of SFE in each field.

While environmental analyses are the widest field of application for SFE, industrial, food and biomedical laboratory applications of this separation technique are also of a high potential. The fact that these analytical fields have so far exploited SFE to a lesser extent is probably the **Environmental analysis**

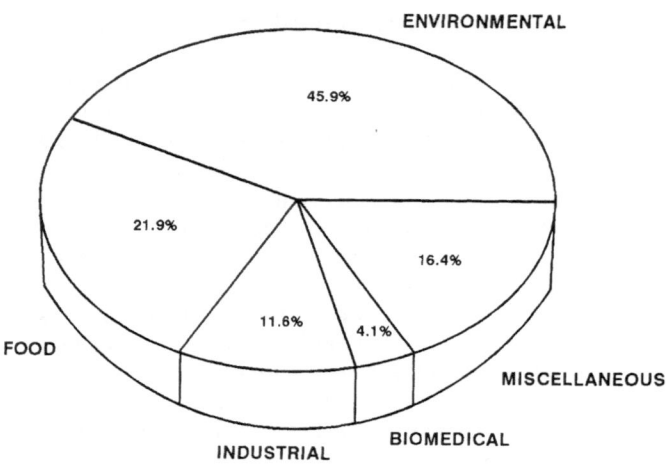

Fig. 5.23. Use of SFE in different fields

result of a reluctance to innovation from conservative workers that stick to solidly established methods and techniques and avoid potential complications arising from experimentation. Carefully structured research showing the advantages of SFE over conventional extraction techniques and a critical comparison between them would probably foster usage of SFE to the extent one would expect from its potential. Skilled personnel with a deep knowledge of the technique could aid in demonstrating it in an easy, affordable way to novices, as could crash or advanced courses, both of which would no doubt help to spread it. A number of official methods are bound to be replaced with SFE alternatives in the next few years on account of their outstanding advantages.

Tables 5.15–5.19 provide a broad picture of the possibilities of SFE. In fact, any type of sample can be conveniently collected for introduction into an SF extractor. Also, any analyte, whether ionic, polar or non-polar, can be isolated from its matrix by effect of an appropriately modified SF.

In addition to the above-discussed applications, which use SF for extraction of the analytes of interest with a view to their subsequent quantification, there are other, scarcely explored, though potentially interesting analytical-scale uses of SFs. Such is the case with the determination of enzyme activities [109] and the reactivity of organic compounds in SFs [135], the use of supercritical fluids to transfer samples to detectors in dynamic systems or even their application on a semi-preparative scale [97].

Enzymes   Former studies on the effect of SFs such as CO_2 on the activity of various enzymes were aimed at adjusting the activities of a number of natural products in order to facilitate (a) denaturation of undesirable enzymes such as those acting on lipid oxidations; (b) the selective action of enzymes that are not deactivated by SC CO_2; and (c) extraction of substances from natural products containing thermolabile enzymes [109].

The unique properties of supercritical water have been used to study the reactivity of some organic substances by using model compounds in order to simulate molecular structures embedded in charcoal. Because the two primary functions of coal processing are separation of heteroatoms and depolymerization of the larger molecules, the reactions of quinoline and isoquinoline have been widely studied in this respect. Zinc chloride, which is typically used as the hydrogenation catalyst, was added to the medium in order to increase the reactivity. Under these conditions, the compounds studied were found to be more reactive and conform to different mechanisms in the presence of supercritical water. In addition, 70% of nitrogen in the quinolines and aniline consumed in the reactions was separated as ammonium ion in the aqueous layer, alkyl side chains in aromatic compounds were more reactive and some carbon atoms were reoxidized by the water, thereby making a source of hydrogen for production of other substances [135].

Using SC CO_2 as carrier for organic solutions of metal complexes from a hot restrictor has proved to be an efficient means for obtaining vapours or aerosols to be introduced into flame atomic absorption spectrophotometers. Calibration data obtained in this way showed an SF to provide 1.2 times higher sensitivity than injection into an aqueous carrier and conventional nebulization. Such a small gain usually does not justify using an SF rather than an aqueous solution. However, future research may reveal other advantages worth exploiting [136].

On-line coupled SFE/SFC is an appealing alternative to obtaining various substances such as vitamins on a semi-preparative scale [97].

5.5 SFE and Other Extraction Techniques

One compulsive step in the development and consolidation of a new technique involves comparing it with those it may compete with in order to clearly establish its advantages and disadvantages so as to provide potential users with clear decision-making arguments. A lack of knowledge in this respect occasionally leads users to make a wrong choice in selecting a new methodology that may pose difficult or even unsurmountable problems.

On the basis of the scheme shown in Fig. 5.24, which outlines the differences between SFE and other separation techniques it can replace, the differential features of the supercritical technique are discussed below in the light of experimental data wherever available.

Differences to other techniques

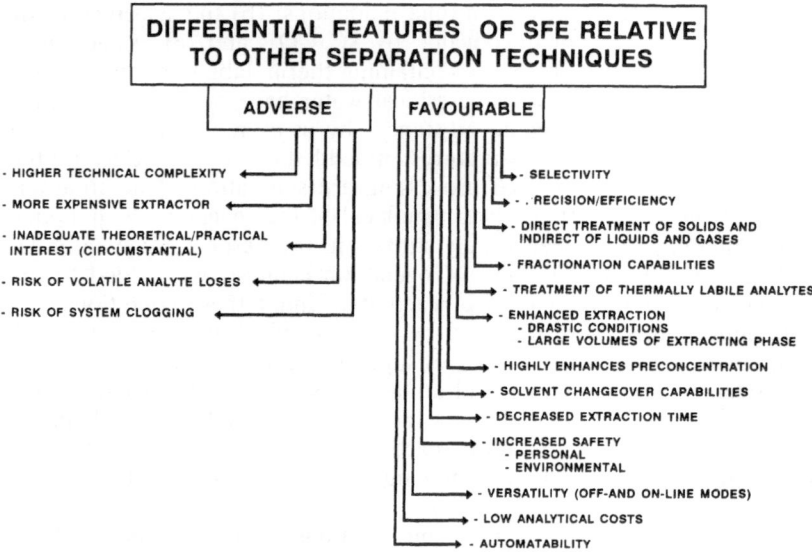

Fig. 5.24. Comparison of SFE with other separation techniques

It should be noted that, since Soxhlet extraction in all of its modes is the most widely used technique for treatment of solid samples, most comparisons of SFE to conventional techniques logically involve Soxhlet extraction. On the other hand, solid sorbent extraction and high-temperature distillation are the usual references for liquid samples.

5.5.1 Advantages of Supercritical Fluid Extraction

Conventional extraction methods do not necessarily rely on complete isolation of the analyte, so increasing the analyte recovery or extracting it quantitatively entails detracting from the extraction precision and reliability. The SF extraction of 2,3,7,8-tetrachlorodibenzo-*p*-dioxin (2,3,7,8-TCDD) from sediment samples [43] is a good example of how quantitative extraction with SFs can be. Virtually 100% of the analyte was extracted within 30 min from a sediment to which 200 µg 2,3,7,8-TCDD/kg was added, using SC CO_2 containing 2% methanol. Standard Soxhlet extraction of the same sample for 18 h provided 65% recovery only. The solvent power, polarity and working temperature of the SF, as well as its high diffusivity – similar to that of a gas – are the chief factors on which complete extraction relies.

Soxhlet extraction

Supercritical fluid extraction :volves a very small number of steps, so it leads to short analysis times and reduced sample transfer, and hence to smaller analytical errors. The simplicity of SFE is clearly reflected in the extraction of atrazine from soil samples. Extraction of the raw soil sample for 15 min with an SF is enough to obtain recoveries comparable to those provided by Soxhlet extraction after sample pretreatment. In addition, the analyte is collected in a small volume of solvent (less than 3 ml), ready for injection into a gas chromatograph with no preconcentration.

Simplicity

As shown in the preceding sections, SFE is preferentially used for treating solid samples though it is also applicable to liquid samples. Obviously, dealing with analytes adsorbed on solid supports (whether by impregnation or circulation of the sample through adsorbing columns) is more convenient and commonplace. Gaseous samples and suspended solids in environmental gaseous samples can also be extracted with SFs after deposition on a solid support – usually a filter. Thus, mebeverine alcohol was isolated from plasma by solid-phase extraction and SFE, which were found to provide comparable recoveries and precision; on the other hand, the selectivity, the ease with which the extracting fluid can be removed, the cost or organic solvents and the time consumed in the process were all much more favourable for the SFE. Table 5.22 shows the recoveries and reproducibility achieved in the extraction of organochlorine pesticides by manual and SF extraction, the latter of which is clearly superior in both respects.

Comparison of metals for pesticides

The fact that SFE includes no preconcentration step is quite significant since the large solvent volumes used in Soxhlet extractions (50–200 ml) must be concentrated to 2–5 ml by using an appropriate device prior to analysis in order to obtain adequate analyte concentrations. This step usually takes between 2 and 24 h, and releases substantial amounts of solvent into the atmosphere. By contrast, SF extracts are usually collected in only 1–2 ml of methanol or another solvent, ready for immediate analysis by GC, HPLC, UV spectroscopy or MS.

Advantage of low volume

Solvent changeovers occasionally required for the isolation or determination of analytes are most readily done in SFE; in fact, after extraction,

Table 5.22. Comparison of the efficiency and precision provided by conventional extraction (EPA method) and SFE in the determination of pesticides in fish

Analyte	Conventional extraction		SFE	
	Mean recovery (μg/g)	rsd (%)	Mean recovery (μg/g)	rsd (%)
DDD	0.70	0.36	1.00	0.33
DDE	1.72	0.32	1.77	0.23
DDT	0.68	0.50	0.62	0.21

the SF is separated by depressurization and the analytes are collected by bubbling through a suitable solvent or deposited on a sorbent or impact solid, and subsequently dissolved in the most appropriate solvent for the intended purpose.

One of the most valuable assets of SFE relative to classical separation techniques is expeditiousness. In fact, it is faster than conventional liquid extraction since (a) the SF penetrates into solid matrices more rapidly than do liquid solvents, (b) no concentration is required after extraction, and (c) the time needed to make the sample ready for measurement is up to one hundred times shorter in many cases. There are a number of applications clearly reflecting this advantage of SFE over Soxhlet extraction. Thus, the SFE of dibenzo-p-dioxins and dibenzofurans in fly ash takes 2 h with CO_2 containing 10% benzene as modifier, at 5880 psi and 40 °C. Comparatively, the Soxhlet extraction of equivalent amounts of these compounds takes 20 h. Also, Soxhlet extraction of anthraquinone from Kraft paper for 1, 2, 3, 4 and 5 h provides 16, 39, 61, 82 and 83% recovery, respectively, while SFE yields over 90% recovery in 15 min [106] and avoids emulsion formation and contact with light and air and hence potential alterations of the analytes. The purity of the products used as SF extractants is usually higher than that of ordinary solvents. As can be seen in Table 5.23, the efficiency achieved with a Soxhlet method applied over an interval twice as long as that required for over 90% extraction with an SF was rather poor [6]. Table 5.24 shows the different steps involved in the extraction/determination of fatty acids in wood pulp by Soxhlet extraction/GC and SFE/SFC. As can be seen, SFE is much more expeditious than its Soxhlet counterpart. However, conventional extraction occasionally takes a similar time to that required for SFE. Such is the case with the determination of pesticides in plant tissues by distillation in a gaseous stream and SFE, where the difference between the time taken by each methodology is negligible [76].

Cleanliness and safety   Supercritical fluid extraction is cleaner and less hazardous than conventional extractions since (a) 90% of the SFE applications developed so far use CO_2, which is non-toxic and uninflammable, and (b) unlike liquid solvents (n-hexane, ether, methylene chloride), SC CO_2 poses no fire risk and leaves no environmentally hazardous wastes.

Cost comparison   The cost of an SF extraction process compares favourably with that of conventional extractions. Thus, the cost per sample of an SF extraction with CO_2 of adequate purity is often only 1–2% of the cost of a liquid–liquid extraction. For example, the SFE of 5 g of sample at a flow-rate of 1 ml/min for 30 min requires around 15 g of CO_2, which costs about $0.10. The solvents used in the Soxhlet extraction of the same amount of sample cost $5–20. In addition, (a) SFE results in no additional waste disposal expenses; (b) staff costs are much lower since analysis times are typically of a few minutes rather than several hours; (c) the need for none of

Table 5.23. Comparison of the phenanthrene extraction efficiency achieved by using a Soxhlet and an SF extractor with the same solvents

Solvent	SFE		Soxhlet	
	Extraction time (min)	Phen extracted (%)	Extraction time (min)	Phen extracted (%)
THF	15	99.5	30	10.2
Pentane	15	91.9	30	8.25
Diethyl ether	15	90.5	30	8.13

Table 5.24. Time required for the determination of fatty acids in wood pulp. Soxhlet extraction followed by GC vs on-line coupled SFE/SFC

Soxhlet/GC	SFE/SFC
– Sample grinding or powdering (10 min) – Conditioning of extraction apparatus (15 min) – Sample weighing (5 min) – Extraction (4 h minimum) – Sample drying at 150 °C (16 h minimum) – Sample weighing and transfer (5 min) – Derivatization (methylation) of acids (1 h) – Sample evaporation in rotavapor and drying (20 min) – Sample weighing and conditioning for GC (5 min) – GC determination (30 min)	– Sample grinding or powdering (10 min) – Sample weighing (5 min) – Extraction for 10 min. Simultaneous collection of the sample in the cryogenic zone – SFC conditioning (6 min) – SFC determination (45 min)
Overall analysis time: 22.5 h	Overall analysis time: 1.5 h

the typical heating or evaporation steps results in substantial savings in electricity and hence in laboratory ventilation costs – in commercial laboratories, these savings are passed onto the clients; and (d) since the extraction time is much shorter, an SF extractor can replace a larger number of Soxhlet extractors and evaporators, thereby saving ample bench space.

Therefore, taking into account the costs involved in implementing Soxhlet and SF extraction of fat from appetizers, it is logical to obtain the results listed in Table 5.25, taken from an SF extractor manufacturer's newsletter. However, more detailed studies such as that summarized in Table 5.26 have led to rather different results. These assessments are usually biassed as manufacturers tend to place special emphasis on some

Table 5.25. Comparison of Soxhlet and SF extraction costs

SF extraction

Component	Cost/extraction ($)	Service life (no. extractions)
Restrictor	2.00	40
Collection vial	1.00	1
Frit	0.50	100
Cartridge	1.70	100
Seals	0.40	200
SC CO_2	0.20	1
SFE system	1.55	10 000
Analyst	7.50	
Waste disposal	–	
Labware cleaning	1.00	
Total cost	$15.85	

Soxhlet extraction

Component	Cost/extraction ($)	Service life (no. extractions)
Miscellaneous	1.60	1 000
Snyder column	0.22	1 000
Flasks and tubes	0.24	1 000
K-D flasks	0.15	1 000
Anhydrous Na_2SO_4	0.20	1
Solvents	0.28	1
Soxhlet and condenser	1.66	1 000
Analyst	10.80	
Waste disposal	5.00	
Labware cleaning	3.00	
Total cost	$22.60	

advantages of newly introduced methodologies and ignore other aspects that also contribute to the overall costs.

Other studies have shown SFE recoveries to compete with those provided by any existing separation technique, as can be seen in Table 5.27, which compares Soxhlet, sonication, batch, exhaustive extraction plus sonication, and acid treatment plus sonication with SFE under different conditions.

The SF extraction of thermolabile samples and/or analytes is sometimes troublesome. Occasionally, labile compounds cannot be extracted with the usual solvents, nor with purging or trapping below their decay temperature. Supercritical fluid extraction overcomes many of the problems

Table 5.26. Comparison of Soxhlet and SF extraction costs

SF extraction

Component (no.)	Total cost ($)	No. extractions	Cost/sample ($)
Extraction cell	410	300	1.35
Silica restrictor	10/m	2	5.00
Internal reducer	20	20	1.00
Ferule	5	2	2.50
4-Way valve rotor	140	140	1.00
Valve head	475	300	1.60
Vials (2)	1.00	1	1.00
Frit (2)	4.90	1	9.80
Stainless steel tube	5.00	1	5.00
Valco bushings and ferules (5/set)	5.00	1	5.00
Valco 4-way valve	1 800	1 000	1.80
CO_2 with/without modifier			1.00
SF cylinder lease	15 000–		0.50
Extractor	50 000	10 000	1.50–5.00
Cost			38.05–41.55
Analyst time 1 h/extraction			15.00–20.00
Total cost			53.05–61.55

Soxhlet extraction

Component (no.)	Total cost ($)	No. extractions	Cost/sample ($)
Soxhlet extractor (10)	960.00		
Condenser (10)	700.00		
Flask (10)	150.00		
K-D flask (5)	150.00		
Snyder column (5)	212.50		
Collection tube (5)	92.50		
Heating mantle (10)	790.00		
Cells (25)	40.00		
Tempearture regulator (2)	232.00		
Hexane (20 1)	105.00		
Acetone	173.00		
Anhydrous Na_2SO_4	195.00		
Total cost	3 800.00	1 000	3.80
Analyst time for 100 extractions at 12 $/h	90 h	100	10.80
Waste disposal			5.00
Labware cleaning			3.20
Total cost			22.80

Table 5.27. Comparison of the pyrene extraction efficiency of various solid–liquid extraction modes

Extraction method	Solvent (volume, ml)	Temperature (°C)	Recovery (%)
SFE/SFC	CO_2 (260)	250	66
SFE	CO_2/methanol (180)	250	63
SFE	CO_2/toluene (150)	250	66
SFE	Isobutane (100)	250	64
SFE	Isobutane/methanol (150)	250	68
SFE	Isobutane/toluene (50)	250	65
Soxhlet	Methanol (10)	60	62
Ultrasounds	Toluene (10)	50	63
Batch	Toluene (10)	25	54
Exhaustive, ultrasounds		50	65
Acid pretreatment/ ultrasounds	Toluene (10)	50	54

posed by thermolabile compounds provided their decay temperatures lie above the critical temperature of the SF, which is close to ambient temperature for many fluids. In this situation, the solvent power of the SF has to be adjusted essentially through the pressure.

Versatility   Supercritical fluid extraction is more versatile than are conventional extraction techniques by virtue of the special features of SFs. For example, depending on the pressure and temperature, the solvent properties of supercritical CO_2 resemble those of a variety of organic solvents, from a typically polar one such as *n*-pentane to the moderately polar ethyl acetate. Addition of small amounts of modifiers expands the solvent action of this fluid. Such a wide solvent power range enables selective multi-extractions of various types of solutes in the same sample.

The high flexibility of SFE is also reflected in its ability to operate off-line or coupled to an appropriate instrument for subsequent ready determination as described in Sect. 4.6. This opens up a door for ready automation of the analytical process since the operator's involvement is limited to weighing the sample in the extraction cell and placing it in the extraction compartment.

5.5.2 Disadvantages of Supercritical Fluid Extraction

The few disadvantages of SFE arise essentially from the following:

a) Handling an SF extractor demands a fair knowledge of its functioning and the behaviour of SFs, which are not commonplace in the analytical **Necessary** laboratory. Therefore, personnel in charge of SFE systems must be **handling skill** trained operators – the skillfulness required can be acquired at courses such as those given by SFE equipment manufacturers or workgroups with technical experience in the topic.

b) The cost of an SF extractor can be a deterrent factor. In fact, adding the initial investment to analytical costs may take these over those of conventional analyses, as shown in Table 5.26.   **Cost**

c) Because SFE started to be used in the analytical laboratory only recently, there is a lack of literature support and theoretical background required to anticipate and solve potential problems analysts may be confronted with in developing new applications.

d) There are other potential shortcomings of lesser consequence, but also worth considering in devising SF extractions. Such is the case with the risk of the restrictor being plugged (and time being wasted as a result), or volatile analytes being lost, which calls for extreme caution in collecting the extract after depressurization of the SF.

5.6 Trends in SFE

Analytical-scale supercritical fluid extraction is a fairly recent technique in a developing state whose present is outlined in Fig. 5.25, which shows the fields of action of this separation technique in order of significance, as well as the fluids used to develop SFE methodologies and the types of analytes and matrices addressed so far.

Like any promising novel technique, analytical-scale SFE still possesses a number of unexplored, scarcely exploited or improvable aspects that are bound to be the target of research in the short to medium run and make a rich seam for interested researchers. Foreseeable trends in SFE can be classified according to whether they are concerned with generic aspects or specific aspects.   **Unexplored areas**

The generic aspects that require and are bound to attract special attention from researchers in the near future include the following: **Research needed**

a) Consolidation of the scientific foundation of the technique through theoretical and practical studies. The latter should be aimed at achieving more rapid and efficient dissolution of species in SFs so as to facilitate the investigation of SF–matrix–analyte interactions in order to provide users with the knowledge needed to anticipate the generic behaviour of any SFE system.

b) Enhancement of available methodologies in the fields shown in Fig. 5.25 and expansion to others. This will not only broaden the scope of application of SFE but also allow previously unresolved problems to be overcome with its help.

c) Improvements in the efficiency and precision levels achieved so far, which are qualitatively depicted in Fig. 5.26. Logically, such levels are   **Efficiency and precision**

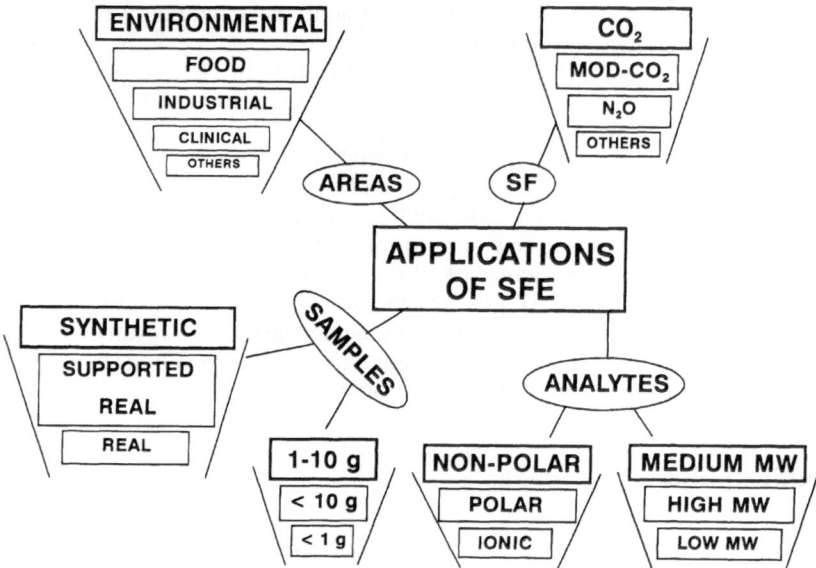

Fig. 5.25. Current uses of SFE

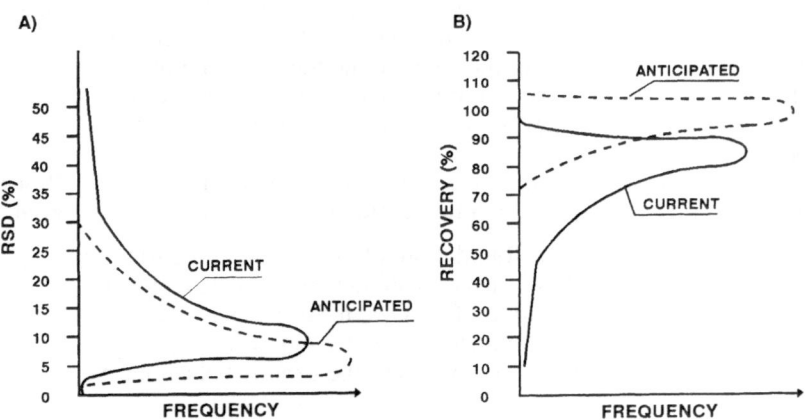

Fig. 5.26A,B. Current and foreseeable precision and efficiency levels provided by SFE

bound to be raised and even surpass those of conventional existing alternative as soon as some R&D endeavours are devoted to this analytical methodology. Because of the difficulties posed by direct extraction and the wide variability of experimental conditions (type and concentration of analytes, nature of matrices, etc.), reaching optimal precision (rsd < 5%) and recovery levels (>95%) in all applications will be next to impossible: there are always bound to be specific situations (e.g. rather complex samples, analytes at very low concentrations) where quality parameters will be underoptimal.

d) Miniaturization, a generic trend in current analytical chemistry, has also started to emerge in SFE, even though somewhat slowly so far. Its growing use in clinical chemistry, where sample size is a major limiting factor, as well as in other areas where samples are scant and/or valuable, makes miniaturization one of the most vital trends; in this sense, it would be desirable to have the classification of techniques according to sample size in Fig. 5.25 reversed and SFE customarily use less than 1 g of sample. **Miniaturization**

e) Automation, another trend in current analytical chemistry, has already emerged in SFE, which now takes part in a number of hyphenated approaches involving various techniques – particularly column chromatographies – and available in customized and commercial forms. Good proofs of the interest in these hyphenated approaches are the monographs recently published on the analytical-scale uses of supercritical fluids [136–138], which deal virtually exclusively with SFE associations and, nearly in passing, with off-line coupled SFE. **Automation**

f) Screening systems are the best choices for expediting routine analyses. The development of units providing a rapid yes/no answer on the extract that emerges from the extraction chamber is pivotal as the answer dictates whether or not an individual separation step will be required and can save time and reagents. This is particularly significant to solid samples since sensors are chiefly applied to liquid and gaseous samples. **Screening systems**

On the other hand, the generic aspects to be dealt with in forthcoming research into SFE should be directed at the different extractor components or the specific step that takes place in each (Fig. 5.27).

a) As far as the supercritical fluid is concerned, research should concentrate on (i) the obtainment of more polar SFs enabling extraction of heavier and more ionic compounds; (ii) the search for new cosolvents allowing the properties of SFs to be finely adjusted in order to expand the number of extractable analytes and also the scope of application of selective extractions through gradual or abrupt changes in the extractant properties; and (iii) the use of mixing systems allowing implementation of various gradients, as well as ternary and even **Supercritical fluid**

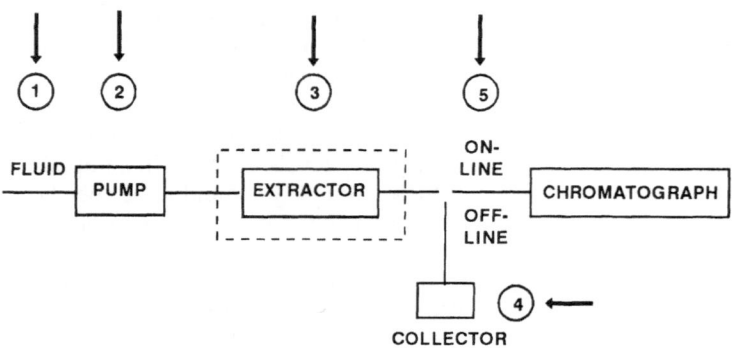

Fig. 5.27. Zones of incidence of SFE trends

quaternary mixtures of SFs and/or modifiers in order to expand application to new analytes and enhance the selectivity of extractions.

Propulsion systems b) Attempts at amelioration of propulsion systems should be aimed at (i) making them more affordable in order to allow usage of several pumps to create gradients in order to enhance the selectivity or expand the scope of application of SFE; (ii) designing systems that can withstand higher pressures in order to broaden the scope of application of this technique; and (iii) developing small units for miniaturized systems.

Extraction chambers c) Improvements in the extraction chamber should be focussed on the cell itself and the process that takes place within. Cell performance could be improved by determining the optimal length to inner diameter ratio and placement (vertical or horizontal) in each instance. This would allow the most suitable designs to be developed in each case, which should be fine-tuned by improving the sealing mechanism in order to enable convenient, expeditious changeovers so as to permit automatic operation controlled by a robotic station. These studies should desirably include the development of cells usable for sampling or alternative treatments prior to extraction proper. On the other hand, the process taking place in the cell should be facilitated by using derivatizing reactions rendering less polar and/or more readily extractable analytes, and using alternative sources of energy (e.g. sonication) before or during the extraction process in order to facilitate or expedite it.

Extract collection d) As regards extract collection, efforts should be aimed at solving the problems posed by volatile analytes and those to be expected from the growing number of analytes that will foreseeably be addressed by SFE.

Interfaces e) On-line systems call for new interfaces in order to circumvent the shortcomings of existing choices, and for hyphenated approaches meeting the new demands of an increasing number of extractable analytes.

In this respect, the use of sensors is bound to expedite the step following extraction. Research in this area should travel in two parallel directions: on the other hand, it should be concerned with the development of straightforward sensors responding to generic groups of compounds, which would be of great use in routine analyses in order to rapidly ascertain the presence or absence of a given compound family in the extracts. On the other, it is also desirable to have multi-sensors responding selectively to individual analytes or providing a composite signal that can readily be discriminated by using a suitable processing method (e.g. deconvolution).

Intelligent, systematic developments in the above-described generic and specific aspects of SFE is bound to endow it with the support required to establish it as an advantageous alternative to conventional solid-liquid extractions, thereby fully developing its still incompletely explored potential.

References

1. Campbell E (1992) Optimization of operating parameters used in supercritical fluid extraction. Pittcon'92
2. Shishikura A, Fujimoto K, Kameda T, Arai K, Saito S (1988) J. Jpn. Oil Chem. Soc. 37:8
3. Hawthorne SB, Miller DJ, Langenfeld JJ (1990) J. Chromatogr. Sci. 28:2
4. McNally MEP, Wheeler JR (1988) J. Chromatogr. 447:53
5. Knipe CR, Smith J (1990) Hewlett Packard Catalogue
6. Ndiomu DP, Simpson CF (1989) Anal. Proc. 26:393
7. Wright BW, Frye SR, McMinn DG, Smith RD (1987) Anal. Chem. 59:640
8. Ikushima Y, Hatakeda K, Ito S, Saito N, Asano T, Goto T (1988) Ind. Eng. Chem. Res. 27:818
9. Liu H, Cooper LM, Raynie DE, Pinkston JD, Wehmeyer KR (1992) Anal. Chem. 64:802
10. Wong JM, Kado NY, Kuzmicky PA, Ning HS, Woodrow JE, Hsieh DPH, Seiber JN (1991) Anal. Chem. 63:1644
11. Schneiderman MA, Sharma AK, Locke DC (1988) J. Chromatogr. Sci. 26:458
12. Pinkston JD, Delaney TE, Bowling DJ, Chester TL (1991) J. High Resolut. Chromatogr. 14:401
13. Taguchi M, Hobo T, Maeda T (1991) J. High Resol. Chromatogr. 14:140
14. Hawthorne SB, Miller DJ (1987) Anal. Chem. 59:1705
15. Messer DC, Taylor LT (1991) Hewlett Packard Catalogue
16. Bartle KD, Cliffors AA, Hawthorne SB, Langenfeld JJ, Miller DJ, Robinson R (1990) Supercrit. Fluids 3:17
17. Hunt TP, Dowle CJ, Greenway G (1991) Analyst 116:1299
18. Schmidt S, Blomberg L, Wannnman T (1989) Chromatographia 28:400
19. Furton KG, Lin Q (1991) Chromatographia 34:185
20. Tena MT, Luque De Castro MD, Valcarcel M (1993) Laborat. Robot. Autom. 5:255

21. Hawthorne SB, Miller DJ, Langenfeld J, Burford MD (1992) Physicochemical and instrumental factors that can control SFE recoveries of neutral and ionic pollutants from environmental samples. Pittcon'92
22. Hawthorne SB, Krieger MS, Miller DJ (1988) Anal. Chem. 60:472
23. Hawthorne SB, Miller DJ, Nivens DE, White DC (1992) Anal. Chem. 64:405
24. Rochette EA, Harsh JB, Hill HH Jr. (1993) Talanta 40:147
25. Field JA, Miller DJ, Field TM, Hawthorne SB, Giger W (1992) Anal. Chem. 64:3161
26. Laintz KE, Wai CM, Yonker CR, Smith RD (1991) Supercrit. Fluids 4:194
27. Laintz KE, Yu JJ, Wai CM (1992) Anal. Chem. 69:60
28. Laintz KE, Wai CM, Yonker CR, Smith RD (1992) Anal. Chem. 64:2875
29. Wright BW, Fulton JL, Kopriva AJ, Smith RD (1989) Supercritical fluid extraction and chromatography, techniques and applications. ACS Symposium
30. Ramsey ED, Perkins JR, Games DE, Startin JR (1989) J. Chromatogr. 464:353
31. Lopez-Avila V, Dodhiwala NS, Benedicto J (1991) Evaluation of four supercritical fluid extraction systems for extracting organics from environmental samples. Short Course on SFE, New Orleans
32. You X, Wang X, Bartha R, Rosen JD (1990) Evironm. Sci. Technol. 24:1732
33. Ashraf-Khorassani M, Taylor LT, Zimmerman P (1990) Anal. Chem. 62:1177
34. Hengstmann R, Hamann R, Weber H, Kettrup A (1989) Fresenius Z. Anal. Chem. 335:982
35. Houben RJ, Janssen HM, Leclercq PA, Rijks JA, Cramers CA (1990) J. High Resol. Chromatogr. 13:669
36. Lohleit M, Hillmann R, Bachmann K (1991) Fresenius J. Anal. Chem. 339:470
37. Tena MT, Luque De Castro MD, Valcarcel M (in press) Lab. Robot. Autom.
38. Hawthorne SB, Krieger MS, Miller DJ (1989) Anal. Chem. 61:736
39. Hawthorne SB, Miller DJ, Krieger MS, Fresenius Z (1988) Anal. Chem. 330:211
40. Levy JM, Cavalier RA, Bosch TN, Rynask AM, Huhak WE (1989) J. Chromatogr. Sci. 27:341
41. Levy JM, Guzowski JP, Fresenius Z (1989) Anal. Chem. 330:207
42. Ryan TW, Yocklovich SG, Watrins JC, Levy EJ (1990) J. Chromatogr. 505:273
43. Sugiyama K (1985) J. Chromatogr. 332:107
44. Suprex Newsletter, Application note SFE 101 (1991)
45. Suprex Newsletter, Application note SFE 102 (1991)
46. Suprex Newsletter, Application note SFE 118 (1991)
47. Suprex Newsletter, Application note SFE 119 (1991)
48. Suprex Newsletter, Application note SFE 121 (1991)
49. Suprex Newsletter, Application note SFE 124 (1991)
50. Jenkins TJ, Kaplan M, Simmonds MR, Davidson G, Healy MA, Poliakoff M (1991) Analyst 116:1305
51. Smith RD, Udseth HR (1983) Sep. Sci. Technol. 18:245
52. Levy JM, Rosselli AC, Boyer DS, Cross K (1990) J. High Resol. Chromatogr. 13:418
53. Schantz MM, Chesler SN (1986) J. Chromatogr. 363:397
54. Hutchenson KW, Roebers JR, Thies MC (1991) Carbon 29:215
55. Capriel P, Haisch A, Khan SU (1986) J. Agric. Food Chem. 34:70
56. McNally MEP, Wheeler JR (1988) J. Chromatogr. 435:63
57. Lohleit M, Bachmann K (1990) J. Chromatogr. 505:227
58. Hawthorne SB, Miller DJ (1986) J. Chromatogr. Sci. 24:258
59. Onuska FI, Terry KA (1989) J. High Resol. Chromatogr. 12:527
60. Hawthorne SB, Miller DJ (1987) J. Chromatogr. 403:63
61. Smith RD, Udseth HR, Hazlett RN (1985) Fuel 64:810
62. Mauldin RF, Vienneau JM, Wehry EL, Mamanton G (1990) Talanta 37:1031
63. Wong JM, Li QX, Hammock BD, Seiber JN (1991) J. Agric. Food Chem. 39:1802

64. Alexandrou N, Pawliszyn J (1989) Anal. Chem. 61:2770
65. Hewlett–Packard Newsletter, Application 228–119
66. Hewlett–Packard Newsletter, Application 228–120
67. Hewlett–Packard Newsletter, Application 228–116
68. Hewlett–Packard Newsletter, Application 228–113
69. Hewlett–Packard Newsletter, Application 228–154
70. Hewlett–Packard Newsletter, Application 228–133
71. Brady BO, Kao CPC, Dooley DM, Knopf FC, Gambrell RP (1987) Ind. Eng. Chem. Res. 26:261
72. Janda V, Steenbeke G, Sandra P (1989) J. Chromatogr. 479:200
73. Hewlett–Packard Newsletter, Application 228–143
74. ISCO Newsletter, Application SFE 101
75. Deroos FL, Bicking KL (1990) Chemosphere 20:1355
76. Thomson CA, Chesney DJ (1992) Anal. Chem. 64:848
77. Dooley KM, Kao CP, Gambrell RP, Knopf FC (1987) Ind. Eng. Chem. Res. 26:2058
78. Cortes J, Green LS, Campbell RM (1991) Anal. Chem. 63:2719
79. ISCO Newsletter, Application 137
80. Hawthorne SB, Miller JD, Krieger MS (1989) J. High Resol. Chromatogr., 12:714
81. Wheeler JR, McNally ME (1989) J. Chromatogr. Sci. 27:534
82. Hawthorne SB, Lagenfeld JJ, Miller DJ, Burford MD (1992) Anal. Chem. 64:2614
83. Wuchner K, Ghijsen RT, Brinkman UAT, Grob R, Mathieu J (1993) Analyst 118:11
84. Oudsema JW, Poole F (1992) Fresenius J. Anal. Chem. 344:426
85. Lee H, Chung BH, Park YH (1991) JAOCS 68:571
86. Gonçalves M, Vasconcelos AMP, Gomes De Azevedo EJS, Chaves Das Neves HJ, Nunes Da Ponte M (1991) JAOCS 68:474
87. Djarmati Z, Jankov RM, Schwirtlich E, Kjulinac B, Djordjevic A (1991) JAOCS 68:731
88. Ondarza M, Sanchez A (1990) Chromatographia 30:16
89. King JW (1989) J. Chromatogr. Sci. 27:355
90. Sugiyama K (1985) J. Chromatogr. 332:107
91. Sugiyama K, Saito M (1988) J. Chromatogr. 442:121
92. Ong CP, Ong HM, Yau Li SF, Lee HK (1990) J. Microcol. 2:69
93. Hewlett–Packard Newsletter, Application 228–130
94. Hewlett–Packard Newsletter, Application 228–115
95. Sandra P, David F, Stottmeister E (1990) J. High Resol. Chromatogr. 13:284
96. Kaliniski HT, Udseth HR, Wright BW, Smith RD (1988) Anal. Chem. 56:2421
97. Saito M, Yamauchi Y, Inomata K, Kottkamp W (1989) J. Chromatogr. Sci. 27:79
98. ISCO Newsletter, Application 156
99. Suprex Newsletter, Application 1063
100. ISCO Newsletter, Application 154
101. Kijppers S (1992) Chromatographia 33:434
102. Bartle KD, Boddington T, Clifford AA, Cotton NJ, Dowle CJ (1991) Anal. Chem. 63:2371
103. Braybrook JH, Mackay GA (1992) Polymer Intern. 27:157
104. Hewlett–Packard Newsletter, Application 228–138
105. Hewlett–Packard Newsletter, Application 228–110
106. Schneiderman MA, Sharma AK, Locke DC (1987) J. Chromatogr. 409:343
107. Ashraf-Knorassani M, Levy JM (1990) J. High Resol. Chromatogr. 13:742
108. Thiebaut D, Chervet JP, Vannoort RW, De Jong GJ, Brinkman UATh, Frei RW (1989) J. Chromatogr. 477:151
109. Taniguchi M, Kamihira M, Kobayaski T (1987) Agric. Biol. Chem. 51:593
110. Li SFY, Ong CP, Lee ML, Lee HK (1990) J. Chromatogr. 315:515

111. Mulcabey LJ, Taylor LT (1992) Anal. Chem. 64:981
112. Schafer K, Baumann W (1989) Fresenius Z. Anal. Chem. 332:884
113. Alexandrou N, Lawrence MJ, Pawliszyn J (1992) Anal. Chem. 64:301
114. Hedrick JL, Taylor LT (1990) J. High Resol. Chromatogr. 13:312
115. Mulcahey LJ, Taylor LT (1990) J. High Resol. Chromatogr. 13:393
116. Jahn KR, Wenclawiak B (1988) Chromatographia 26:345
117. Furton KG, Rein J (1991) Chromatographia 31:297
118. Sharma AK, Prokopczyk B, Hoffmann D (1991) J. Agric. Food Chem. 39:508
119. Hedrick J, Taylor LT (1989) Anal. Chem. 61:1986
120. Hewlett–Packard Newsletter, Application 228–109
121. Hewlett–Packard Newsletter, Application 228–132
122. Hewlett–Packard Newsletter, Application 228–144
123. Nielen MWF, Sanderson JT, Frei RW, Brinkman UATh (1989) J. Chromatogr. 474:388
124. Ndiomu DP, Simpson CF (1988) Anal. Chim. Acta. 213:237
125. Nam KS, Kapila S, Yanders AF, Puri RK (1990) Chemosphere 20:873
126. Damian J, Myer L, Liescdheski P, Tehrani J (1992) Supercritical fluid extraction of organic analytes from aqueous media and wet matrices. Pittcon'92
127. Michael RS, White LD, Hill DR (1992) Optimization of supercritical fluid extraction of additives from polyolefins. Pittcon'92
128. Ezzell JL, Richter BD (1992) Suprecritical fluid extraction and chromatography following the solid phase extraction of analytes from water. Pittcon'92
129. Di Maso M, L'Archeveque B, Vas A, McClintock SA (1992) The use of supercritical flui extraction for samples of pharmaceutical importance. Pittcon
130. Howard AL, Taylor LT (1992) Supercritical fluid extraction of chlorsulfuron and sulfemeturon methyl from aqueous matrices. Pittcon'92
131. Niessen WMA, Bergers PJM, Tjaden UR, van der Greef J (1988) J. Chromatogr. 454:243
132. Rynaski AF, Richter BE, Graig CA, Baumgartner G, Cunningham J (1992) Using supercritical fluids to separate dyes from soil matrices. Pittcon'92
133. Liebman SA, Levy EJ, Lurcott S, O'Neill S, Guthrie J, Ryan T, Tocklovich S (1989) J. Chromatogr. Sci. 27:118
134. Houser TJ, Tiffany DM, Li Z, McCarville ME, Houghton ME (1986) Fuel 65:827
135. Bysouth SR, Tyson JF (1992) Anal. Chim. Acta. 258:55
136. Jinno K (1992) Hyphenated techniques in supercritical fluid chromatography and extraction. Elsevier, Amsterdam
137. Wenclawiak B (ed.) (1992) Analysis with supercritical fluids: extraction and chromatography. Springer-Verlag, Berlin
138. Westwood SA (ed.) (1993) Supercritical fluid extraction and its use in chromatographic sample preparation. Chapman & Hall, London

Subject Index

Springer-Verlag
and the Environment

We at Springer-Verlag firmly believe that an international science publisher has a special obligation to the environment, and our corporate policies consistently reflect this conviction.

We also expect our business partners – paper mills, printers, packaging manufacturers, etc. – to commit themselves to using environmentally friendly materials and production processes.

The paper in this book is made from low- or no-chlorine pulp and is acid free, in conformance with international standards for paper permanency.